Sound Authorities

Sound Authorities

SCIENTIFIC AND MUSICAL KNOWLEDGE
IN NINETEENTH-CENTURY BRITAIN

Edward J. Gillin

The University of Chicago Press   CHICAGO AND LONDON

The University of Chicago Press, Chicago 60637
The University of Chicago Press, Ltd., London
© 2021 by The University of Chicago
All rights reserved. No part of this book may be used or reproduced in any manner whatsoever without written permission, except in the case of brief quotations in critical articles and reviews. For more information, contact the University of Chicago Press, 1427 E. 60th St., Chicago, IL 60637.
Published 2021
Printed in the United States of America

30 29 28 27 26 25 24 23 22 21   1 2 3 4 5

ISBN-13: 978-0-226-78777-0 (cloth)
ISBN-13: 978-0-226-80917-5 (e-book)
DOI: https://doi.org/10.7208/chicago/9780226809175.001.0001

This monograph is part of a project that has received funding from the European Research Council (ERC) under the European Union's Horizon 2020 research and innovation program (grant agreement No. 638241).

Library of Congress Cataloging-in-Publication Data

Names: Gillin, Edward John, 1990– author.
Title: Sound authorities : scientific and musical knowledge in nineteenth-century Britain / Edward J. Gillin.
Description: Chicago : The University of Chicago Press, 2021. | Includes bibliographical references and index.
Identifiers: LCCN 2021027261 | ISBN 9780226787770 (cloth) | ISBN 9780226809175 (e-book)
Subjects: LCSH: Music—Acoustics and physics—England—History—19th century. | Music and science—History—19th century. | Mathematics—England—History—19th century. | Sound.
Classification: LCC ML3805.G53 2021 | DDC 781.2/30942—dc23
LC record available at https://lccn.loc.gov/2021027261

♾ This paper meets the requirements of ANSI/NISO Z39.48-1992 (Permanence of Paper).

To my Nan, Estelle

Contents

List of Figures and Tables ∗ ix

INTRODUCTION
Sounds and Sweet Airs ∗ 1
Science, Sound, and Music in Britain, 1815–1914

PART I
Experiments and Mathematics
The Making of Sound as a Scientific Object

CHAPTER 1
The Laboratory of Harmony ∗ 25
The Transformation of Sound within British Science, 1815–46

CHAPTER 2
A Harmonious Universe ∗ 83
Herschel, Whewell, Somerville, and the Place of Sound in British Mathematics, 1830–70

PART II
Contesting Knowledge
Mathematicians, Musicians, and Sound Measurements

CHAPTER 3
The Problem of Pitch ∗ 123
Mathematical Authority and the Mid-Victorian Search for a Musical Standard

CHAPTER 4
Accuracy and Audibility ∗ 165
Mathematics, Musical Consensus, and the Unreliability of Sound, 1835–81

PART III
Materialism and Morality
Religious Authority and the Science of Sound

CHAPTER 5
Musical Matter * 201
*Religious Authority, John Tyndall, and the Challenge
of Materialism, 1859–1914*

EPILOGUE
Musical Spiders and Sounds Scientific in the Modern Age * 231

*Acknowledgments * 241
Notes * 245
Select Bibliography * 271
Index * 297*

Figures and Tables

Figures

Figure 1.1	The arrangement of four trumpets to project the voice of the "Invisible Girl"	29
Figure 1.2	Brewster's illustration of the "Invisible Girl"	30
Figure 1.3	Ernst Chladni (1756–1827)	31
Figure 1.4	Wheatstone's arrangement for the Enchanted Lyre	36
Figure 1.5	Wheatstone's kaleidophone	40
Figure 1.6	The kaleidophone on the front cover of the *Mechanics' Magazine* (1827)	41
Figure 1.7	The production of an acoustic figure through Chladni's experimental method	58
Figure 1.8	Wheatstone's acoustic figures, published in *Philosophical Transactions* (1833)	64
Figure 2.1	Herschel's cornfield analogy	84
Figure 3.1	The 1859–60 pitch inquiry	137
Figure 3.2	Various pitches (for both C and A) considered during the 1859–60 pitch inquiry	139
Figure 3.3	Herschel's intervals for various keys, each divided into a thousand parts	148
Figure 3.4	Pole's diagram to show the differences between small intervals	150
Figure 3.5	Hullah's "ladder" diagram of the chromatic scale	151
Figure 3.6a and 3.6b	Pole's diagram of the musical scale	152
Figure 3.7	The 1886 Society of Arts pitch inquiry's circular	158

Figure 4.1   Airy's comparison of sound, light, and water waves   171

Figure 4.2   The recasting of the Great Bell for the Palace of Westminster's clock tower   174

Figure 4.3   Airy's drawing of the vibrating circumference of Big Ben   176

Figure 4.4   Robert Werner's first report on the notes of the Westminster bells   177

Figure 4.5   Hopkins's report on the chimes of the bells of Great St. Mary's Church, Cambridge   179

Figure 4.6   Airy's figures to show the proportion of vibrations in "God Save the Queen"   184

Figure 4.7   The Calton Hill time-ball   188

Figure 4.8   Observations made on the propagation of the Edinburgh time-gun's sound signal   191

Figure 4.9   Hislop's time-gun map of Edinburgh and Leith   192

Figure 5.1   Tyndall's example of a game of solitaire showing sonorous impulses moving through air particles   210

Figure 5.2   Tyndall's second demonstration of the communication of a sonorous impulse   210

Figure 5.3   Tyndall's use of paper "riders" to illustrate the nodal points of a vibrating string   211

Figure 5.4   The figures produced by the movement of the vibrating rods of Wheatstone's kaleidophone   212

Tables

Table 1   Royal Institution Friday lectures on sound-related subjects, 1825–54   46

Table 2   Royal Institution lecture season of 1829   55

Table 3   Royal Institution lecture season of 1831   56

Table 4   Subscribers to separate courses of lectures, 1835–41   57

※ INTRODUCTION ※

Sounds and Sweet Airs
Science, Sound, and Music in Britain, 1815–1914

> *Be not afeard. The isle is full of noises, Sounds and sweet airs that give delight and hurt not. Sometimes a thousand twangling instruments will hum about mine ears, and sometimes voices That, if I then had waked after long sleep, will make me sleep again. And then, in dreaming, the clouds methought would open and show riches Ready to drop upon me, that when I waked I cried to dream again.*
>
> SHAKESPEARE, *The Tempest*

Why does music sound good? Or, more precisely, why is it that some sounds are pleasing and some sounds are not? For hundreds of years this question troubled natural philosophers, instrument makers, musicians, and theologians. Reconciling sound's physical properties with musical experiences was difficult. Traditional understandings of music taught that it was metaphysical, with its almost magical ability to affect the human senses attributed to its spiritual nature. Its power was divinely ordered and its aesthetic character was evidence of a benevolent Creator. The potential of a musical performance to determine a listener's emotional state was not so much a material process as a sacred encounter that touched the human soul. But amid the nineteenth-century expansion of industry, scientific culture, and imperialism, all this changed. New machines raised questions about how music was manufactured, radical accounts of nature and scientific practices revolutionized ideas about what it was, and diverse instruments and audiences problematized preconceptions about how it was heard. At stake were broader questions of music's role in nature and the place of sound in the universe. Nowhere was this more apparent than in Britain, the first industrial nation and home to a booming scientific culture. Yet these questions were not merely philosophical but social. Attempts to define, explain, utilize, control, measure, and regulate sound, and specifically music, were, above all, about cultural authority. They engendered tensions over the sort of knowledge that society could trust and the kind of individual who could be relied on to deliver it.

From 1815 until the outbreak of the First World War in 1914, sound and music took on a distinctive role in the natural sciences. *Sound Authorities* examines this relationship and argues that the connections between the scientific and the sonorous are crucial to our understanding of the place of science within nineteenth-century British society. Arguably, no other question in natural philosophy implicated such a diverse range of social audiences. As a science, acoustics was subject to experimental and mathematical investigation, but sound was never an exclusively scientific concern. Music was so central to human existence and culture that any scientific observation made on sonorous phenomena raised questions over who was an authority to instruct society in matters of nature. Musicians, composers, and instrument makers all offered rival sources of sonorous knowledge to those scientific, while theological commentators provided philosophical instruction through sermons that reached beyond the relatively elite audiences to which leading scientific practitioners lectured. Clergymen were eager to inform listeners on nature, including the sonorous, and they frequently drew on scientific publications in sermons. Nevertheless, music and sound generally remained beyond material explanation, instead appearing as divinely ordered phenomena. Throughout this book, it becomes clear that debates over the materiality of music and its medium, sound, engendered crucial concerns over man's place in the universe and the divine origins of all nature.

Above all, it was sound's transcendental character that makes it such a rich subject for historical investigation. Sound, and specifically music, had an unrivaled capability to permeate social orders, class boundaries, and hierarchies of knowledge.[1] As anyone with a functioning ear could appreciate music's beauty, recognize harmonic principles, and experience sonorous phenomena, sound provided ways of engaging with natural philosophy that did not demand higher education or even basic literacy. Similarly, by witnessing sound waves moving through water or vibrating sand, audiences could comprehend how natural forces like light and heat operated, without specialist expertise. Across the British Isles, scientific popularizers mobilized musical resources to attract diverse new audiences to lectures, mathematicians invoked harmonious analogies in their portrayals of an ordered universe, natural philosophers employed the ear to scrutinize nature, and musical instruments provided valuable apparatus for experimentalists. This was an exciting time for the natural sciences in Britain, and sound was a central part of this.

However, sound's ubiquitous character made it a troubling scientific subject: the sonorous was difficult to control. From their acoustic observations, scientific practitioners endeavored to extend their work to the nature of music, bringing them into conflict with traditional sources of musical knowledge, including musicians and theologians. Musical sounds were not confined to any one social group but experienced in churches, theaters, opera houses, music halls, taverns, and streets. Debates over the nature of sound and music were, therefore, difficult to restrict to elite scientific communities. More than any other natural phenomena, philosophical discussions over the sonorous were open to a wide range of individuals with sharply contrasting credentials for social authority. It is this socially diverse quality that makes sound and music so insightful for historians of science and demonstrates what musicologists, so often maligned among historians, can contribute to historical analysis. Indeed, this book argues that there is no better subject for understanding questions of nineteenth-century scientific authority than that of sound.

Religion and Romanticism: The Rival Authorities of Sonorous Knowledge

The close relationship between music and the study of nature was not new to the nineteenth century. Along with arithmetic and geometry, music was an important part of ancient Greek philosophy, with Pythagoras's conception that the musical relations between different pitches were explainable in terms of ratios illustrated through the production of notes on strings of varying lengths. By showing that a string sounding one note could, on being halved in length, produce the same note an octave higher, Pythagoras's work was pivotal to the classical study of harmony, which was subsequently extended through Ptolemy's *Harmonics*. This science of harmony was the basis of musical speculation throughout the Middle Ages and Renaissance. Increasingly, music provided powerful analogies for the study of light, most famously in Isaac Newton's undergraduate manuscript "Of Musick," written between 1664 and 1666, where he compared the seven notes of the diatonic scale to the seven perceived colors of the spectrum. These connections between light and sound subsequently dominated sonorous discussions throughout the eighteenth century, especially in the mathematical writings of Leonhard Euler and Daniel Bernoulli.[2]

During the nineteenth century, however, amid the contexts of industrialization, mechanization, growing atheism, and scientific investigation, this relationship between the natural sciences and sound underwent a profound change.[3] The French Revolution of 1789 marked the start of a quarter-century of almost continual conflict between Britain and France, first through the revolutionary wars of the 1790s and then the Napoleonic wars of the 1800s. Along with war came isolation, not just in trade and culture but in science and industry. On the Continent, new methods of French mathematical analysis replaced the Newtonian calculus which remained dominant in Britain until well after the peace of 1815.[4] These differences were equally evident in the science of acoustics, especially through the works of the German natural philosopher Ernst Chladni (1756–1827). Published in his *Entdeckungen über die Theorie des Klanges* (1787) and *Die Akustik* (1802), Chladni's experiments to make invisible sound waves observable to the eye were the most significant acoustic inquiries of the late eighteenth and early nineteenth centuries. By sprinkling sand on a glass plate and then putting this into vibration with a violin bow, he produced patterns, known as "Chladni figures," which varied in relation to the frequency of the pitch sounded, as sand collected at the points of least vibration to reveal the wave motions of musical sounds. Although securing attention with European audiences, including Emperor Napoléon in 1808, these experiments did not come under serious scrutiny from British audiences until well after 1815, as part of a gradually increasing international exchange of scientific and industrial ideas and practices.[5]

Chladni's experiments were important because they represented one of the first philosophical frameworks for visualizing the relationship between frequency and musical tones as a mechanical process. This was not the only way in which understandings of sonorous phenomena were redefined. Nineteenth-century Britain was a place of expanding industry, rapid urbanization, spiraling population growth, and increasing mechanization. With this came strange sounds, both musical and disruptive, natural and manmade. Booming railways and steamships, music-making automata, and factory machines transformed Britain's soundscape, as did novel devices, including microphones and early recording instruments, which shaped unprecedented listening practices.[6] New technologies encouraged radical interpretations of the sonorous, engendering a reconception of what separated music from sound generally. Naval architect John Scott Russell (1808–82), for example, drew on steam transport to illustrate musical phenomena. Most famous for collaborating with Isambard Kingdom Brunel to build the enormous *Great Eastern* steamship during the 1850s, Russell's theoretical work on fluid dynamics shaped the design of his ships. He drew comparisons between water waves and the movement of sound through air. The more "rapid and uniform speed of railways," he alleged, enabled engineers and natural philosophers "to test the nature of sounds still more accurately," providing new means of measuring the relationship between mathematics and musical tone. Russell asked his audiences to imagine a straight line of track, running parallel to wooden fence posts set one foot apart, before explaining that a train traveling at twenty miles per hour would pass sixty-four of these posts per second, creating a musical tone corresponding to a C tuning fork. Were the train then to pass fence posts placed two feet apart, equaling thirty-two posts per second, the "musical effect is an octave lower than it was before."[7] Likewise, in a lecture on sound delivered at London's Royal Institution, John Tyndall drew on new industrial mechanisms to distinguish between music and dissonance. The "puffs of a locomotive" created noise, which increased in frequency as the train accelerated. If this continued "until the puffs numbered 50 or 60 a second," explained Tyndall, "the approach of the engine would be heralded by an organ peal of tremendous power."[8] For both Tyndall and Russell, industrial technologies provided resources both for new methods of listening and for discussing how their audiences thought about what they heard.

During the second half of the nineteenth century, physicists, psychologists, and physiologists united to construct new understandings of the sensory perception of sound. This was a moment when acoustic experiments were combined with physiological and psychological accounts of sound to explain the sensory perception of music. Alexandra Hui has shown that this was a particularly pertinent matter among liberal middle- and upper-class Germans and Austrians, and her study rightly focuses on those working within these societies. Hermann Helmholtz (1821–94), Gustav Fechner (1801–87), Ernst Mach (1838–1916), Carl Stumpf (1848–1936), and Wilhelm Wundt (1832–1920) were all concerned with examining how humans experienced music.[9] This investigation was of undoubted urgency in the musically cultivated German lands, but the place of sound and music within the natural sciences also came under scrutiny in industrial Britain. However, to understand the relationship between science and music, we have to move our analysis away from scientific elites.[10] Those professing experimental or mathematical skill subjected sound to physical investigation, but they did not have inherent authority to define and comment on sonorous phenomena.[11] New experimental techniques and apparatus induced fresh hope that physical accounts of sound would eventually provide knowledge of why music was pleasing and how it evoked human emotions. Yet in moving from sonorous to musical phenomena, scientific practitioners found themselves in competition with well-established rival authorities.

Musicians, critics, composers, conductors, and vocalists all boasted musical knowledge, as did instrument makers who had working experience of how to produce different sounds, be it through violins, flutes, pianos, or tuning forks. Innovative manufacturers constructed new contrivances to improve musical performances, mobilizing skills and know-how which took time to acquire. And both the armed forces and religious commentators were deeply invested in questions of music. The nation's largest organized musical bodies were the bands of the British Army and Royal Navy, while church worship, in which music was often prominent, represented society's most extensively shared cultural experience throughout the century. In Victorian Britain, the pulpit was a space of immense authority, providing spiritual guidance for audiences, including on questions of music and nature.[12] The 1851 religious census found that 3.8 million Anglicans attended church every week, along with a further 3.2 million who participated in nonestablished dissenting services, including Baptists, Quakers, and Congregationalists.[13] This religious diversity stimulated contrasting musical cultures. In dissenting communities, music took on a particularly important function. For instance, from its formation in 1832, London's Sacred Harmonic Society cultivated sacred music as a means to Christian unity and moral improvement.[14] Likewise, the city's dissenting churches encouraged hymn singing as an intensely emotional form of worship. Yet the spiritualism and individual religious feeling that such musical practices emphasized troubled many Anglicans, who were suspicious of such zealous Protestantism, so much at odds with established Church of England order and decorum.[15] The degree to which music should feature in the performance of Anglican prayer book liturgies was equally controversial. Some worried about the frivolity of music, with extreme Protestants fearing chanting to be popish. Others stressed the unifying quality of song, echoing the endorsements of sixteenth-century theologians John Calvin (1509–64) and Martin Luther (1483–1546).[16] Although the Church of England witnessed much evangelical influence that stressed the value of scripture and the Atonement, High Church Anglicanism emphasized the importance of the Church itself as an institution, tracing its authority back to the medieval Catholic Church. High Churchmanship involved a middle path between radical Protestantism and Roman Catholicism in which rituals, including those musical, were central.[17]

It was not just the place of music in worship that concerned religious commentators but the nature of music itself and the sonorous phenomena that comprised it. Clergymen were just as eager as natural philosophers to provide knowledge of sound and mobilize it within a vision of a divinely ordered universe. In sermons, religious commentators exerted authority over what music was and why it was aesthetically pleasing.[18]

Spiritual accounts of sonorous phenomena portrayed music, as well as human speech, as a God-given gift. Nineteenth-century theology shaped understandings over musical principles, the voice, the organization of the ear, and sound's place in nature. For example, few Victorian sermonizers were as eager to cast judgment on matters scientific as the minister of the Kingsland Congregational Church, Thomas Aveling (1815–84). Renowned for his entertaining sermons which drew moral lessons from recent technological and industrial spectacles, such as the Great Exhibition in 1851, Aveling employed the Bible as a prophetic history which offered lessons for contemporary society.[19] In 1849, he defined music as "modulated sound, regulated by certain laws, called the laws of harmony, which, like mathematical truths, are fixed, unaltered, and eternal." This constant law of harmony, with its fixed, permanent measures of vibrations, exerted a mysterious influence over the minds of men. Aveling warned his listeners that science could not explain music's power. As he put it, music was "capable of producing very powerful effects upon the mind, exciting the liveliest sensations," despite being but the "result of harmonic combinations." This "wondrous effect of music" transcended scientific explanation. Aveling declared that while the "materialist asserts that it is in consequence of the fine construction of the organs for the transmission of sound" and the "psychologist will tell us, that the soul, in its spiritual mechanism, is marvellously formed, and most singularly susceptible of external influence," neither truly offered an "explanation of the mystery." Instead, Aveling argued that there was no accounting for music beyond that the divine Architect of the universe had made it so, and in doing so, provided humans with a "species of spiritual electricity." It was a sacred gift, for the benefit of mankind, to sooth and calm, excite passions, and sustain moral improvement. Aveling recalled one clergyman who attributed the gentle temperament of his three daughters to piano and singing recitals. Nevertheless, he contended that music at the theater and opera was an immoral waste of time, while he found oratorios particularly irreligious. Although popular with many Christians, these performances offended Aveling, representing a "profanity in the exhibition, when a number of people,—not one in fifty of whom, perhaps, has the fear of God in the heart,—get up and sing the most solemn words . . . with no intention of worshipping Him."[20] These performances were not for the glory of God but the pleasure of man. Aveling not only provided a religiously informed framework for knowing nature, including musical phenomena, but outlined how it should be regulated.

When scientific practitioners attempted to provide physical explanations for sonorous phenomena, they found themselves competing with religious figures like Aveling for authority. This is certainly not to say that scientific and religious accounts of sound and music were always in opposition. Sermonizers and clergymen frequently drew on scientific publications in their portrayals of natural phenomena, while scientific authorities regularly invoked theological interpretations of nature within their own work. Although nineteenth-century discussions over sound were not divided between the scientific and the religious, there were escalating tensions over the material nature of euphonic sounds. Physical explanations for musical aesthetics and the scientific study of sound increasingly threatened to demystify music, contradicting religious understandings that it was something spiritual and divine. And yet the Church proved a remarkably resilient bastion of social authority. Anglican clergymen and dissenting ministers spoke to socially diverse congregations, including the rich and poor, literate and uneducated, from the well-to-do gentry and aspirational middle classes to farm hands, artisans, factory workers, and unskilled laborers. Crucially, while scientific writers depended on literate audiences having some degree of mathematical understanding, sermons were accessible to all and generally comprehensible. Sermons on music and sound delivered powerful lessons and instruction on nature that transcended social, political, and economic boundaries. Theological knowledge of sonorous phenomena was a robust rival to scientific authority, but these interactions were, in fact, just one element within the relationship between science and faith that was becoming ever more strained.

Until the mid-nineteenth century, leading scientific authorities often invoked God within their accounts of nature. During the 1820s and 1830s, leading science writers like John Herschel, Mary Somerville, and William Whewell worked within a religious framework of divine creation in which God was central. Within these interpretations of nature, sound contributed to grand unifying visions of the universe. Mathematicians and sermonizers alike mobilized the sonorous within their broad portrayals of nature in which all phenomena, both celestial and terrestrial, were connected and governed by a few divinely created laws. Both music and the universe appeared to reveal the harmonious order of God's creation. However, throughout the century the natural sciences became increasingly troubling, both socially and theologically.[21] As something championed in Revolutionary and Napoleonic France, science had long carried potent connotations: this knowledge, proclaimed as rational, had underpinned French efforts to reorder society, such as through the metric system, decimal time, and the Republican Calendar which divided months into ten-day *décades*. French revolutionary science had been in direct confrontation to the established authority of the Catholic Church. Apocryphally, when Napoléon asked Pierre-Simon Laplace (1749–1827), the doyen of French mathematics, where God fit in his account of the universe, Laplace allegedly replied that he had "no need of that hypothesis."[22] Although science became more fashionable in Britain during the 1820s and 1830s, it continued to present society with worrying implications. Mathematical accounts of a mechanical universe, working without divine interference, threatened the traditional role of God within Creation, as did chemistry's suggestion that all in nature might be mere matter in motion. Geology provided further religious concern, with new theories over the age of the Earth and its ongoing geological processes challenging Genesis. Astronomy offered little comfort, with new nebulae observations suggesting stars were not divine creations but products of gaseous bodies.[23] Most dramatic of all were the socially disturbing theories of evolution that became ever more prominent during the 1850s and 1860s. Following Robert Chambers's *Vestiges of the Natural History of Creation* (1844), Herbert Spencer's "Progress: Its Law and Cause" (1857), and Charles Darwin's *On the Origin of Species* (1859), proevolution writers like Tyndall and Thomas Huxley challenged what they perceived to be the clerical dominance of science, arguing that it should be scientists rather than clergymen who informed society on questions of nature.[24]

Nevertheless, God's creating role persisted in British natural philosophy, especially within physics, well into the twentieth century. At the 1868 meeting of the British Association for the Advancement of Science (BAAS), the Scottish mathematician James Clerk Maxwell (1831–79) famously asserted that molecules provided evidence of a Designer, claiming that

> the exact quality of each molecule to all others of the same kind gives it . . . the essential character of a manufactured article, and precludes the idea of its being eternal and self-existent. . . . Science is incompetent to reason upon the creation of matter itself out of nothing. We have reached the utmost limit of our thinking faculties when we have admitted that because matter cannot be eternal and self-existent it must have been created.[25]

This was a crucial concept in the science of thermodynamics, which taught that the amount of useful energy in nature was constantly diminishing, leading to the inevitable cold death of the universe. This law suggested that, as there would one day be an end, so there must have been a beginning. These were notions that were highly compatible with Biblical accounts of the Creation and Apocalypse in Genesis and Revelation. As much as the nineteenth century witnessed growing tensions between the Church and science, sermons continued to address questions of natural philosophy until the eve of the First World War, while scientists often placed their work within a theological framework. Within these escalating debates over the physical world and anxieties that all within the universe might be matter, music and sound took a central part.

It was not just religious understandings of music that presented a challenge to those seeking to exert scientific authority. New mathematical and experimental accounts of sound were not the only responses to industrialization and the rise of mechanized labor. Along with the cultivation of sound as a scientific object, the early nineteenth century witnessed the development of increasingly Romantic understandings of music: religious frameworks were not alone in ascribing metaphysical value to the musical. Largely in response to industrial expansion and French mechanical philosophy, such Romantic thinking was often in conflict with empirical approaches to sonorous phenomena. At the beginning of the century, music was already widely conceived of as metaphysical.[26] Such idealism became more entrenched in the 1800s, with German Romantic philosophy rejecting accounts of nature in which all was reducible to "physiochemical principles." On the contrary, the German writer and statesman Johann Wolfgang von Goethe (1749–1832) argued that organic living creatures were not the products of mechanical principles. Late-eighteenth- and early-nineteenth-century German Romantic philosophers preferred to celebrate the spiritual; though chemical and physical science could enhance knowledge of the organic, this sort of Romanticism emphasized the existence of an immortal human soul, with music idealized for its ability to connect with the metaphysical, typifying man's creative powers.[27] Georg Wilhelm Friedrich Hegel (1770–1831), for instance, proposed that music was beyond mathematical understanding.[28] For the German music critic and Romantic author Ernst Theodor Amadeus Hoffman (1776–1822), music's metaphysical power was indisputable and produced transcendental experiences. He described how Mozart's compositions led audiences "deep into the realm of spirits": such music "irresistibly sweeps the listener into the wonderful spirit-realm of the infinite."[29] Similarly, in 1820 the Italian philosopher Giacomo Leopardi (1798–1837) asserted that music was unique among the arts due to its power over human emotions. The effect that music had was not due to harmony, "but from sound, which electrifies and shakes us from the very first note. . . . This is what makes music special compared with the other arts, although a fine, bright color does affect us, but to a much lesser extent."[30] While harmony modified sounds, Leopardi maintained that it was sound itself that commanded so much influence over the soul.

These conceptions developed throughout the nineteenth century. As late as 1854, the German music critic Eduard Hanslick (1825–1904) warned that music, as "the most ethereal of the arts," could not be reduced to mechanical questions of sensation. While philosophers were eager to break down the arts to their technical characteristics and analyze them objectively, music alone was inconceivable from "this objective standpoint." Disinterested inquiry was the definition of "scientific knowledge," but Hanslick felt that music was above such investigation.[31] New scientific accounts of music that sought to reduce it to mechanisms of sensation and physical phenomena were clearly contrary to such Romanticism. As Friedrich Lange showed in his *History of Materialism and Critique of Its Present Importance* (1866), the attributing of everything in nature, including that considered spiritual, to molecular processes, was in direct contradiction to accounts of music as a spiritual entity that produced transcendental experiences.[32] These tensions remained throughout the century, with Richard Wagner (1813–83) warning in 1878 that "once chemistry has finally laid hold of logic . . . [e]very mystery of being [would be exposed] as mere imaginary secrets." Wagner resented the scientific assault on music as an idealistic art. In particular, he condemned scientific practitioners whom he perceived to "haughtily look down upon us artists, poets, and musicians, as the belated progeny of an absolute worldview."[33] As Daniel Chua has argued, this reduction of music to mechanical principles can be seen as part of what Max Weber termed the "disenchantment of the world," in which phenomena once regarded as supernatural were rationalized and refashioned as natural.[34]

Accompanying scientific attempts to reduce musical phenomena to physical laws were schemes for mechanically reproducing aesthetically pleasing sounds. Machines that could speak or play music were not new but grew increasingly common and sophisticated. Away from the Romantic philosophy of Goethe and Hegel, artisans and mechanics produced musical automata capable of humanlike performances, but these lifeless devices raised further questions over the ethereal nature of music. Could music carry the same emotional value if it lacked the human expression of a vocalist or instrumentalist? Just as machines replaced laborers in factories, new industrially produced music threatened to destroy the spiritual element of the art.[35] As Hoffman put it in his *Die Automate* (1814) when describing a music-making automaton, "Could you look at such a sight, for an instant, without horror? At all events, all mechanical music seems monstrous and abominable."[36]

For all this, the dichotomy between materialism and Romanticism needs careful qualification. Music and sound were certainly part of a wider nineteenth-century discourse surrounding the Romantic and the mechanic but, as John Tresch has demonstrated in *The Romantic Machine* (2012), Romanticism and mechanism were not always opposed. Well into the nineteenth century, aesthetic and emotional values shaped the mechanical sciences: it remained possible to interpret machines and instruments as part of, rather than opposed to, nature.[37] This was equally true of scientific accounts of sound and music; though attributing musical phenomena to physical causes, this did not inevitably equate to materialism. Amid industrial, colonial, and technological expansion, Britain's soundscape changed dramatically throughout the nineteenth century, as did understandings about the nature of sound and music. Within this transformation, it was unclear who should instruct society on the sonorous. Religious and Romantic frameworks accompanied, informed, and rivaled those scientific. Throughout this volume, it will be demonstrated how sound was at the center of a reorganization of the natural sciences, which became increasingly disciplined and defined after 1815.

Sound and Vision in the History of Science

The place of the sonorous within nineteenth-century British science matters because this was a crucial period of transformation within European scientific culture. More than ever before, empirical, experimental, and mathematical approaches to knowledge production were looked to for understandings of nature. It was an era of definition and discipline, and at the heart of this was the question of authority: To what extent should the scientific guide society in what it thought and how it behaved? In the early nineteenth century, "science" was a vague term. Who should be doing it, what it was, and how it should be produced were all under discussion. There was nothing self-evident about the value of experimental practices, what these might involve, or even what it was to be "empirical." These concerns remained in British natural philosophy well into the century. The 1850s and 1860s saw the emergence of physics laboratories in Glasgow, Edinburgh, London, Oxford, and Cambridge, largely for the production of accurate new measurement standards.[38] Along with growing specialization in the natural sciences and an expanding laboratory culture, physics ascended from something hardly existent in 1800 to the most prestigious of all sciences by 1900.[39] As science became increasingly defined, it seemed to exert a growing social authority: this expanding influence was particularly felt and contested in questions of sound and music.

Despite this, in histories of science the eye has customarily been construed as the dominant organ for knowledge production, and it is to the visual that historians have traditionally looked for insights into the development of scientific culture during the nineteenth century. This is something Bruno Latour has exemplified in his emphasis of the role of visual practices within laboratory culture, examined in his "Visualization and Cognition: Thinking with Eyes and Hands" (1986). Here he asserted that most scholars would agree "that writing, printing and visualizing are important *asides* of scientific revolution." Inscription, whether as signs, symbols, prints, images, diagrams, lists, formulae, drawings, files, or equations, provides essential visual resources for presenting science and making it readable. This is vital for mobilizing knowledge and shaping its reception.[40] In many respects, the power of knowledge to convince has been historically analyzed in terms of its visual representation. It is not surprising that histories of science are so occupied with the ocular, especially since Michel Foucault's *Discipline and Punish* (1975) employed eighteenth-century philosopher Jeremy Bentham's Panopticon design for a prison to explore the visual power relations between dominant and subordinate subjects. In the Panopticon, architectural techniques sustained disciplining regimes which operated in terms of what and who could be seen; power, in this sense, is completely visual.[41] As Chris Otter surmised, "vision and power are symbiotic" and "*Who* could see *what, whom, when, where,* and *how* was, and remains, an integral dimension of the everyday operation and experience of power." Otter went further, arguing that the "cultural history of light and vision . . . becomes inseparable from two political histories, those of discipline and of capital."[42] In a post-Foucauldian world, it seems obvious that a study of political power would entail questions about vision and surveillance rather than hearing and listening. If we are to examine nineteenth-century scientific authority, then we are effectively dealing in problems of political and social power, but, as this book makes clear, these extend to the sonorous. While Otter's *Victorian Eye* (2008) illustrates the historic relationship between liberal politics and the eye, the "Victorian Ear" has yet to be written.

Thinking about the visual cultures of science has been productive in histories investigating the manufacturing and disseminating of philosophical knowledge. The importance of visual techniques in knowledge production are especially relevant for the study of nineteenth-century Britain. Elizabeth Musselman asserted in her 2006 *Nervous Conditions* that "sight seemed to provide the natural philosopher with most, and the most accurate, information." Visual inconsistencies like hallucinations or sights that few could see, such as directly observing the sun, raised epistemological problems but, Musselman argued, eyesight provided superior evidence to alternative senses. Chemists, for example, could feel temperature but preferred the visuality of a thermometer. The same was true of sound.[43] Visual culture was not just central to displaying new scientific knowledge but to its construction. Frequently the techniques of showing science were inseparable from the ways in which it was made. The work of visualization was as important in the theater and lecture room as it was in the laboratory. As Simon Schaffer has shown, late-Victorian cinematic technology was closely integrated with the image machinery used in scientific practice.[44] For Tyndall, physics was a discipline qualified in terms of what could be seen; it was a subject defined by the eye. The subjects of physics were "those which lie nearest to human perception," observed Tyndall in 1855.[45] This science was about making natural phenomena visible, and it was built on new optical devices, like microscopes and photographic cameras, that revealed nature to the eye.

Just as seeing was important for believing, showing was essential for persuading. Bernard Lightman's study of Victorian scientific visual culture, *Victorian Popularizers of Science* (2007), argued that the eye was inseparable from efforts to popularize and promote new bodies of knowledge. "Popularizers responded to the pictorial turn in Victorian culture by increasing the use of illustrations and vivid literary images in their books" as well as in their lectures and public performances. Aesthetically satisfying images revealed the beauty of nature to audiences. Spectacles made science accessible to broad society, and here it was eye-witnessing an experiment, demonstration, or technological wonder that mattered.[46] In London, with its array of venues offering the latest scientific wonders to popular audiences, this was very evident. From the workshops of the West End, with the newest mechanical contrivances, to the Royal Institution's experimental demonstrations of spectacular natural phenomena for fashionable metropolitan elites, scientific knowledge was accessible to more diverse audiences than ever before.[47] The Adelaide Gallery, for instance, charged a shilling for entry to exhibitions and specialized in machines and amusements employing electricity. Away from public institutions of science, new forms of knowledge were experienced within domestic spaces, with cheap, industrially produced publications available from the 1820s, aimed at new nonspecialist reading audiences and often excitingly illustrated.[48]

Similarly, Iwan Morus has emphasized the epistemological centrality of vision within nineteenth-century science in his aptly titled "Seeing and Believing Science" (2006) and subsequent "Illuminating Illusions, or, the Victorian Art of Seeing" (2012). As he has shown, the management of optical illusions was an important practice in the legitimization of new scientific knowledge. Visual displays helped scientific performers to delineate the relation between experience and judgment for their audiences; illusions which revealed the mind's inability to make sense of an experience showed the limitations of human judgment. As Morus argues, how a scientific practitioner mobilized visual resources was crucial to persuading audiences that they were seeing nature itself rather than artificial, manmade, tricks.[49] Sight was the fundamental sense for making knowledge; as he put it, seeing was "the model for knowing."[50] However, as much as historians of science treat hearing as secondary to seeing, *Sound Authorities* reveals that Victorian commentators were, in fact, very aware of the ear's epistemological value.[51] As an organ capable of measuring vibratory frequency, distinguishing between different notes, and recognizing musical relations, the ear had unique claims to mathematical precision.

Until recently, when sound has appeared in histories of science, it has generally been for its analogous qualities within the broader study of natural phenomena, usually light.[52] The notable exception has been the sustained interest in Helmholtz's work on musical consonance and dissonance in an effort to unite the physical nature of sound with musical aesthetics.[53] Predictably, the visual emphasis within the history of science has attracted criticism from scholars specializing in aural culture. Most famously, in 2003 Jonathan Sterne argued that notions of modernity are implicitly associated with vision; the Enlightenment appears grounded in the use of light and sight as metaphors for truth and knowing. Rejecting what he identified as "the pervasive narrative that says that, in becoming modern, Western culture moved away from a culture of hearing to a culture of seeing," Sterne maintains that although sight might sometimes be "the privileged sense in European philosophical discourse," sound is also central to modernity.[54] Sterne advises historians to rethink the traditional privileging of vision and to "take seriously the role of sound and hearing in modern life," which would "trouble the visualist definition of *modernity*."[55] With new technologies for recording and disseminating sound, between 1750 and 1925 "sound itself became an object and domain of thought and practice, where it had previously been conceptualized in terms of particular idealized instances like voice or music." As much as an Enlightenment, arguably, there was also an "Ensoniment."[56]

Although there is still a prioritizing of seeing over hearing in histories of knowledge production, recent studies have done much to redress this visual hegemony in scientific culture and situate sound within the broader history of science. In 2013, the *Osiris* special edition "Music, Sound, and Laboratory from 1750–1980" offered direction on the interconnections between sound and science, arguing that listening and hearing were in fact crucial philosophical practices. While scientific practitioners often labored to make sound visible to the eye, the ear had a distinct role in making and disseminating science. As Alexandra Hui, Julia Kursell, and Myles Jackson have argued, you need all your senses to fully comprehend nature: within the laboratory, sound offered valuable means of investigation.[57] Beyond the laboratory, Graeme Gooday has asserted the importance of architectural practices and knowledge in the formation of new acoustic knowledge.[58] At the same time, there are clear parallels between musical and scientific instruments. While both involved highly skilled labor to manufacture, musical instruments provided new ways to investigate nature. The eighteenth-century aeolian harp, for example, produced eerie sounds if exposed to the wind, making nature appear to be playing the instrument.[59] In his *Harmonious Triads* (2006), Jackson has explored how in nineteenth-century Germany, musicians, instrument makers, and physicists all shared in the broad construction of a science of acoustics, while Peter Pesic's *Music and the Making of Modern Science* (2014) has demonstrated how useful music has been in the study of nature; both works have argued that music is not a supplementary body of knowledge to, but inseparable from, the natural sciences.[60] More recently, James Davies and Ellen Lockhart's *Sound Knowledge: Music and Science in London, 1789–1851* (2016) explored how London's early nineteenth-century scientific culture provided a multisensory experience. The city's scientific and musical cultures were intertwined and music was often central to public displays of new knowledge of nature. Beyond visual techniques for exhibiting science, London audiences were encouraged to listen and acquire knowledge aurally. In conjunction with what could be seen, what could be heard was a valuable epistemological resource for scientific practitioners to mobilize.[61] In a similar fashion, Viktoria Tkaczyk's account of Chladni's acoustic figure experiments in the 1800s reveals that the performative quality of sonorous phenomena was invaluable to securing new popular audiences and political patronage.[62] Since Sterne's call to consider the ear, as much as the eye, historians and philosophers of science have shown how nineteenth-century knowledge of nature was heard, just as much as it was seen.

Building on these studies, *Sound Authorities* extends our understanding of the nineteenth-century intersections between science, sound, and music by showing how central this relationship was to the formation of broader scientific culture. It does this by placing sound knowledge within unfamiliar social, political, religious, and musical contexts, thereby shifting our historical analysis away from traditional scientific elites: it is to broader, often popular, communities and audiences that we need to look to understand the development of nineteenth-century science. In many respects, this was a period of what Thomas Gieryn has described as "demarcation," or "boundary work," over what constituted "science."[63] Between 1815 and 1914, natural philosophers and scientists looked to fashion sound as a scientific object and claim it firmly as a subject within the natural sciences. However, contemporaries attached varying meanings to "sound" and "music." Often, natural philosophers, experimentalists, and mathematicians, when examining physical questions of sonorous phenomena, allowed their discussions to move into musical concerns. A new theory on resonance or harmonic principles might initially be part of the science of sound but could also have ramifications for musical knowledge and practice. There was no clear demarcation between "sound" and "music," though these were clearly different concepts, contingent on the context in which they were used. Although music theorists sometimes referred to a "science" of musical composition, scientific practitioners did not, generally, deal with this element of the art, but they did comment on phenomena which carried urgency for musical practices, such as harmonic divisions and the frequency of pitch. At times, music was implicitly included within discussions on sound, but on other occasions, contemporaries considered music as a discipline within its own right. Much depended on who was using the term: a natural philosopher employing musical instruments in experimental inquiries might regard "music" as part of acoustic science, but a composer or performer could regard "musical knowledge" as something quite distinctive from the study of sonorous phenomena.

Entangled with these unstable terms, the social distinctions between different scientific and musical interlocutors were but loosely defined. The boundaries between "gentlemanly," "amateur," and "professional" within nineteenth-century scientific practice were flexible. One of this book's challenges is to identify the complexities of these interchangeable terms and social positions. The tensions over sound and music involved a contest over what sonorous and musical phenomena fell within various intellectual traditions and the credentials an individual required to be regarded as an authority within these disciplines. Charles Wheatstone, for example, was an instrument maker but also fashioned himself as an experimentalist. His scientific credentials enhanced the popularity of his novel musical instruments, but this did not mean orchestras or professional musicians adopted such devices in their performances. On the other hand, John Herschel was an astronomer but felt compelled to enter into debates over accurate musical tuning. His mathematical skill earned him respect within Britain's philosophical networks but failed to secure the approval of musical performers. New sonorous experiences and understandings exacerbated this struggle for authority between scientific, musical, and religious communities.

Central to this problem of authority was the question of audience. Religious guides speaking to their congregations exerted influence over listeners different from those at scientific lectures, while publications in elite scientific journals like the *Philosophical Transactions of the Royal Society* sought credibility from an alternate readership to those that purchased cheap popular treatises or musical and religious periodicals. Sometimes, however, these audiences overlapped. While entertaining scientific exhibitions of sonorous phenomena appealed to diverse social groups, the same arrangements could be reworked as experimental demonstrations for Britain's scientific elites. Similarly, religiously informed accounts of music could secure sympathy from scientific and nonscientific audiences alike. What was important was the context in which scientific observations were made, and this was very much a matter of audience. Michael Faraday, for instance, so celebrated for his electrical research, was by no means an expert on music, and yet within the Royal Institution and Royal Society was happy to reflect on the aesthetic beauty of musical tones. But it was hard for him to exert a similar authority in different contexts, such as with church musicians and composers. Throughout *Sound Authorities*, it is evident that the role of sound within the natural sciences was inseparable from more extensive concerns over who could be trusted with providing knowledge of nature to society.

Hearing Is Believing: Outline of Analysis

It is often said that "seeing is believing." Of course, this Biblical cliché actually emphasizes that seeing is not the most privileged form of knowing. Christ warns Thomas that "because thou hast seen me, thou hast believed; blessed are they that have not seen, and yet have believed" (John 20:29). Theologically and philosophically, visual epistemological techniques have never been sufficient to obtain truths, be it of a Creator or of nature. Throughout this book it becomes clear that in the nineteenth century, to see was not, in itself, to know. Natural philosophers relied on the ear, as well as the eye, in the construction and promotion of empirical knowledge, while musical experiences and religious convictions could also be important in the production of scientific understandings of nature.

Chapter 1 explores sound's role within the natural sciences between 1815 and 1846. In London, scientific exhibitions and spectacles competed with rival entertainments, including music halls and concerts. Popular science appealed not only to traditional social elites and the educated middle classes but to laborers, artisans, and women. Sonorous phenomena provided dramatic marvels of nature and helped inform diverse audiences of the latest philosophical findings. Experimentalists like Wheatstone and Faraday combined musical and scientific knowledge seamlessly in London's workshops, laboratories, lecture theaters, and showrooms. In particular, it was sound's transcendental character which made it such a valuable scientific resource: audiences did not require mathematical training or a scientific education to be capable of appreciating the sonorous effects and musical sounds through which nature could be investigated. This experiential quality also made the ear an invaluable organ in constructing new knowledge, especially for natural phenomena that could not be directly seen, such as electromagnetism, heat, and light. Sonorous experiments helped natural philosophers to share new ideas on how these forces operated, but they also shaped their inquiries into the workings of nature.

Sound played a part in explaining natural phenomena but, as **chapter 2** argues, it also contributed to grand unifying visions of the universe throughout the midcentury. John Herschel, Mary Somerville, and William Whewell all mobilized sound within their broad portrayals of nature in which all phenomena, both celestial and terrestrial, were connected and governed by a few permanent laws that were divinely created. Similarly, there were clear parallels between such accounts of nature and the beauty of music, which appeared to reveal the harmonious order of God's creation. It was not only that Britain's leading science writers worked within a religious framework but that religious authorities mobilized scientific knowledge within their own accounts of nature. These opening chapters illustrate how, by the middle of the century, sound was prominent in British science, and that scientific accounts of sonorous phenomena were, until the 1850s, highly compatible with broader spiritual understandings of music.

However, as becomes clear in **chapter 3**, there were growing tensions between those professing scientific and musical expertise. By investigating the attempt to regulate Britain's musical pitch, conducted in 1859 and 1860, it becomes clear how difficult it was for natural philosophers to exert influence within the nation's musical communities. Different orchestras across Europe performed music to different note pitches and, concurrently, there was a widespread fear that the frequency of these notes was increasing to such an extent that it altered historical compositions and endangered the voices of singers. When Herschel led a mathematical campaign for musical performers to adopt a unified national pitch based on what he asserted to be a natural standard, musicians, instrument makers, and composers combined to reject his proposal. They preferred a pitch that was aesthetically pleasing rather than one invested with scientific credentials. It was through these discussions over standardization that the limits of scientific authority in British musical society became startlingly evident.

The pitch controversy of 1859 and 1860 marked the realization that scientific and musical frameworks might not always be compatible, but it was not just pitch that raised concerns over science's role within music. **Chapter 4** examines how one of Britain's most respected mathematicians, the Astronomer Royal George Biddell Airy, struggled to apply mathematical solutions to musical problems. Airy wrote extensively on the theory of sound and encouraged students to engage with such practical phenomena but found it hard to build a consensus with musicians and instrument manufacturers. Music practitioners often could not agree among themselves over the note sounded by various instruments or how best to tune devices: Airy found music a frustrating business to work in. Unlike the ordered discipline of astronomy that he was accustomed to, music appeared chaotic, with the ear a difficult organ to manage. As a result, Airy came increasingly to rely on the eye for disseminating information from the Royal Observatory at Greenwich. Airy's experiences contrasted sharply with the romantic notion that music and science were united by a shared sense of order.

By the 1860s there were disagreements between mathematicians and musicians over the place of scientific knowledge within the organization and practice of music. But, as **chapter 5** demonstrates, these philosophical concerns surrounding sound and music engendered far greater theological questions over Creation and man's place in nature. Sound and music were very much part of wider controversies over biological evolution and God's role within nature. Idealized accounts of music as something ethereal and spiritual came into conflict with scientific investigations that maintained that all sonorous phenomena could be reduced to physical and physiological laws. In Britain, it was Tyndall who did most to promote such interpretations, placing them within his arguments that knowledge of nature should be removed from the purview of the Church and entrusted to scientists. Nevertheless, religious commentators continued to deliver accounts of sonorous phenomena and music to broad social audiences. Beyond the educated readers to whom leading scientific authorities spoke, the pulpit provided much more diverse social guidance. These were places of immense cultural authority, and it is here that we should look to understand the relationship between nineteenth-century science and music. While often drawing on scientific publications in sermons, for sermonizers, music and sound remained beyond material explanation: they were divinely created. Such understandings endured into the twentieth century.

As much as Britain's soundscape was profoundly different at the start of the First World War from what it had been on the eve of Waterloo, sound's place in the natural sciences had also changed dramatically. *Sound Authorities* examines and explains this change. Natural philosophers, scientists, experimentalists, and mathematicians all worked to refashion sound as a scientific product: they endeavored to take what was a theological, musical concern and transform it into a subject of empirical, physical investigation. In doing so, they raised social and political questions over authority, as they looked to reconceptualize music in terms of sensory experience and sonorous phenomena. Few questions of natural philosophy entailed such socially diverse audiences as those regarding sound, from theologians and composers to instrument makers and musicians. Without placing the sonorous within the history of nineteenth-century natural philosophy, it is difficult to fully appreciate science's troubled social value. Examined through their religious and artistic contexts, sound and music demonstrate the limits of scientific authority between 1815 and 1914.

PART I

Experiments and Mathematics
The Making of Sound as a Scientific Object

✱ CHAPTER 1 ✱

The Laboratory of Harmony
The Transformation of Sound within British Science, 1815–46

There is no truer truth obtainable by man than comes of music.
ROBERT BROWNING, *Parleying with Charles Avison*

During the 1820s and 1830s, British natural philosophers transformed sound's place in the natural sciences, refashioning the sonorous into a scientific object. Traditionally confined to the opera, concert halls, theaters, churches, streets, and taverns, harmonious sounds were now increasingly scrutinized within the laboratory and lecture theater. At once, this blurred distinctions between musical amusement and scientific instruction. London's fashionable West End was at the center of this transformation, where sonorous displays, especially those of a musical nature combined with philosophical inquiry, provided mystifying and wondrous entertainment.[1] Among the Westminster theaters and showrooms, scientific displays were popular and competed with rival sources of leisure. A new musical instrument or acoustic illusion could cause a sensation, and promoters of these performances were eager to show the philosophical implications of their work: such sonorous exhibitions helped to bring new scientific understandings to specialist and nonspecialist audiences. No one understood the value of these sonic resources better than the instrument-maker Charles Wheatstone (1802–75). During the 1820s, his mechanical creations astonished and bemused London audiences at his shop on Pall Mall. Yet by 1828 Wheatstone had changed the nature of this work. In collaboration with Michael Faraday (1791–1867), Wheatstone worked between 1828 and 1832 to turn his acoustic research into something much more philosophically significant. Together, Faraday and Wheatstone transformed London's sonorous spectacles into lectures at the Royal Institution. Employing exotic instruments, new acoustic sensations, and wonderful applications for sound, they mobilized musical resources to help popularize science. In doing so, they challenged distinctions between experiment and entertainment, transcended boundaries between scientific and popular audiences, and raised questions over the social order between natural philosophers, musicians, and instrument makers. By tracing the story of musical performances from parties and theaters into scientific lecture rooms, laboratories, and scientific journals, this chapter explores how music provided a powerful cultural resource for scientific practitioners to mobilize. In analyzing the transition of what were, originally, forms of musical entertainment, into philosophical subjects, it becomes apparent how vague the boundaries were between the popular and the scientific. Often, the audiences of West End shops and showrooms were eager for scientific spectacles and mysterious natural phenomena, while those of the fashionable scientific venues did not just want scientific discourse, but to be entertained. For both, music provided valuable resources, both marketable and educational. Likewise, the journals reporting on these exhibitions included specialist and broadly cultural publications, from *The Times* and *Literary Gazette* to the more scientifically focused *Mechanic's Magazine* and *Philosophical Magazine*.

Melissa Dickson has argued that while the construction of new acoustical knowledge depended on sight, this could be combined with sonorous experiences: to be convinced, audiences were encouraged both to see and hear sound.[2] This chapter looks at the role of sound, specifically music, in popularizing and substantiating new philosophical claims and scientific authorities. Rather than treating science simply as a sort of knowledge, this chapter also considers it as a competing form of entertainment. Wheatstone and Faraday mobilized fashionable musical resources in what was a highly competitive entertainment market, as the Royal Institution had to compete with a broad range of alternative amusements, including operas and theaters.[3] Between 1828 and 1832, this elite scientific venue was the site of a transformation in the nature of sound within British science. Yet this was not just a reorganization of how sonorous phenomena were presented. Through publications in the Royal Society's *Philosophical Transactions*, university lectures, and the experimental investigation of heat, electricity, light, and magnetism, the ear took on new significance as an organ of scientific analysis. It featured prominently in Faraday and Wheatstone's work and, increasingly, within the natural sciences across Britain. While popular accounts of acoustic experiments appealed to a general readership, these same inquiries could be refashioned for elite philosophical journals to recast sound as a scientific object. Sound exposed the vague boundaries between contrasting readerships and authorities: efforts to make the subject either popular or empirical revealed the nuances between amateur experimentalists, learned natural philosophers, instrument makers, and gentlemanly specialists. These were, of course, very fluid terms, with individuals like Faraday and Wheatstone respected both in specialist scientific networks and with polite metropolitan audiences.

This chapter explores Wheatstone's efforts to disseminate acoustic knowledge, especially that of vibrating plate experiments, as something that was valuable for the study of nature more broadly. In Britain's post-1815 experimental culture, Chladni's acoustic plates provided a rich example of how specialist and nonspecialist audiences alike could engage with new scientific ideas from the Continent. Above all, it was sound's experiential quality which made it so epistemologically valuable. It was not enough to read about sonorous phenomena on the pages of scientific treatises, no matter how accessibly written: they had to be experienced in show rooms and lecture theaters to be fully appreciated. This meant that sound was especially adept at transcending the boundaries between popular and elite science, between gentlemanly amateurs and specialist experimentalists, and between amusement and philosophical instruction. It did not require experimental or mathematical training to experience or conduct sonorous investigations. This character also made hearing valuable to the production of new knowledge: it was through both the eye and ear that nature could be known. Wheatstone and Faraday emphasized the connections between the study of sound and other phenomena, principally light and electromagnetism, conscripting their experiences of acoustics into later works on electricity and telegraphy. From metals to electricity and heat, acoustic knowledge and practices informed radical new approaches to natural philosophy, effecting a revolution in experimental techniques in which hearing was often as important as seeing.

Musical Marvels and Philosophical Exhibitions

Writing in 1832, the Edinburgh experimentalist David Brewster (1781–1868) reflected that sound was in the midst of a transformation. Historically, he observed, the manipulation of sonorous phenomena had been a powerful means of social control. The "science of *Acoustics* furnished the ancient sorcerers with some of their best deceptions," with the imitation of thunder in temples indicating the presence of a "supernatural agent" and "the vocal statue of Memnon, which began at the break of day to accost the rising sun" being a similar sonic deception. Sound's ability to conjure up spiritual feeling provided ancient governments with "influence over the human mind."[4] This metaphysical character of the sonorous had endured, with music's ethereal quality a valuable resource for religious authorities, not least within Christian traditions. Chanting and melody were central to medieval worship, including the singing of Psalms and music for mass. In 1832 such performances were still prominent within the Roman Catholic Church, while instrumental music, especially with church organs, and hymns had flourished in the Protestant Church of England during the eighteenth century. Be it for ancient sorcerers or nineteenth-century clergymen, Brewster demonstrated how sonorous phenomena had provided audiences throughout history with apparently metaphysical experiences.

Brewster was familiar with sound's religious implications. Although most famous for his invention of the kaleidoscope in 1817 and his commitment to the Newtonian conception of light as projected corpuscles, rather than undulating waves, Brewster had intended to become a minister in the Presbyterian Church of Scotland. After attending the University of Edinburgh between 1794 and 1800, he was forced to abandon his religious aspirations due to a chronic fear of public speaking.[5] Yet at the heart of Brewster's science was a Calvinistic conviction that to experiment was to reveal the truth of God's creation. An understanding of the material world which was grounded in sensual experiences, including those sonorous, exposed the divine power of its author. Brewster, however, was aware that the place of sound in natural philosophy was undergoing a rapid transformation: it was becoming a scientific object.

Sound had long been a subject of philosophical inquiry but, as Brewster asserted, sonorous knowledge was permeating the boundaries between scientific discourse and popular entertainment. Natural philosophers were using acoustic marvels to captivate broad audiences and impart new knowledge of nature. No example of this was of greater celebrity than the French inventor M. Charles's "Invisible Girl."[6] As early as 1807, spectators had been amazed by the voice of a small girl, able to answer questions, sing beautifully, converse in several languages, and make observations on members of the audience, emanating from a series of four trumpets and a ball, suspended by ribbons from a frame made of railings and posts (fig. 1.1). The faint sound of the girl appeared to come out of the trumpets which were suspended in midair, leaving audiences "in utter amazement." Brewster explained that this illusion went beyond all other acoustic marvels because, unlike the talking heads of ancient times, the sound-producing ball and trumpets "communicated with nothing through which sound could be conveyed." Yet this was a simple deception. In two of the horizontal railings opposite the trumpet mouths there was an aperture joined to a pipe connected to a vertical post which then ran down through the floor and into a room next door,

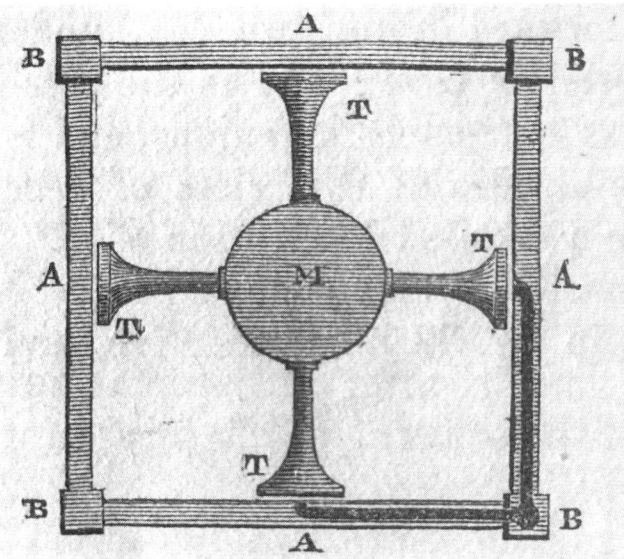

FIGURE 1.1 The arrangement of four trumpets to project the voice of the "Invisible Girl." In the bottom right corner, a hidden tube was positioned through which sound was conveyed to and from a woman hidden in the next room. Brewster, *Letters on Natural Magic*, 162. Image by permission of Edinburgh University Library Centre for Special Collections, 2019.

where the "invisible lady sat" (fig. 1.2). From there she could speak and listen through the trumpet. Sound, then, was capable of arousing superstition within even the most rational of minds.[7]

Yet of all sounds, Brewster argued that it was those harmonic which produced the most entrancing influence over listeners. He explained that increasing knowledge over how a string vibrated to create harmonics, the production of concordant sounds by vibrating columns of air, and the vibration of solid bodies, were all contributing to acoustic effects of increasing wonder. The precise mechanisms of these processes were, Brewster maintained, known only to philosophers; as an example, he invoked Ernst Chladni's (1756–1827) production of nodal lines with sand distributed over vibrating glass plates to demonstrate the oscillations of solid bodies (fig. 1.3). As Brewster concluded, "Among the discoveries of modern science there are few more remarkable than those which relate to the production of harmonic sounds. We are all familiar with the effects of musical instruments, from the deep-toned voice of the organ to the wiry shrill of the Jew's harp. We sit entranced under their magical influence."[8]

In 1820s London, the place of sound within scientific and popular cul-

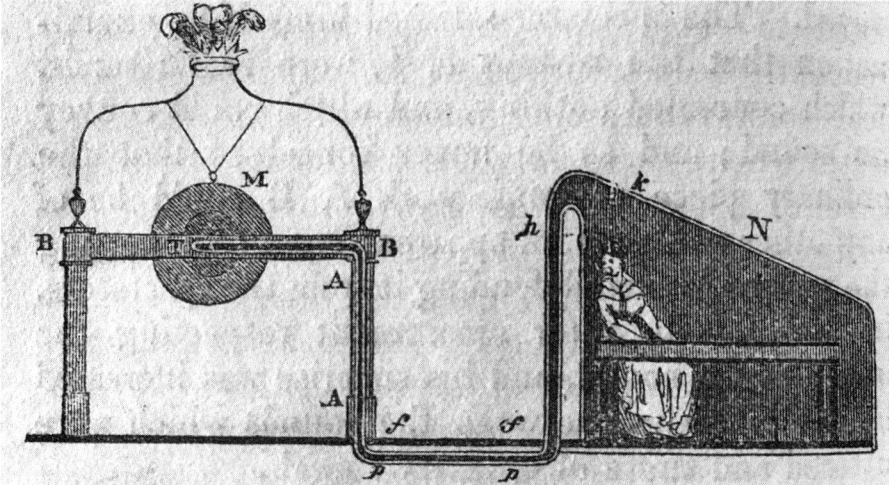

FIGURE 1.2 Brewster's illustration of the "Invisible Girl," showing the route that sound took between a hidden woman and the apparatus that mysteriously projected the voice of a small girl. Brewster, *Letters on Natural Magic*, 164. Image by permission of Edinburgh University Library Centre for Special Collections, 2019.

ture had indeed changed. Brewster invoked two acoustic marvels which had, during the previous decade, aroused sensation. The first of these was Wheatstone's "kaleidophone," named after Brewster's own kaleidoscope. The second was the Jew's harp. Both were typical of an experimental culture in which sound, specifically music, took on prominence both as an investigatory tool and a popularizing medium. In the early nineteenth century, the Jew's harp was rarely considered a musical instrument.[9] About two to three inches in length, consisting of a metal frame of various shapes and with a thin metal tongue suspended through its center, this device was popularly conceived of as a children's toy. To play, the frame had to be brought in contact with the teeth, while a finger put the metal tongue into vibration. The sound produced, Brewster's "wiry shrill," was commonly thought lamentable. So his reference to this as a wonder of harmony is curious and almost certainly made relating to the dramatic musical career of Karl Eulenstein (1802–90), whose transformation of the Jew's harp from irritant to instrument captivated scientific and musical audiences alike between 1826 and 1834.

Initially, Eulenstein's career was not scientific but the subject of popular publications and socially elite musical performances. His biography was one easily romanticized for readers of local and national newspapers. A popular narrative of Eulenstein's life detailed how he was born

FIGURE 1.3 Ernst Chladni (1756–1827). Tyndall, *Sound*, frontispiece. Public domain.

in 1802 in Heilbronn, Württemberg. Aged six, he had inherited a violin and a passion for music from his father, but his mother was loath to see her son pursue a future of "idleness and poverty," while his wider family, haunted by images of "the lean figures of poverty-stricken music-professors," secured him work in a shop. Having a hatred of music, the shop's mistress banned him from playing any instrument. For a while he played his violin secretly in his room, but his downfall came on purchasing a French horn. After much resistance, "one fine evening, the spirit of blowing the French-horn came upon him . . . the temptation was too great."[10] On hearing the instrument, the shopkeeper confiscated both of his instruments. Eulenstein found refuge with the Jew's harp. On sale for a penny in the shop, and popular among local school boys, he took a few to his room for "experiments" and discovered amazing sounds hidden within the toy. He played in secret, hidden in his bed, until one night he fell asleep holding the instrument and woke with his face cut by the harp's sharp metal tongue. To avoid a repeat, he covered the tongue's end with a knob of sealing wax, discovering that he could tune the harp accurately to a note and, by using different amounts of wax, produce a set of harps covering several octaves. The "discovery and the inference were, for a mere boy, alike wonderful; and his skillful and really scientific use of both were not less so," observed one British journalist.[11] Eulenstein learnt to move swiftly between each harp, constructing fine tunes and developing a model of how to perform with the instruments. By first giving a musical performance in the dark, keeping the identity of the instruments secret, and then lighting candles to reveal the harps, he thus provoked wonder that beautiful music could be produced by such devices.[12]

Armed with sixteen waxed harps, Eulenstein set off to make his fortune, finding success in Stuttgart and then before the Queen of Württemberg, who advised him to visit Paris. In February 1826 *London's Morning Post* reported how Eulenstein was "at present astonishing all Paris" with the elegance of his Jew's harp and the speed with which he moved between his sixteen instruments.[13] There he met the celebrated British admiral, Sidney Smith (1764–1840), who gave him letters of introduction and, after astounding King Charles X and the Duc d'Orléans with a performance, Eulenstein traveled to London. After an introduction to the Marchioness of Salisbury, he performed before a party of three hundred guests, but his "tiny instruments with their ethereal softness ... produced no audible sound." Fearing that "all London routs were noisy," he almost returned to Paris before the Duke of Gordon offered him patronage after reading of his exploits in *the Literary Gazette*. The duke secured him silent audiences, including one before King George IV, and Eulenstein soon became a musical celebrity with fashionable London elites. In 1828 his career almost came to an end, as the "iron of the Jew's harp had affected his teeth, and produced a general decay." The vibrating tongue caused such pain that it was only thanks to a dentist who "ingeniously contrived a glutinous case or covering" that his career could continue.[14]

Accounts of Eulenstein's music were full of hyperbole and romanticizing rhetoric, portraying his art as something nearing magic and pandering to the British public's hunger for sensation. His biographer described how the sound was reminiscent of a violin, combined with "a tiny bell," as "fairy strains creep about the room in undulating harmony, now nearer, now more distant, as if a moving chorus of any little beings, in their tenderest mood, were delighting themselves in song."[15] Another of his contemporaries, on hearing his harps, captured this sense of wonderment: "These tones are not of this world, They are the sounds of gentle spirits Which God entrusted for your consecration, Harmonies from the tent of stars."[16] Characterizing these romanticized depictions of Eulenstein's music was a shared sense of mystery regarding the sounds emanating from the harps. Although such accounts of Eulenstein's art were aimed at a broad readership with little or no scientific expertise, the nature of his performances began to take on new value following a demonstration at the Royal Institution. It was here in 1828 that "scientific men investigated the theory of his instrument, the effects of which were quite new to them." In front of more than five hundred attendees, Eulenstein's music was incorporated into a scientific paper; his credibility and fame boomed after playing at this venue, before "all the *savans* of London."[17] Eulenstein increasingly emphasized that his performances contributed to making music scientific. At a party, one guest inquired how Eulenstein was able to get music from two instruments at the same time, while another admired that he "must have been indefatigable . . . to have acquired such perfection on so unpromising an instrument." After appeals for an explanation, Eulenstein attributed his skill to "the remarkable order and uniformity which exist among harmonic bodies," which placed music "among the exact sciences."[18]

In 1830 Eulenstein toured from Cheltenham to Gloucester, Worcester, Birmingham, and Edinburgh. The *Caledonian Mercury* anticipated his arrival in Scotland with excitement, reporting that it was "impossible to imagine anything so beautiful from such means; and when we also find that science is deeply indebted to these charming vibrations for the advancement of a very interesting inquiry, we are the more inclined to point out the subject for public notice."[19] The timing of this performance in Edinburgh, less than two years before the publication of Brewster's *Letters*, was significant: Eulenstein's art appealed to scientific and nonscientific audiences alike. Wherever Eulenstein traveled, he evoked exactly the sort of wonder which Brewster argued was so important to the cultivation of natural philosophy. The *Liverpool Mercury* informed readers in 1832 that Eulenstein was in the city, bringing with him his "magic strains." The Liverpool reporter confessed to being skeptical until witnessing the musician execute tunes "as perfectly as Paganini himself could execute them upon his violin." As much as fine music, this was a philosophical demonstration because, as the *Liverpool Mercury* put it, "the German musicians are musical philosophers."[20] Eulenstein understood harmonics, the action of air, and the chromatic scale, therefore combining musical and philosophical knowledge to produce a magical acoustical effect. A large part of the wonder of the Jew's harp was that Eulenstein had employed philosophical knowledge, rather than mechanical craft, to transform a grotesque toy into a beautiful instrument. This transcending of any distinctions between philosopher and musician appealed to the popular press's readership, which was eager for both amusement and improvement.

A year later, in 1833, Eulenstein visited Bristol where audiences were again left satisfied, with the *Bristol Mercury* providing a euphoric account of his first concert in Clifton. Eulenstein displayed an "astonishing facility of execution," producing chords and harmony of unparalleled beauty which united the softness of a flute with the tone of a trumpet and the richness of an organ. This effect was, the journal admitted, impossible to comprehend without witnessing.[21] As one Bristolian correspondent put it, this was a heroic story of a musician, denied his violin, "who yet found an instrument for himself, one all but tuneless in other hands, and, by stealth, acquired the power on that instrument a music of such excellence, science, taste, and beauty, that princes and nobles came to be his patrons."[22] That this instrument was apparently "tuneless" to all others very much contributed to Eulenstein's success. Combined with his heroic biography and the philosophical implications of his performances, this made for a sensation. It was, however, as a demonstration within the lectures of Charles Wheatstone that Eulenstein's music would take on its most significant philosophical value. Few were so instrumental in refashioning sound's position in popular and scientific culture as Wheatstone, and, within this work, he would radically change the nature of Eulenstein's musical performances into empirical exhibitions of an almost experimental character.

Born in Gloucester, Wheatstone had worked in his father's musical instrument shop at 128 Pall Mall, in the center of London's West End, since 1806. There he made what his sister described as "contrivances and experiments . . . to amuse and puzzle the gaping groups that gathered at the shop window."[23] Despite such entertaining projects, Wheatstone harbored serious philosophical ambitions. After beginning an apprenticeship in his uncle's musical instrument manufacturing business at 436 the Strand in 1816, Wheatstone purchased a book on Alessandro Volta's (1745–1827) electrical experiments. He subsequently began replicating Chladni's sand experiments with water and, in 1819, exhibited "his first practical experiments in acoustics in Pall Mall."[24] Wheatstone's first scientific publication, "New Experiments on Sound," appeared in *Thomson's Annals of Philosophy* in 1823 and combined his musical knowledge with philosophical inquiry. This paper divided sound, produced from oscillating bodies, into transversal vibrations, moving at right angles to the axis of a body, such as wires or forks, and longitudinal vibrations, moving in the direction of the axis, such as in liquids and aeriform fluids. The calculation of the movement of waves was made by finding the "nodal lines" of a vibrating body; that is, the part where the oscillation was least. The evidence for this was to be found through an adaption of Chladni's experiments. A plate glass covered with water and vibrated by a violin bow created beautiful patterns of oscillating particles. As the sound became more acute, so the centers of vibration became more numerous. While Chladni had used sand, Wheatstone preferred water, believing it showed vibrations with greater accuracy, but he also reported trying Chladni's experiments with mercury and oil in a glass vessel. Longitudinal oscillations could be examined by blowing a flute on to a surface of water to produce what he termed "crispations." Together, he felt that these experiments established the existence of molecular vibrations and offered ways of exploring the physical properties of different sounds. Though Chladni was the first to observe vibrations by these techniques, Wheatstone claimed to have applied a more rigorous experimental method to the phenomenon. He communicated his results to the Danish chemist Hans Christian Ørsted (1777–1851), who had conducted similar trials with a fine powder, publishing the results in Lieber's *History of Natural Philosophy* in 1813.[25]

Despite these philosophical contributions, Wheatstone remained primarily occupied with running his Pall Mall shop, having inherited it on his uncle's death in 1823, along with the family workshop at 128 Pall Mall. At the workshop, he designed mechanical and musical devices for fashionable West End audiences, eager for popular entertainment. Wheatstone's earliest celebrity came during the early 1820s with the exhibition of his "Enchanted Lyre." In 1821 Wheatstone displayed an antique lyre in his shop, which seemed to have no wires or strings attached, apparently floating below the ceiling, above a small surrounding fence. In the spring of 1822 these demonstrations were repeated in a shop along the Royal Opera Arcade where audiences could enjoy an hour's orchestral entertainment for five shillings. In the same room, the Enchanted Lyre was followed by a performance of the Invisible Girl.[26] *The literary New Monthly Magazine* announced that the exhibition had "excited considerable sensation among the lovers of music, and the curious in matters relating to natural philosophy." Suspended in air, the antique lyre appeared to perform beautiful music of its own intuition (fig. 1.4). The journal's anonymous writer joined other observers in being

> convinced at once that it is not the lyre which gives us the musical treat, but that a skillful player is somewhere else occupied in entertain-

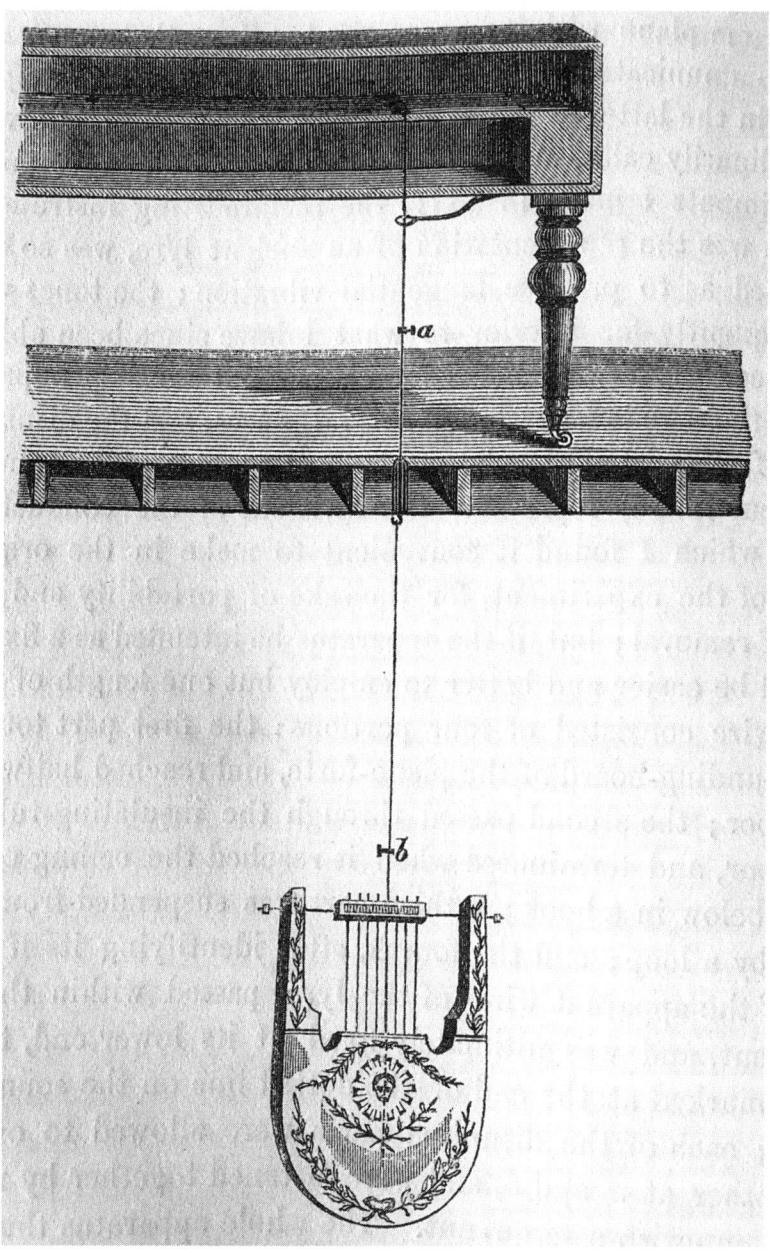

FIGURE 1.4 Wheatstone's arrangement for the Enchanted Lyre, which appeared to project music while suspended from a ceiling. The music was in fact played on an instrument, such as a piano, the soundboard of which was in contact with the lyre by either a wire or deal rod. Wheatstone, *Scientific Papers of Sir Charles Wheatstone*, 56. Public domain.

ing and puzzling us. Nevertheless, on approaching the lyre, and holding the ear close to it, we are equally assured that the sound proceeds from the belly of the lyre itself . . . we are left to conclude that the sound is conducted into the lyre; but the means of this harmonic introduction have as yet eluded the most minute investigation.[27]

This was not an exhibition for acoustic specialists and gentlemen philosophers; it was aimed at popular audiences, eager for mystery and amusement. Nevertheless, it drew on Wheatstone's inquiries into the transmission of vibrations through solid bodies, particularly their movement through rods which, though usually inaudible, could be heard if connected to a soundboard. This principle was behind his Enchanted Lyre experiments, which Wheatstone boasted were "universally imagined to be one of the highest efforts of ingenuity in musical mechanism."[28]

Wheatstone did not publish an account of how the Enchanted Lyre worked until 1823. For two years, this amusement captivated fashionable London viewers, including composer Thomas Busby (1754–1838). A leading musical journalist, Busby received a doctorate in music from Cambridge in 1801 before launching a series of unsuccessful journals to inform London readers of the latest Continental developments in music, including *the Monthly Musical Journal*, England's first musical periodical. Busby promoted the scientific study of music, believing that harmonics was a subject comparable to astronomy or mechanics, asserting in 1818 that "Music is a science having sounds for its element. . . . By the theory of Music, we are enabled to measure the intervals between sonorous impressions; and so to understand the principles of *melody* and *harmony*."[29] Wheatstone's unification of natural philosophy and music was exactly the sort of work which appealed to Busby.

Although ostensibly a commercial stunt, Busby portrayed Wheatstone's demonstration more as a scientific experiment than frivolous exhibition. After witnessing the Enchanted Lyre in 1822, he reported how London's musical society was all in excitement because of "one of the most beautiful and striking experiments that has ever been witnessed, in the philosophy of sound," reflecting "much credit on the science and talents of Mr C. Wheatstone." The hour-long performance included several famous compositions and began with Wheatstone applying a key to the instrument, "giving it a few turns representative of the act of winding it up."[30] Indeed, this theatrical winding up contributed to the device's mystery, with the process "evidently intended for the sake of illusion." Wheatstone's lyre was, in short, "an experiment . . . of high interest, both in a musical and scientific point of view."[31] Another journal reviewer detailed how Wheatstone invited skeptical observers

> to stoop under the fence, and hold their ear close to the belly of the lyre; and they, including ourselves, are compelled to admit, that the sound appears to be within the instrument.... the attentive auditor is instantly convinced that the music is not the effect of mechanism (a fact indeed which Mr Wheatstone not only concedes, but openly avows, even in his notice).[32]

Importantly, the music was of great beauty, including a performance of Daniel Steibelt's (1765–1823) battle piece, with audiences seemingly enjoying the sounds produced.

Yet how the lyre operated remained unknown. A correspondent from *the Literary Gazette* explained that this secrecy added pleasure to the mystery but was also convinced that the display was a mechanical trick rather than the operation of natural phenomena. Nevertheless, he reported that Wheatstone

> disclaimed mechanism altogether... and asserts that the performance of the Enchanted Lyre is entirely the result of a new combination of powers. Be that as it may, its execution is both brilliant and beautiful. The music *seems to proceed from it*; the tones are very soft; the expression soft or powerful.[33]

Although these descriptions featured in popular journals, aimed at audiences with broad artistic and literary interests, it is clear that there was a market for scientific curiosities.

Wheatstone's Enchanted Lyre demonstrations continued throughout the 1820s, becoming a staple of London's entertainment scene and fueling visions of a revolution in public music. In 1825 one commentator, having witnessed the lyre, speculated that before long a single performer would be able to command fifty instruments across the city. With such capabilities, he expected "ere long, to have music laid on as the gas and water-pipes are now furnished through the streets, and at a probably still lower rate."[34] Arguably it was the wonderful nature of his performances which persuaded viewers of the practicality of Wheatstone's communication promises. As one journalist put it,

> Water, earth, fire are already meandering under our footsteps in every street in the metropolis. Air was only wanting to complete the circu-

lation and conduit of all the four elements. Mr Wheatstone's conductors of sound may be considered in the latter light. Who knows but by this means the music of an opera performed at the King's Theatre may ere long be simultaneously enjoyed at the Hanover-square Rooms, the City of London Tavern, and even at the Horns Tavern in Kennington, the sound travelling, like the gas, through snug conductors, from the main laboratory of harmony in the Haymarket, to distant parts of the metropolis.[35]

This was a grand vision which promised to mobilize elite philosophical knowledge for mass entertainment. Wheatstone combined music and natural philosophy, regulated from a "laboratory of harmony," to deliver a communications revolution.

Wheatstone's Enchanted Lyre was not his only popular contrivance. Equally sensational was his "kaleidophone," consisting of a nine-inch circular board with three perpendicular steel rods (fig. 1.5). The first rod had a spherical bead head to reflect light, while the second had a moving plate instead of a bead, and the third suspended a four-sided prism. A fourth rod, bent at a right angle, carried another bead. These could be put into vibration with a bow or hammer and the reflection of the different media displayed its orbit. Each rod's pitch could be increased if vibrated at a higher speed, revealing smaller excursions of reflected light. The best effects were achieved by ensuring that just a single light ray focused on the rod's vibrating head. If several of the rods vibrated concurrently, Wheatstone observed that he could create "symmetrical figures, resembling elegant specimens of engine-turning." This was, Wheatstone declared, an "application of the principles of science to ornamental and amusing purposes" which rendered "them extensively popular; for the exhibition of striking experiments induces the observer to investigate their causes." He boasted that this amusing device made "obvious to the common observer what has hitherto been confined to the calculations of the mathematician; it presents another proof, that however remote from common observation the operations of nature may be, the most beautiful order and symmetry prevail through all."[36] The English natural philosopher Thomas Young (1773–1829) had first shown this principle in 1800 with the strings of a piano, around which he wound wire, and then contracted light from a window on to the wire. When the piano was played, the reflected light made visible the wire's movements. In this way, the vibrations which Chladni had shown through his acoustic figures were made observable in a musical instrument.

Although Wheatstone intended his kaleidophone as a contrivance for

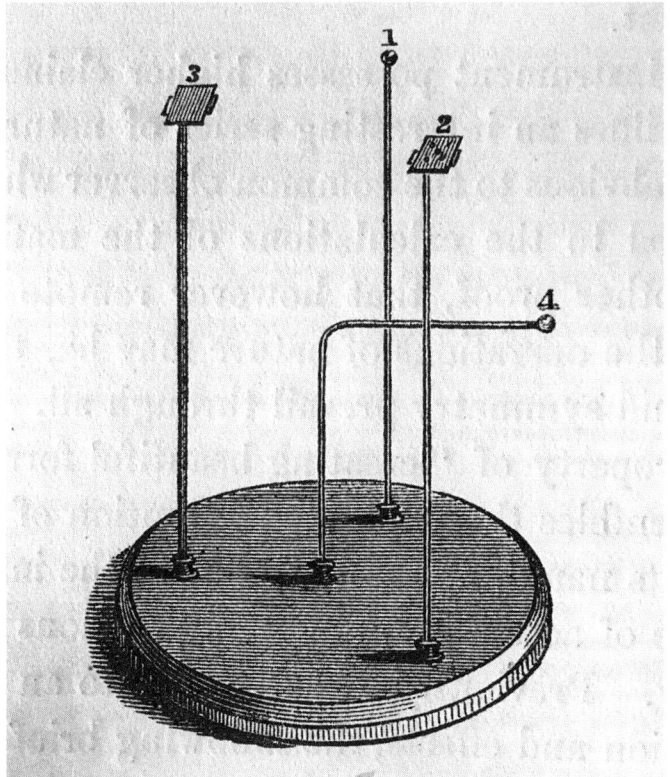

FIGURE 1.5 Wheatstone's kaleidophone, exhibiting a selection of rods suspending various media, including a bead, prism, and mirror, which reflected the oscillations of the rods when put into vibration. Wheatstone, *Scientific Papers of Sir Charles Wheatstone*, 22. Public domain.

popularizing acoustic knowledge beyond scientific audiences, the technical press picked up on the instrument's philosophical implications. In August 1827, images of the kaleidophone's acoustic figures made the front cover of the *Mechanic's Magazine* (fig. 1.6). The specialist journal reported that Wheatstone's instrument made visible the immense diversity of vibrations, revealing the beauty and order of nature. This ocular demonstration of "the orbits of paths" of vibrating rods promised to open up many new principles in science.[37] Through his Enchanted Lyre and kaleidophone, Wheatstone was at the forefront of bringing new forms of sonorous analysis to popular audiences in 1820s London: such demonstrations were aimed both at fashionable West End audiences and learned scientific specialists. His replication of Chladni's vibrating sand experiments and novel instruments entertained listeners while simultaneously

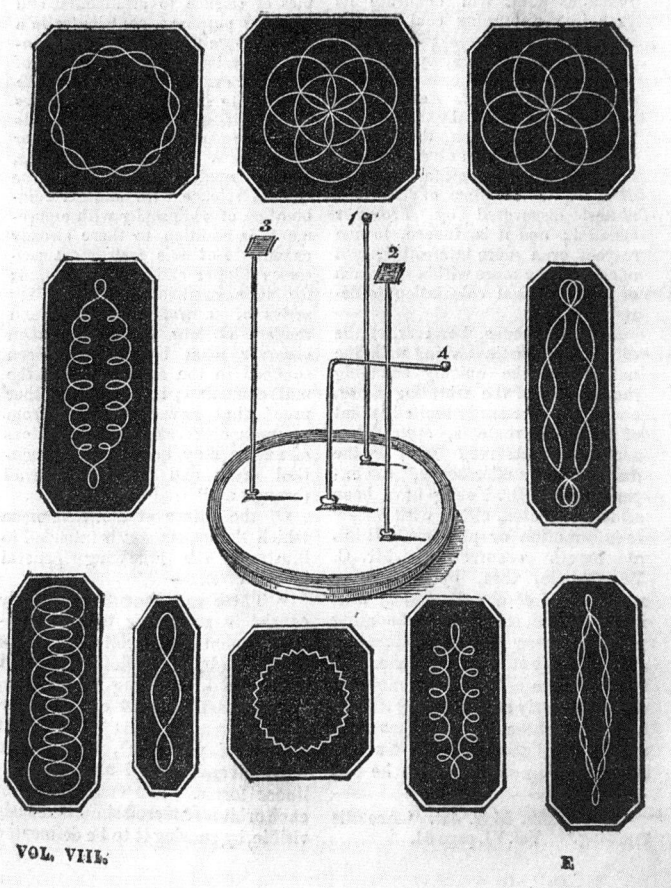

FIGURE 1.6 The kaleidophone on the front cover of the *Mechanics' Magazine*, displaying both the instrument and its figures. "The Kaleidophone, or Phonic Kaleidoscope," *Mechanics' Magazine* 8 (August 11, 1827): 49–52, 49. Image by permission of Edinburgh University Library Centre for Special Collections, 2019.

encouraging them to think of sound as a movement of waves.[38] Wheatstone was actively refashioning sound from something aesthetically pleasing to a scientific object.

A Sound Revolution at the Royal Institution

Wheatstone brought new acoustical phenomena to London audiences, but it was through his collaboration with Michael Faraday that he was to orchestrate a transformation of sonorous phenomena intended for popular amusement, including those musical, into objects of scientific discourse. Together, Faraday and Wheatstone would move the sounds of West End shops and showrooms to London's elite laboratories and lecture theaters. With Wheatstone working at Pall Mall and, eventually, King's College London, he was never far from Faraday, who lived and worked at the Royal Institution on Albemarle Street and was, from its foundation in 1824, secretary of the Athenaeum Club at 107 Pall Mall. Faraday and Wheatstone's association began sometime around 1825; they probably met initially at one of Wheatstone's musical performances.[39] Between 1828 and 1832, Wheatstone and Faraday refashioned the musical demonstrations of West End showrooms and popular journals, translating them to the Royal Institution's lecture theater and the pages of the Royal Society's *Philosophical Transactions*.

Originally apprenticed to a bookbinder, Faraday became Humphry Davy's (1778–1829) chemical assistant at the Royal Institution in 1813. Davy had worked at the Royal Institution since 1801, but in October 1812 he damaged his eye when a mixture of nitrate and chlorine exploded in his face, and so he temporarily employed Faraday as his amanuensis. After gaining valuable experience of experimental practices in this position, Faraday became superintendent of the laboratory at the Royal Institution in 1821 and director in 1825. Faraday secured fame for his celebrated observation of electromagnetic rotation in 1821, building on Ørsted's recent discovery that when electricity passed through a wire parallel to a compass needle, the needle turned at an angle toward the wire. Faraday found that by passing a current through a wire suspended about a magnet, he could get the wire to oscillate around the magnet's pole.[40] However, he devised his most renowned experiment in 1831. While the effect of an electrical impulse on a magnet was well known, how to combine movement and magnetism to induce electricity was a mystery. Faraday eventually produced the phenomenon by passing a magnet through a coil of wire, marking the start of over twenty-five years of research into electricity.

A flautist since childhood, music was always important to Faraday.

Throughout the 1810s and 1820s he attended "a small party for the practice of vocal music once a week at a friend's house" where he apparently "sang bass correctly, both as to time and tune." He even took music into the laboratory. In 1813, while assisting Davy with some highly dangerous experiments on explosive chloride of nitrogen, in which both experimentalists were injured, Faraday offered moments of solace, "occasionally with a song, and in the evening playing tunes on his flute."[41] In the same year, Faraday reflected on the importance of music to his well-being, writing to Benjamin Abbott that,

> "He that hath not music in his heart &c" confound the music say I.— it turns my thoughts quite round or rather half way round from the letter[.] You must know sir that there is a grand party dinner at Jaques hotel which immediately faces the back of the institution and the music is so excellent that I cannot for the life of me help running at every new piece they play to the window to hear them—I shall do no good at this letter tonight and so will get to bed and "listen listen to the voice of" bassoons violins clarinets trumpets serpents and all the other accessories to good music.[42]

Faraday was especially keen on the opera. When he accompanied Davy and his wife, Jane, on their Continental tour between 1813 and 1815, he went to the opera in Paris at the Salle Louvois but recorded that he was "not particularly gratified" owing to his lack of French and "want of goût."[43] He was more impressed in Genoa, where he witnessed "the performance of a difficult piece of singing," followed by an eccentric dramatic audience response in which

> a shower of printed papers descended from the top of the theatre amongst the audience. Some of them were copies of the piece just sung and others were verses in praise of the actress. . . . After the shower of papers several pigeons were thrown one by one from the top of the theatre into the pit and some of them suffered cruel deaths.[44]

Throughout his career, Faraday remained a regular at the Royal Opera House in Covent Garden, where his acquaintance with its manager, Frederick Gye (1810–78), provided a reliable source of tickets. Throughout the 1850s he enjoyed performances of Beethoven's *Fidelio*, Rossini's *Barber of Seville*, and Meyerbeer's *Les Huguenots* and *L'Etoile du Nord*, before taking his niece to see the great Italian soprano Giulia Grisi in Donizetti's *Lucrezia Borgia* in June 1858.[45]

Nature's sounds were also of interest, with thunderstorms making frequent subjects for correspondence between Faraday and his scientific associates. John Tyndall (1820–93) wrote to Faraday in 1855, reporting a storm of such power that it could be heard all over London. Tyndall described an acoustical wonder, noting that the "flashes and peals grew nearer. . . . Louder rolled the cannonade and one crash broke above us which almost seemed intended to shiver the building to atoms."[46] Six years later, George Biddell Airy (1801–92), the Astronomer Royal, wrote to Faraday recalling two recent thunderstorms. In the first, he reported a "thunder-clap which commenced by an instantaneous very heavy singing sound which gave the idea that an Armstrong shot of enormous size had struck." In the second, each peal of thunder began with a low noise, followed by a heavy echo; an effect Airy found hard to explain and requested Faraday's opinion on the matter.[47] Thunderstorms were intriguing acoustic spectacles for natural philosophers.

It is unsurprising then that Wheatstone's sonorous exhibitions piqued Faraday's curiosity. Yet before his association with Wheatstone there is little to suggest Faraday regarded acoustic phenomena as a subject worthy of experimental attention. The single exception was Faraday's inquiry into the sounds produced from flames in glass tubes during late 1817. At an evening meeting of members and friends of the Royal Institution, Faraday had been persuaded to try to explain the sounds produced by a burning jet of hydrogen gas in an open-ended glass tube. This phenomenon had been observed in England and Italy during the eighteenth century, amid speculation that the effect was unique to hydrogen, with sound caused by the release of an aqueous vapor. Instead, Faraday contended that the cause was the vibration of the flame and could result from all combustible gases. To undermine notions of the mysterious vapor, Faraday heated the whole tube above 212°, at which temperature sound was still heard without the chance of any aqueous substance remaining. He then employed a cracked tube, producing a similar sound, from which he deduced that the sound was not due to the vibration of the tube itself. By producing sound from other gases he also established that hydrogen was not unique in creating a "singing" flame.[48] He concluded that the noise resulted from the continuing explosion of gas and air.

These singing flame experiments represented Faraday's only substantial acoustic investigations prior to his association with Wheatstone: he had little experimental interest in acoustics in the early 1820s. This was confirmed when, in early 1826, he privately listed all the subjects he felt suitable for lectures, "to illustrate popular subjects" and connect "facts generally separated." These included papers on the atmosphere, water, gases, acids, alkalines, electricity, steam, metals, magnetism, and heat, but omitted any reference to sound. He also considered "future experiments and illustrations," but once again with little sonorous interest aside from a few references to a demonstration on the "singing of a sugar pan furnace" and vibration experiments in connection with the production of high-quality optical glass.[49] In this context, Faraday's collaboration with Wheatstone is significant: it was Wheatstone who raised Faraday's sonorous concerns, but it was Faraday who proved crucial to refashioning Wheatstone's popular acoustic amusements for London's elite scientific audiences.

In 1825 Faraday had initiated the Royal Institution's Friday Evening Discourses, which novelist George Eliot later described as becoming "as fashionable an amusement as the opera."[50] It was through these fashionable gatherings that Wheatstone and Faraday oversaw a brief, but intense, focus on sound. Between 1828 and 1832, no fewer than thirteen Friday lectures were focused on sound-related subjects, with Faraday and Wheatstone delivering twelve of these (table 1). The Royal Institution had occasionally held lectures on acoustics before Wheatstone's demonstrations. In March 1820, for instance, John Millington (1779–1868) lectured on "Pneumatics and Acoustics; of the Nature and Cause of Winds, and Their Variations; Theory of Sounds and Echoes; Description of the Air; of Acoustic Deceptions."[51] However, under Faraday and Wheatstone, interest in sound intensified. Over ten years, from 1828 until 1838, there were seventeen papers on sound. This is quite startling considering that before 1828, and from 1838 until 1849, not a single Friday Lecture analyzed an acoustic subject. For four years, the Royal Institution was home to a radical reappraisal of sound's philosophical significance.

Usually including thirty to forty experimental demonstrations, Friday Evening Discourses were not just scientific but fashionable social occasions for metropolitan elites: music and musical instruments made an obviously entertaining subject for these audiences.[52] The Royal Institution provided Faraday with the city's premier laboratory space, a lecture theater, and access to expensive materials. It was not until the late 1820s that this exclusivity changed with the emergence of University College London and King's College London, both eventually building laboratories of their own. The Royal Institution's eighty-square-meter basement laboratory, with a furnace, gas supply, and mineralogical collection, became Faraday's own private space, while he lived above the theater with his wife Sarah.[53] Faraday's skill in making experimental displays appear to reveal natural truths ensured Royal Institution lectures were popular. Clearly, experimental displays had to be organized and rehearsed, but

TABLE 1. Royal Institution Friday lectures on sound-related subjects, 1825–54

Speaker	Title	Date
Michael Faraday	Resonance or the Reciprocation of Sound Illustrated by Experiments and Some Novel Musical Performance	February 15, 1828
Faraday	Supplementary Remarks on the Reciprocation of Sound	March 7, 1828
Faraday	On the Production of Musical Sounds and on a New Musical Instrument	May 9, 1828
Faraday	Phonics - Resonances of Volumes of Air	April 3, 1829
Benjamin Humphrey Smart	The Audible Qualities of Speech	May 1, 1829
Faraday	On the Nodal Figures of Vibrating Surfaces	May 22, 1829
Faraday	Transmission of Musical Sounds through Solid Conductors and Their Subsequent Reciprocation	March 5, 1830
Faraday	The Application of a New Principle in the Construction of Musical Instruments	May 21, 1830
Faraday	The Laws of Co-Existing Vibrations in Strings and Rods	June 11, 1830
Faraday	Mr. Trevellyan's [sic] Recent Experiments on the Production of Sound during the Conduction of Heat	April 29, 1831
Faraday	The Arrangements Assumed by Particles on the Surfaces of Vibrating Elastic Bodies	June 10, 1831
Charles Wheatstone	On the Vibrations of the Air in Cylindrical and Conical Tubes	March 16, 1832
Faraday	The Crispations of Fluids Lying on Vibrating Surfaces	May 18, 1832
Wheatstone	The Construction of Speaking Machines	May 8, 1835
Wheatstone	The Means of Investigating the Structure of Crystallised Bodies by Their Sonorous Vibrations	March 18, 1836
William Ritchie	The Velocity of Sound and the Discrepancy between Theory and Observation	February 10, 1837
William Edward Hickson	The Introduction of Vocal Music as a Branch of National Education with Illustrations	May 4, 1838
Thomas Joseph Pettigrew	The Mechanism and Functions of the Organ of Voice in Man, with the Introduction of a Case of Double Utterance	May 11, 1849
John Tyndall	On the Vibrations and Tones Produced by the Contact of Bodies Having Different Temperatures	January 27, 1854

Source: Royal Institution lists.

Faraday was especially good at making phenomena appear to exist independent of laboratory contrivances, transferring basement experiments to the first-floor theater. In contrast, Wheatstone was a shy man, who struggled to speak in public despite being lucid in private company. It was therefore convenient for Faraday to act as Wheatstone's "voice" by delivering his papers.[54]

Wheatstone's Royal Institution lectures included musical performances, brought straight from his Pall Mall shop and replicated in the style of experimental demonstrations, to substantiate his scientific claims. In February 1828, Faraday read Wheatstone's first of these, on the reciprocated vibrations of columns of air, outlining how elastic bodies vibrated to produce sound. His paper investigated these laws of resonance or, rather, the phonic reciprocation of columns of air. The first experiment involved the sounding of a tuning fork over the embouchure of a flute; if the note selected on the flute was the same as, or a multiple of, the fork's fundamental note, this augmented the sound. The demonstration was repeated with resonating bottles, tuned to correspond to different forks by containing different volumes of water. Faraday then exhibited several novel instruments. First, he showed a "Gender," a Javanese musical instrument which used resonant columns of air to augment the sounds of its vibrating metallic plates. There were eleven plates, suspended by strings, under which each had its own tube of bamboo, of a length to reciprocate the lowest sound of each individual plate.[55] Though there was no such instrument in Europe, similar devices had been found over Asia and Africa.

It was not only the exoticism of the gender which impressed the audience. One reviewer noted how this lecture was "illustrated by some striking experiments, by many curious instruments of music . . . and by some very novel and curious musical performances on the Jew's-harp."[56] Wheatstone employed Eulenstein and his Jew's harp to demonstrate the relationship between vibrating bodies and their potential to put columns of air into reciprocating vibration. The harp's tongue produced vibrations of a low sound which could be made to correspond with the cavity of the mouth by altering the form of the tongue and lips. As Wheatstone explained, "when the number of vibrations of the contained volume of air is any multiple of the original vibrations of the tongue, a sound is produced, corresponding to the modification of the oral cavity." He went on to explain how the instrument could produce sounds relative to multiples of its original note.[57] By explaining the sonorous phenomena behind Eulenstein's music, Wheatstone transformed what had been a form of musical recreation into a scientific demonstration.

The observation of Eulenstein's music helped inform Wheatstone's understanding of how columns of air could be made to vibrate in reciprocation of the sound of an instrument. For the Royal Institution audience, he showed exactly how the Jew's harp worked and why Eulenstein had been able to produce such fine harmonies on the device. He devised a demonstration in which the harp's tongue was tuned with sealing wax to sound a C, corresponding to the sound of a closed four-foot-long tube. Faraday, who performed the experiment for the audience, then

> placed before the tongue the open end of a closed tube, containing a column of air two feet in length and one inch in diameter, and furnished with a movable piston, by which the column could be shortened to any required length. On striking the tongue, the octave of the fundamental sound was produced, being the sound 2 of the scale in the preceding section; by shortening the column of air successively to one third, one fourth, one fifth, one sixth, one seventh, &c. . . . all the notes of the series were correctly obtained.[58]

This analogous illustration of the contraction and expansion of the mouth showed that no audible sounds could be produced but those reciprocating from columns of air identical with the multiple of vibrations of the Jew's harp. Therefore, Wheatstone defined Eulenstein's skill as the ability to accurately control the size of vibrating columns of air within his mouth and throat. As for Eulenstein's creation of chords with several simultaneously reciprocating sounds, Wheatstone explained that, to produce a major C chord,

> three jew's harps, incapable of producing lower sounds than the fourth of the series, are selected; and the mouth being made to correspond with the C, the two other sounds, E and G, are likewise reciprocated, though faintly, because these sounds are nearer that to which the mouth is adapted, than any other multiples of the original vibrations.[59]

The same column of air could thus give several differing sounds. Eulenstein's ability to produce several notes simultaneously was because a column of air vibrated not only with bodies whose vibrations were isochronous with its own, "but also *when the number of its own vibrations is any multiple of those of the original sounding body.*"[60]

Faraday concluded Wheatstone's discussion of the Jew's harp in his subsequent lecture on March 7.[61] Eulenstein's performances represented just the sort of acoustical phenomena that appealed to London audiences who were eager for musical entertainment but also craved scientific explanation. Wheatstone and Faraday made full use of this: the recruitment of Eulenstein and his Jew's harp into their scientific discourse represents a very powerful example of how closely musical performances could be refashioned into scientific exhibitions. Expanding on Wheatstone's initial lectures, Faraday discussed Wheatstone's phonic communications in April, showing new experiments to illustrate the principles of resonance. Once again, Faraday and Wheatstone orchestrated a musical marvel. There was, it seems, a fine line between serious demonstrations and mere parlor tricks but, by linking philosophical explanations to musical acts, they provided both information and entertainment. The *Arcana of Science and Art* reported that

> a very singular illustration was given by Mr Mannin, of the power of producing two simultaneous sounds from the mouth. It was strikingly shown in some airs which he whistled first as solos, and then as duets. From the command possessed over this power by Mr Mannin, so as to produce the simultaneous sounds either quickly or slowly at pleasure, no possible doubt could be entertained of the fact. It appeared that at the same time, the mouth is divided into two parts by the tongue, and that each portion of air, is thrown into a separate state of vibration by the embouchure formed at the lips.[62]

In their subsequent discourse, "on the laws of the coexisting vibrations of strings and rods," Wheatstone and Faraday continued to refashion objects of amusement as "matter[s] of philosophical investigation." After showing the possibility of coexisting vibrations in the same length of string, Faraday displayed a kaleidophone, amazing the audience with visual and musical wonder. The *Dublin Literary Gazette* described the "extraordinary curves, circles in every variety of combination and concatenation, ellipses, spirals 'in wandering mazes lost,' . . . description fails to convey any thing like an idea of their dazzling brilliancy and effect!"[63] Faraday continued to draw out the kaleidophone's philosophical implications. By applying the laws of vibrating strings and rods, he suggested that

> it might be possible ere long, by further attention to the measurement of minute particles of time, so to procure a kind of microscopic method of magnifying them, as that we might ultimately be able to determine, by observations, in the transmission of the electric fluid, whether there were two electricities or not.[64]

The kaleidophone here appeared less as an amusing toy and more as a serious experimental apparatus. To add weight to this analysis, Wheatstone placed his experiments within the broader context of Baconian natural philosophy. Having established that sound was purely vibratory and could be conveyed by solid bodies alone, Wheatstone attacked seventeenth-century natural philosopher Francis Bacon's assertion that "sound is propagated by spirits contained within the pores of the body."[65] All sonorous solid bodies, according to Bacon, had spirits which communicated with each other, including bone and wood. Wheatstone, however, showed that sound was not spiritual but material.[66]

Musical wonders did more than substantiate Wheatstone's claims — they also helped ensure that later acoustical lectures were well attended. Reviewing the female attendance of Faraday's lecture on the coexisting vibrations of strings and rods, the *Dublin Literary Gazette* noted that there

> were some fair listeners in the gallery, attracted no doubt, by the hope of enjoying some more of those musical performances on strange instruments, which sent them home so delighted this night three weeks. If they failed, however, to be so much *amused* by the evening's entertainment, they must certainly have reaped greater advantage from these more serious illustrations of the philosophy of sound.[67]

This reflection revealed how the emphasis of these musical displays had changed from the recreational to the philosophical and asserted that this was of particular interest to female audiences.

Based on Wheatstone's music shop sales, the fashionable women who attended Royal Institution lectures found the exhibition of new musical instruments especially intriguing. Seizing on this market, Wheatstone incorporated his own musical wares as objects of scientific investigation and brought his commercial products into the lecture theater. The aeolina created a particular sensation in 1828, with the *Times* noting that before featuring in Wheatstone's paper the instrument had been unknown. However, thanks to Faraday, "a simple mode of producing very delightful musical sounds, has lately been brought before the public in a variety of amusing and useful applications." The principle applied to the aeolina was of a series of small springs fitted in apertures in a metallic plate, which vibrated when breathed over. Each spring oscillated at a different frequency and opened or closed the apertures. After the success of the aeolina at the Royal Institution, such instruments went on sale in shops across London and Europe. However, these were, the *Times* asserted, "much less perfect in their construction . . . and have not the richness or variety of tone of those made by Mr Wheatstone."[68] He used the same principle to manufacture springs intended as an alternative to tuning forks for violins and guitars. Sets of these springs were on sale, including for harps and pianos, being very portable and maintaining an accurate tone. For London's many piano tuners, these were precise, transportable alternatives to forks. Evidently Faraday's delivery of Wheatstone's papers at the Royal Institution had commercial value.

Similarly, in a lecture on May 21, 1830, on new musical instruments, Faraday explained how vibrating springs produced the sounds of Wheatstone's symphonium and concertina, both patented in 1829.[69] To demonstrate how the mechanical regulation of air on the oscillating tongues of these devices controlled their sounds, the audience witnessed Mr. Godbé, a talented musician, perform on Wheatstone's symphonium.[70] This was followed by renditions on an accordion, and then on an organ and pianoforte, which illustrated the same principle. While Faraday lectured, Godbè provided philosophically informing musical demonstrations on reed instruments. Indicating the success these instrumental exhibitions had with female audiences, Wheatstone's concertina subsequently became popular with gentrified women. Production of this instrument began in the 1830s and was a direct result of Wheatstone's experimental investigations into mouth-blown reed instruments.[71] Between 1835 and 1870, 88 percent of the 15,056 recorded transactions for this instrument were to men's accounts, but women dominated its purchase within aristocratic circles. From the 1830s until 1859, of 186 titled customers, 129 were women.[72] The sociable nature of Royal Institution lectures assisted with the promotion of these instruments. After Friday night discourses, it was customary for the heiress Angela Burdett-Coutts (1814–1906) to host parties with Wheatstone a regular guest. There he was known for his amusing party pieces and acoustical tricks which enchanted guests, including Florence Nightingale and Jane Webb, author of *The Mummy!* (1827).[73]

Though Wheatstone conceived of these lectures, Faraday played a crucial role in transforming these popular musical amusements into philosophical demonstrations. Faraday personally arranged for some of the more exotic instruments to be transferred to the Royal Institution. In February 1828, he applied to the librarian of the East India Trading Company for the loan of a China gong, supposedly in the company's museum, but was informed that the museum did not possess such a device. The late Sir Stamford Raffles had, however, owned several, kept at the Duke of Somerset's residence on Park Lane. Faraday subsequently applied to Somerset, who had kept them since Raffles's final voyage to India, but Somerset had returned them to Raffles's widow, Lady Sophia Raffles.

Through this correspondence, Faraday nevertheless secured the loan of several exotic instruments, including the gender from Java.[74]

Of equal importance was Faraday's delivery and panache in presenting Wheatstone's work. As the *Dublin Literary Gazette* put it,

> What an extraordinary being is this Michael Faraday! Only give him a subject, no matter what, and he will descant upon it . . . [including] the music of the spheres. . . . He is the improvisatore of the scientific world—the organ of new inventers—the mouth-piece of dumb mechanic geniuses. Do you wish to make known your discoveries in any art or science? Talk to Mr Faraday for ten minutes, and he will announce your labours to the public ear in the most captivating form imaginable.[75]

In contrast, the journal portrayed Wheatstone as "a young man of anxious, spectacle-on-nose appearance," who remained hidden behind the scenes of the lecture until entering the theater at the end to offer some concluding remarks.

Likewise, Faraday's lecture notes emphasize his considerable input into preparing sound demonstrations for Royal Institution audiences: these were in fact experimental notes to accompany Wheatstone's explanatory script on sonorous phenomena. Here, then, was the written conversion of West End amusements into philosophical trials. In the lecture of February 1828 Faraday devised no less than twenty-five arrangements, including the "Sound of a jar or bell," a "Tambourine & voice," "Sand on paper," and "Tuning forks or guitar," as well as the Enchanted Lyre, "Whistling," a Jew's harp, and "Mr Eulenstein." A few weeks later, on March 7, the second lecture on sound adopted a similar structure, this time with thirteen demonstrations, including a tuning fork vibrating over a bottle-mouth, two reciprocating flutes, singing hydrogen flames, Eulenstein and his Jew's harp, and an organ. His third lecture, on May 9, included no less than twenty-two experimental illustrations, covering the nature of pitch, melody, and harmony. Here he employed bell flutes, monochords, vibrating copper plates, tuning forks, a violin and bow to be sounded by "jerks," a "stopcock & air," and, as usual, a Jew's harp.[76]

Faraday's preparatory notes reveal how thoroughly Wheatstone's showroom amusements had been transformed into experimental procedures which, though still entertaining, were intended to be illustrative of scientific principles. The sound lectures of 1829 were no different in this respect. To show the formation of nodal figures on vibrating surfaces, Faraday assembled eighteen experiments, including square plates for producing Chladni's figures, followed by "long plates . . . triangles, hexagons . . . [and] circular plates." Although Faraday's notes on the transmission of musical sounds through linear conductors in March 1830 were lost (when a correspondent from the *Athenaeum* took them to provide an account for the journal), his surviving papers reveal that lectures on sound included some of the most extensive experimental displays on offer at the Royal Institution; they represented a model of how to present new knowledge, with experimental support, to dilettante and specialist audiences alike.[77]

The great strength of these sonorous exhibitions was that they could be experienced without specialist knowledge or mathematical training: they were valuable resources for extending scientific understanding to broad audiences. The apogee of Wheatstone and Faraday's transformation of Pall Mall's mechanical and musical marvels to Royal Institution experimental procedure featured in 1830, when they set up the Enchanted Lyre in the theater for a lecture on the transmission of musical sounds through solid linear conductors. On March 5, Faraday replicated the phenomena of these experiments, the sonorous impulses of which "were capable of exciting, in surfaces with which they were in connexion, a quantity of vibratory motion, sufficient to be powerfully audible when communicated through the air." Wheatstone believed his 1821 Enchanted Lyre arrangement had proved this, and he used his 1831 Royal Institution lecture to detail this work to new audiences. Subsequently published as a scientific paper in the *Journal of the Royal Institution of Great Britain*, for the lecture he combined West End spectacle with natural philosophy, uniting his work on Chladni's acoustic figures and resonance with the lyre. Waves transmitted through an elastic body had, he observed, to be connected to a sounding body capable of resonating at the correct frequency so as to augment the sound, such as the transition of vibrations from a violin's strings to its wooden soundboard via its bridge. To substantiate this claim, Wheatstone took a violin bridge and replaced it with a glass rod, showing that the strings still communicated sound in an audible manner. He then provided a demonstration of the Enchanted Lyre, which had been installed in the lecture theater. By boring a hole through the theater's floor, a rod of forty feet was assembled, from the open window of the cupola to the floor. When an assistant sounded a tuning fork at one end, no sound was audible inside the theater, but when he connected the theater end to a soundboard the fork became audible. Wheatstone then connected the rod to a piano at one end and to either the lyre or a guitar in the theater.[78]

Wheatstone stressed the implications of this scientific knowledge for communication, repeating earlier claims from his Pall Mall exhibitions. As the rods were silent without a sounding board, Wheatstone suggested they could pass through different houses across London inside insulated tubes. Inhabitants could have music conveyed to their homes: they would, he alleged, be able to "hear at pleasure the performance in any of these rooms, by merely attaching a reciprocating instrument to the conductor." This Friday Evening Discourse was the culmination of some ten years of experiments which had shown the practicality of communication with solid conductors. Ideally, deal wood rods with a sonorous velocity of 18,000 feet per second would transmit sound at some 200 miles a minute. Although sound decreased in intensity as the square of the distance of its diffusion increased, Wheatstone asserted that by the sonorous impulse in a single rectilinear direction, the possibilities for communication were infinite. The challenge, however, was that conducting bodies were never perfectly homogeneous. Should such a body be produced, Wheatstone asserted, "it would be as easy to transmit sounds . . . from Aberdeen to London, as it is now to establish a communication from one chamber to another."[79] The scientifically focused *Philosophical Magazine* believed this work of considerable significance, while the *Literary Gazette* was convinced that Wheatstone's research proved

> the possibility of telephonic communication. It was stated, that as sound has proved to travel through several conducting substances at the extraordinary velocity of 18,000 feet in a second, were it possible to transmit audible sound so far, a phonic communication might be made between London and Edinburgh—a distance of nearly 400 miles—in less than two minutes.[80]

While Wheatstone's showroom musical performances had been popular, Faraday's replication of these demonstrations in the Royal Institution's lecture room enhanced claims that sound could seemingly annihilate time and space through high-speed telegraphic communication.

Beyond Wheatstone's collaborative papers with Faraday, several invited lecturers contributed to this transformation of the sonorous from the amusing to the philosophical. Benjamin Humphrey Smart (bap. 1787–d. 1872), a leading elocutionist and close associate of Faraday, delivered an 1829 Friday Evening Discourse entitled "The Audible Qualities of Speech."[81] Faraday had evidently been impressed with Smart, who had provided him with elocution lessons and helped him perfect his public speaking.[82] In 1837 Reverend William Ritchie (ca. 1790–1837), the Royal Institution's professor of philosophy, gave the lecture "The Velocity of

TABLE 2. Royal Institution lecture season of 1829

Lecturers	Subject	No. of lectures [in the series]	Sums paid [per series]	Cost for lecture	Mean no. of the audiences	Cost of lectures for hundred of attendants
Brand	Chemistry	X	£150.0.0	£15.0.0	152.4	£9.16.10
Millington	Nat. Phil.	X	£52.10.0	£5.5.0	115.0	£4.11.4
Faraday	Chemistry	VI	£52.10.0	£8.15.0	153.8	£5.13.9½
Webster	Geology	VIII	£47.0.0	£5.17.0	106.4	£5.9.11½
Phillips	Painting	VII	£73.0.0	£10.9.0	105.0	£9.19.0¼
Crotch	Music	VI	£71.0.0	£11.16.0	114.1	£10.6.10
Burnett	Botany	IX	£0.0.0	£0.0.0	91.0	£0.0.0

Source: Royal Institution (RI) Ms F/5/D, "Administrative Notebooks: Duties of Servants," p. 39.

Sound and the Discrepancy between Theory and Observation." A year later the politically radical William Edward Hickson (1803–70) delivered his "The Introduction of Vocal Music as a Branch of National Education with Illustrations." Along with his programs for political reform, including women's rights and national schools, Hickson promoted changes to musical education. Challenging composer John Hullah's (1812–84) method of teaching singing by sight, Hickson instead proposed beginning with melodies; and his 1836 *The Singing Master* defined singing as a moral discipline for families and schools.[83]

Yet the Friday Evening Discourses were, effectively, just the Royal Institution's showpiece, aimed at fashionable metropolitan audiences. Faraday and Wheatstone's introduction of the sonorous was not confined to such ostentatious occasions. On a much deeper level, they gave sound a prominent place in the Royal Institution's more focused, specialist lecture series. In 1829, composer William Crotch's (1775–1847) course of six lectures on music attracted an average of 114 attendants and, after the chemistry lectures, constituted the most expensive series at £11 and 16 shillings per class (table 2). While music did not feature in the 1830 lecture season, the mechanic and mathematician Robert Willis (1800–1875), who eventually became Cambridge University's Jacksonian Professor of Natural and Experimental Philosophy, provided a course on acoustics in 1831, averaging 135 attendants per session (table 3).[84] Willis had given the first of two papers, both entitled "On Vowel Sounds," to the Cambridge Philosophical Society in April 1828, treating the replication of the human voice as a problem of fluid mechanics, before presenting his acoustic machine, the lyophone, to the Royal Institution two years later.[85] The records of

TABLE 3. Royal Institution lecture season of 1831

Lecturers	Subject	No. of lectures [in the series]	Sums paid [per series]	Cost for lecture	Mean no. of the audiences	Cost of lectures for hundred of attendants
Webster	Geology	VI	£42.0.0	£7.7.0	155	£4.14.10
Brande	Metals	IX	£150.0.0	£16.13.4	167	£10.0.0
Ritchie	Nat. Phil.	VII	£50.0.0	£7.2.10	108	£6.11.5½
Vigas	Anthology	VI	£0.0.0	£0.0.0	118	£0.0.0
Faraday	Miscellaneous	IV	£52.10.0	£13.2.6	164	£8.0.0
Webster	Geology	VI	£40.0.0	£6.13.4	146	£4.11.4
Montgomery	Poetry	VI	£94.10.0	£15.15.0	133	£11.16.10
Willis	Acoustics	VIII	£73.10.0	£9.3.9	135	£6.16.1
Lundly	Botany	IV	£21.0.0	£5.5.0	117	£4.9.9
Friday Evenings	Miscellaneous	XIX	———	———	276½	———

Source: Royal Institution (RI) Ms F/5/D, "Administrative Notebooks: Duties of Servants," p. 40.

subscribers to Royal Institution lectures reveal a high proportion of attendants to music lectures were female. Although Royal Institution members were not recorded individually, twelve out of sixteen of those attending who were not full members were women. Clearly nonmembers represented only a small number of attendants, but this suggests that music was an incentive for attracting female audiences to scientific subjects. For example, to the musician and music writer Edward Taylor's (1784–1863) 1836 lectures on music, all four non–Royal Institution subscribers were women (table 4). Wheatstone's 1836 lectures on sound attracted two male and one female non–Royal Institution attendants, while in the popular press the Royal Institution's musical lectures were specifically advertised as open to ladies.[86]

Although the Royal Institution was at the center of sound's transformed scientific character, Wheatstone and Faraday's work was very much a national endeavor. They not only kept their fashionable West End viewers entertained, but they repeated these lectures to national audiences. On March 16, 1832, Wheatstone read his own paper "On the Vibrations of Columns of Air in Cylindrical and Conical Tubes" at the Royal Institution. This examined wind instruments and subjected them to experimental analysis, defining the relationship between vibratory movements and an instrument's harmonics.[87] In 1835 Wheatstone presented a similar paper to the BAAS in Dublin and, in doing so, adapted his musical

TABLE 4. Subscribers to separate courses of lectures, 1835–41

1836—To Mr Taylor's Lectures on Music	
Miss Neal	
Miss M. Neale	
Miss Horne	
Mrs Day	

1836—Professor Wheatstone's Lectures on Sound	
Mrs Fisher	Mr Verini
	Mr Bishop

1837—To Edward Taylor's Lectures on Vocal Harmony	
Miss Neal	Mr Jones
Miss A. Neal	
Miss Kendall	
Mrs Marshall	

1838—Mr Taylor's Lectures on Vocal Harmony	
Miss Gillies	
Mrs Fladgate	

1840—Mr Taylor's Lecture on Vocal Harmony	
Miss Jardine	Mr Jardine

Source: From Royal Institution (RI) Ms F/5/H.

performances for a national audience. In this, he exhibited experimental evidence in support of the Swiss mathematician Daniel Bernoulli's (1700–1782) theory of wind instruments, that the sound of a tube, open at both ends, moved from its center in opposite directions. By taking a circular pipe and placing a glass plate between where the two pipe ends met, and then putting the plate into vibration, he found the vibrations passing through the tubes neutralized each other. This was effectively the adaptation of Chladni's figures to measure the internal vibrations of wind instruments.

Indeed, Chladni's figures provided a particularly productive means of disseminating acoustic knowledge beyond London's West End (fig. 1.7). Chladni had himself recognized that the manner of presenting an experiment and of retaining some degree of mystery around his work was crucial to securing favorable reviews from audiences. As Viktoria Tkaczyk has argued, Chladni worked to perfect performative techniques for showing his acoustic figures. Most famously, in 1808 he had traveled to Paris and exhibited his experimental arrangement to Napoléon Bonaparte, who

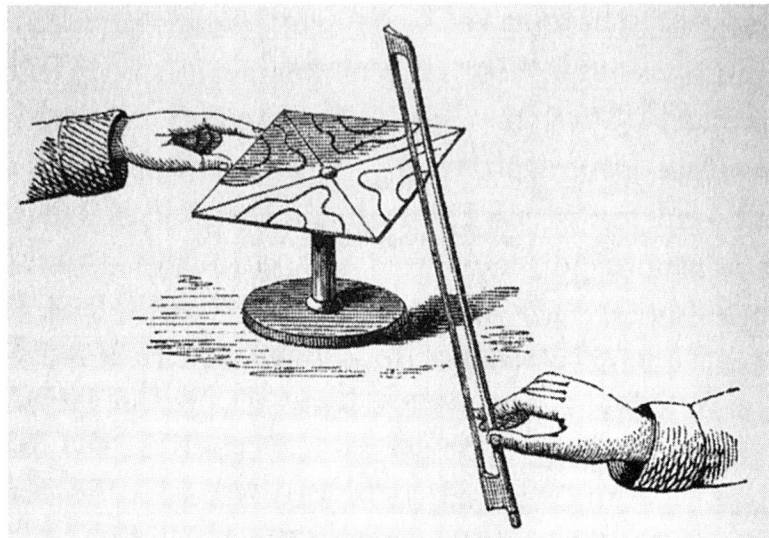

FIGURE 1.7 The production of an acoustic figure through Chladni's experimental method of putting a plate covered in sand into vibration. Tyndall, *Sound*, 142. Public domain.

was captivated with the natural philosopher's work. Chladni dedicated his 1809 *Traité d'acoustique* to the French emperor and later reflected that his audience was less impressed by his scientific competence than by the novelty of the experiments and musical instruments displayed.[88] Wheatstone was equally well aware that for British audiences to appreciate the value of Chladni's figures, they had to actually witness the experiments and, while the BAAS offered one venue for achieving this, it was crucial that other practitioners reproduced these vibratory demonstrations.[89] It was not enough for audiences to read about sonorous experiments in printed texts: these were investigations that had to be experienced.

In Oxford it was Joseph Blanco White (1775–1841) who first brought Chladni's experiments before the city's audiences. Born in Seville, White had been ordained a Catholic priest in 1798 but soon abandoned this vocation, before fleeing to London during the French invasion of Spain in 1810. From 1814, he studied Anglican theology at Oxford, where he became a member of Oriel College in 1826. A close friend of the controversial High Church Anglican John Henry Newman, with whom he played the violin and shared a love of Beethoven, White was very active in the university's High Church spiritual and musical life.[90] Interested in both science and music, White attended the earliest meetings of the newly established Ashmolean Society, formed to promote scientific research in Oxford. In

1829 he was invited to present a paper, "On Music Sounds."[91] Although lacking experience of experimental practices, White had heard of Faraday's Royal Institution lectures on sound. In December 1829 he wrote to the experimentalist informing him that he had been "engaged to write a paper on Music for the Ashmolean Society" and asking if Faraday had "published any thing on the vibrations of Musical sounds," having heard that he had "lectured on that subject." White requested assistance for his own Ashmolean paper, either through recommended readings or the loan of Faraday's lecture notes.[92] Faraday forwarded details of Wheatstone's work on acoustic figures and the kaleidophone. White also consulted Herschel's 1830 *Encyclopaedia Metropolitana* entry on sound. From this material, White prepared an elaborate volume of notes on sound, including details of the ratios of vibrations involved in harmonics and the oscillation of strings, rods, plates, and air in tubes. Of these he noted that the "vibrations of most of these bodies were unexamined until Chladni submitted them to experiments."[93]

When White delivered his paper in 1830, he declared that his aim was "not so much to advance as to recommend science" and illustrate the pleasure derived from its study. To do this, he had chosen music because this had been his "favourite recreation from the very dawn of life" and he desired a "practical knowledge" by collecting "a few facts which persons incomparably better acquainted with physical and mathematical science, would probably find obscure." Along with little mathematical training, he also acknowledged that "the doctrine of sounds demands the exhibition of experiments, in which I must confess myself totally unpractised: as the ground of the Art of Music, it requires a perpetual reference to principles involved in the science of harmony."[94] Although acting as a scientific authority, offering instruction in the natural sciences, he was by his own admission an amateur, with little more than a general interest in the subject.

White began by defining musical sound as "equal vibrations sufficiently rapid to be perceived in continuity," before moving on to describe Chladni's investigation of vibrating plates. Chladni, "the leader of the distinguished philosophers who have lately devoted their attention to Acoustics," had measured the vibrations of a plate in comparison to the "oscillations of an astronomical pendulum." To demonstrate how vibrating bodies had to be homogeneous to convey musical sounds, White performed experiments comparing the transmission of sound through water and effervescent (and therefore heterogeneous) liquid, which obstructed the sound's passage. Wheatstone's influence was very evident throughout, with White observing that he would praise his "knowledge of the present subject and great mechanical ingenuity with which he has illustrated it . . . were it not, that . . . Wheatstone himself happens to be present."[95] White subsequently exhibited Wheatstone's kaleidophone: although amateurish, he performed a role similar to London's elite experimentalists.

White's discussion of sound's propagation was largely based on Herschel's recent "Sound" treatise, published in the *Encyclopaedia Metropolitana*, but he regretted that he had no images from this work to display as they were not yet available to the public. White had evidently been granted a very early viewing of this treatise, from which he presented details concerning the pitch and quality of sounds as well as harmonic theory. While he believed that "the power of perceiving the harmonic relation of one note to another, is a faculty very generally given by nature," he felt that physiologists had been completely unable to explain what constituted a musical ear: it was a "most mysterious combination of organic means."[96] Hinting at music's metaphysical character, White observed that there were limits to scientific reasoning.

However, the highlight of White's paper was his replication of Chladni's acoustic figures. The centerpiece of the evening's entertainment was this illustration, by metal and glass plates strewed with sand, of how a vibrating body subdivided itself. White's performance of Chladni's figures evidently impressed his viewers. Some sixty years later, Francis William Newman (1805–97) recalled being in the audience. The younger brother of John Henry Newman and a fellow of Balliol College, Francis remembered how White

> sat before a table on which fine sand was strewed, and he held his violin so close that he could press any of the strings against the edge of the table. . . . The object of his lecture was to show the relations between musical sounds and mathematical curves. From my chair I could see everything on the table: so I suppose could every visitor, in the moderate-sized room.[97]

Newman's account of how the audience experienced Chladni's figures detailed how White proceeded to draw

> his bow along a string, its vibration on the edge of the table made the sand leap about, and, by the contra continuance of the smooth pure note, quickly form itself into a set of curves, each bearing relation to the other. By changing the note he made the particles of sand rearrange themselves into a new set of curves, all of them, however complex, commending themselves to the eye as harmonious and elegant.

Every note that he drew was by a long and steady bow. I cannot add any of his comments on the process, & only remember what every one without mathematical training would accept by mere sight.[98]

These experiments, then, were not only astonishing but conveyed something of nature's order, without requiring any mathematical skill from the audience. This experience was both visual and sonorous.

Another Oxford undergraduate and future archdeacon of Bath, Robert William Brown (1809–95), also enjoyed White's experiments, which he could still recall in 1892.[99] Erroneously, Brown believed that "Blanco White had anticipated the discovery of Chladni's curves, which Sir Charles Wheatstone exhibited at one of the early meetings of the British Association." Brown had evidently forgotten the references to Chladni and Wheatstone, and Brown's assertion that "Blanco White was an accomplished violin-player, and probably arrived at this scientific fact by a happy accident" was mistaken.[100] But this example shows the impact of White's lecture. For Wheatstone and Chladni's work to travel, the acoustic plate experiments had to be performed before audiences. When done, as in Oxford, they certainly left an impression, but without disseminators such as White, news of the acoustic figures was difficult to propagate.

Wheatstone and Faraday's support of Blanco White's Oxford demonstrations highlights the experiential nature of sonorous scientific exhibitions: to be fully comprehended, audiences had to witness acoustic effects, such as those of Chladni's vibrating plate experiments. It was not enough for descriptions of these visually and aurally spectacular exhibitions to be read aloud. These were more than lectures to be passively heard; they were performances to be actively absorbed on multiple levels. Although of aesthetic value, such displays were epistemologically important, encouraging spectators to use their ears as well as their eyes in the interrogation of nature. Yet acoustic science and the act of hearing were not just valuable for disseminating knowledge of sound, but for examining nature more broadly. Faraday and Wheatstone's refashioning of sound as a scientific object reconfigured the amusing as the philosophical illustrative by employing acoustic experimental techniques in the study of extensive natural phenomena.

Hearing Nature and the Laboratory Ear

New understandings of sound increasingly provided novel methods for scientific investigation and shaped experimental work on other invisible forces, especially heat and electromagnetism. Acoustical experiments and musical instruments provided valuable resources for popularizing science, but they were also conscripted into investigations of nature. Elsewhere in Continental Europe, where Chladni's acoustic figures had gained celebrity during the early 1800s, research into sound had paralleled that of electricity and magnetism. While Ørsted and Johann Wilhelm Ritter worked on acoustics and electromagnetism, French natural philosopher Felix Savart (1791–1841) extended Chladni's acoustic plate experiments to the study of solid materials.[101] Long interested in sound, designing his own trapezoidal violin and giving his name to the unit of measurement for musical intervals, the "Savart," in 1829 Savart explained to Paris's Académie des sciences how he had investigated the elasticity and structure of crystallized bodies by cutting flat plates out of various materials, strewing sand over them, and then recording "the sound which they emit while vibrating, and the modes in which they divide themselves." Through this analysis "by means of sonorous vibrations," Savart argued that the acoustic figures formed from vibrating crystal revealed the direction of the force of cohesion which had originally formed them. Savart boasted that, though not complete, this experimental method "may yet become a powerful means of studying the structure of solid bodies" and extended his inquiry to industrially manufactured metals the following year.[102] No one was more excited by Savart's work than Wheatstone, who instantly foresaw such acoustic techniques for material investigation to be of economic and philosophical value.[103] Wheatstone himself presented a Friday Evening Lecture on the subject in March 1836, emphasizing "the importance to science of the discovery of the connection of Acoustics with Crystallography."[104]

The use of sound to investigate natural phenomena and physical materials was, just as much as the refashioning of popular amusements into philosophical discourses, central to the changing place of the sonorous within British science. Sound was not just a useful means of popularizing science but also of actually producing new knowledge of nature. In the early 1830s, Wheatstone and Faraday did more than present acoustic phenomena to fashionable audiences, they used sound as a means of inquiry. This was made evident by their mutual decision to publish accounts of sonorous experiments within the eminent scientific journal, the *Philosophical Transactions of the Royal Society*. In many ways, this marked the culmination of sound's transformation, first from a subject of entertainment, then to experimental exhibition, and finally to serious scientific investigation.

In 1833, Wheatstone published a paper exploring Chladni's vibrating plate experiments on sound in *Philosophical Transactions*. These experiments had first been published in 1787 in Chladni's *Entdeckungen über die Theorie des Klanges*, before being developed in his later works, *Akustik* (1802) and *Neve Beytrage zur Akustik* (1817), but as these appeared only in German, they were little known in Britain.[105] Wheatstone's article, featured in the illustrious journal of the venerable Royal Society, marked a considerable escalation of the standing of Chladni's acoustics with British audiences.[106] Wheatstone identified different modes of vibration passing through the plates and cataloged these figures by noting the relations between their parallel and perpendicular lines. As the pitch of sound produced intensified, so these relations became more complex (fig. 1.8). Identifying the significance of his *Philosophical Transactions* paper a year later, Wheatstone described that Chladni "produced all these figures on plates of different forms and sizes, he delineated them, and he ascertained the series of sounds which corresponded with the series of figures of each plate of different form," but regarding "the causes of these phenomena," he had said nothing. "Desirous of obtaining the key to these hieroglyphics," Wheatstone determined to explain the phenomena and "at last succeeded in deciphering them . . . and the results of which are in such strict accordance with experiment that there is not the slightest bend in a line which it does not fully account for."[107] His publication with the Royal Society marked what Wheatstone believed to be the completion of Chladni's research.

Wheatstone's *Philosophical Transactions* paper on Chladni's vibrating plates actually marked the second study of acoustic figures to feature in the Royal Society's elite journal. Two years earlier, in 1831, Faraday had published the results of his own extensive examination of sonorous vibrations, having himself taken up acoustics as a research subject. He brought a new body of experimental skills to the science of sound, investigating sonorous vibrations between February and July of 1831 following his reading of John Herschel's *A Preliminary Discourse on the Study of Natural Philosophy* (1830). In 1832 Faraday wrote to Herschel congratulating the author on his work and reflecting that the paper had improved his own philosophy: he had found Herschel's analogies between sound and light impressive.[108] Faraday's views on the unity of laws of electromagnetism and acoustics in fact carried theological implications, with his Sandemanian religious values central to his study of nature.[109] Sandemanianism envisaged a literal interpretation of the scripture and taught that life should be lived through Christ's example. Faraday believed that God had laid out laws of nature for man's own good and that it was up to man to utilize these productively. To uncover natural laws was to reveal evidence of God and his benevolence, and to find consistencies between these

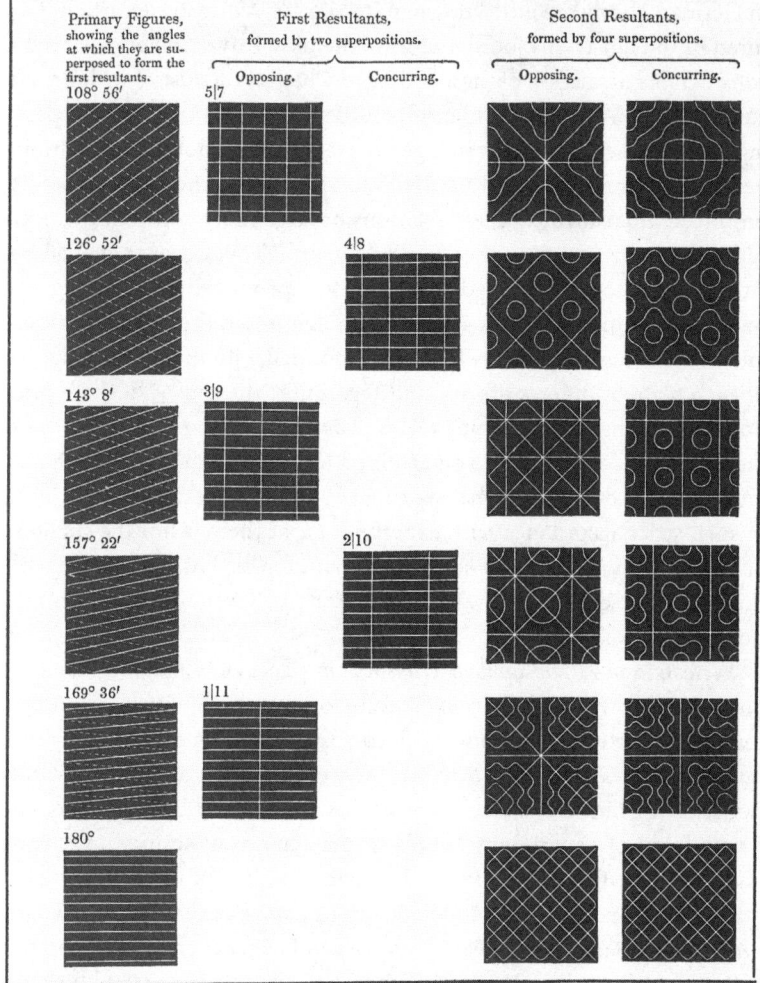

FIGURE 1.8 Wheatstone's acoustic figures, published in *Philosophical Transactions* in 1833, showing the ratios between transverse and longitudinal nodal lines. Public domain.

laws, as with acoustics and electromagnetism, confirmed this theology of nature. The observation of patterns, contingent on the sounds produced through vibrating plates, enhanced Faraday's vision of a unified universe of beautiful order.[110] In the early 1830s, sound was the most observable force which could be investigated through experiment: acoustic plate experiments were not simply aesthetically pleasing but, for Faraday, carried religious implications for the unity and order of God's creation.

In early 1831 Faraday began experimenting on acoustic vibrations in solid and fluid bodies. He adopted Chladni's techniques but used silica, powdered glass, phosphorus of lime, and red lead instead of sand. On February 2 he conducted twenty-eight experiments, resuming these the next day, including the observation of powder vibrating on a plate within an air pump. After systematic trials and comparing different substances for their ability to display the movements of a vibrating plate, he examined how plates vibrated in water.[111] By February 18, Faraday had completed at least sixty-seven experiments, and he conducted several more on March 14. Although initially impressed with the beauty of Chladni's experiments, it was Savart's observation of the accumulation of particles on centers of vibration, when using powder instead of sand, which captured Faraday's subsequent attention. He wanted to examine Savart's explanation that these resulted from secondary modes of vibration, which coarser sand did not show, instead believing that the principal vibrations, which sand revealed, were the only movements at work on the plate. The problem then was to account for how and why powder accumulated on areas in motion. The answer, according to Faraday, was that vibrations caused the dust to be thrown into the air and "form clouds." To produce this phenomenon, an experiment required a very fine powder, such as silica, lycopodium, peroxide of tin, red lead, or vermilion. To show how dust piles resulted from air currents, Faraday repeated his experiments on vibrating plates, but this time employed a series of cards, which he placed at the edge of a plate covered with lycopodium and put into vibration. By observing where the powder accumulated, Faraday was able to determine the movement of the lycopodium, finding that the dust always accumulated at an angle "in accordance with the idea of currents intercepted more or less by the card." To confirm the presence of air currents, Faraday applied a layer of gold leaf to the plate, which when vibrated caused the leaf to form "a blister." He subsequently rejected Savart's theory of coexisting secondary vibrations, presenting his results to the Royal Institution on March 21, 1831, and then at the Royal Society in May.[112] Publication followed in the *Philosophical Transactions*, confirming the transformation of these sonorous experimental techniques from trivial amusement to serious scientific discourse. Wheatstone was impressed, lauding the "time and ingenuity that you have bestowed upon the task," the results of which concurred with the "fact that we do not hear these presumed co-existence sounds" and confirmed their "non-existence."[113]

In June, Faraday returned to sonorous phenomena, this time experimenting on what he termed "crispations," these being the forms assumed by fluids when in contact with vibrating surfaces. He found these forms enchantingly beautiful but believed that past work on the subject did not provide a complete understanding of the phenomenon, which was, he asserted, "an obstacle to the progress of acoustical philosophy." Water with ink made an adequate vibratory fluid, but Faraday preferred mercury, while milk, alcohol, or egg whites all worked well. Mercury vibrating over tin or copper was, he felt, especially beautiful. When in vibration, Faraday found the mercury quickly covered with film, but he prevented this by adding a layer of diluted nitric acid. On replacing this with ink, the summits of vibrating mercury appeared very visible, popping up through the dark ink. This was staggering, "producing the appearance of pearls of equal size beautifully arranged in a black medium."[114] In this way, Faraday stumbled on a particularly aesthetic method of observing vibrations.

On June 17, 1831, Faraday constructed a circular plate of glass with six cork feet attached in a nodal line on which the plate could stand, and then added a glass rod that caused water on the plate's surface to vibrate. Faraday observed how, at first, circular waves moved outward from the center, but on increasing the vibration intensity these changed into hexagonal and then parallel lines. Three days later, Faraday worked on long rectangular vibrating glass plates, employing a candle to illuminate the formation of fluid wave heaps over the plates' surfaces in a manner similar to Wheatstone's kaleidophone. Faraday then carried out some very large experiments on June 25 with an eighteen-foot-long deal board positioned on two stools and put into vibration. By scaling up smaller vibratory experiments, he attempted to make the cycles of rising and falling heaps more observable. He concluded his experiments on July 5 by suspending a wooden lath with cork into a bowl of water, connected to a glass rod which caused vibrations on touch. He observed how "as the cork touched the water a beautiful store of ridges formed all round it, running out 2, 3, or even 4 inches. These were more or less numerous according to number of vibrations."[115] If the vibration was increased, linear heaps began dividing the circles, eventually forming quadrangles and increasingly complex patterns as the frequency rose.[116]

Faraday surmised that a vibrating "fluid may be considered as a pendulum vibrating to and fro under a given impulse." This relating of the motions of acoustic plates to an oscillating pendulum provided what Faraday believed to be an adequate explanation of sonorous vibrations. But while Faraday's experiments focused on the vibration of fluids, he was eager to draw out broader implications, reflecting that "the phenomena as yet described are such as take place on the *surfaces* of those fluids" but this inferred "that analogous effects should take place in gases and vapour." Faraday went further, suggesting that, "by analogy these views extend to the undulatory theory of light, and especially to that theory as modified by M. Fresnel . . . [who] conceived it necessary to admit that the vibrations of the ether take place transversely to the ray of light, or to the direction of the wave causing its phenomena."[117]

French engineer Augustin Fresnel (1788–1827) had argued that light behaved as a wave rather than a projection of particles, so Faraday's crispations appeared consistent with such optical understanding, as he explained to the Royal Institution on July 30, 1831.[118] This work on crispations and acoustic figures had consumed about five months of experimental labor, with Faraday later regretting that his resulting "paper on vibrating surfaces was too heavily laden with experiments."[119] However, Faraday's work was important in that he brought unprecedented experimental resources to sonorous investigation, including the Royal Institution's laboratory and his own experimental experience.

The dissemination and publication of these experiments very much fit within the broader exhibition of acoustic phenomena to public audiences. Faraday's acoustical researches made a fitting Royal Institution lecture, as he impressed viewers with beautiful acoustic figures on June 10, 1831.[120] For Faraday, however, this paper carried epistemological and ideological implications, as it differentiated between what he perceived to be the Royal Institution's and the Royal Society's varying approaches to experimental evidence. The paper was originally read before the Royal Society, before being repeated at the Royal Institution, where Faraday stated that

> his principal reasons for bringing it forward on the present occasion arose from a desire to illustrate the characteristic differences between the Royal Society and the Royal Institution, in their modes of putting forth scientific truths; and his conviction that every thing, whether small or great, originating in the latter establishment, should be placed, as soon as possible, in the possession of the members at large.[121]

According to Faraday, while his paper had been read and listened to at the Royal Society, the Royal Institution engaged observers with a more performative style of scientific demonstration. Acoustic experiments, such as vibrating plate demonstrations were, in this way, a powerful epistemological example of how experimental performances validated new "scientific truths." Wheatstone shared in this appreciation of Faraday's inquiries, asserting that the real value of your experiments is that of assigning the true cause of phenomena, which was so little obvious as to have escaped the observation of such experienced philosophers and successful experimentalists as Chladni, Oersted and Savart.... It will render the investigation of the residual phenomena of elastic surfaces less complicated, and it promises some further valuable information from the application of similar considerations to other phenomena.[122]

Wheatstone appreciated the methodical practices Faraday had employed in developing Chladni's experiments and believed such a systematically produced body of experimental evidence provided a solid basis for relating acoustic knowledge to the study of other transitory forces.

Wheatstone's connection of Faraday's experiments to broader natural phenomena was preemptive: Faraday's acoustic work finished in July 1831 and the very next entry in his experimental diary concerned electromagnetism. On August 29 he experimented on the production of electricity from magnetism, employing an iron ring with coiled copper wire wound about one half. By placing this within a circuit, he found that on making and breaking a connection he could cause a needle to oscillate.[123] He continued working on this phenomenon throughout October and into 1832. Faraday subsequently published the results from his initial electromagnetism experiments on November 24, 1831, outlining the principles of induction. He announced his second series of experiments on this phenomenon in his Bakerian lecture in January 1832.[124] In March, when considering the duration of a traveling magnet's impulse and the time of impulse of electric induction, he made a note that this might be related to the "theory of vibrations."[125] Faraday's earliest understanding of electricity was that it was a state of tension between particles and that it could be conducted through different bodies, just as sound traveled through conductors.[126]

Past histories of Faraday have trivialized his acoustic experiments, with James Hamilton concluding that the "six or so months of crispation experiments had amused, entertained and gratified Faraday," offering him "some kind of occupational therapy ... before he undertook the next major step in extending human understanding of nature."[127] In contrast, Ryan Tweney has argued that Faraday's visualization of sound was essential to constructing theories of electricity and magnetism as fields, in which invisible forces moved in curved lines.[128] He asserted that Faraday's acoustic experiments provided crucial skills for his subsequent observation of electromagnetic induction. After all, both were transient phenomena, under experimental investigation, with the object to make small effects over short spaces of time increasingly visible.[129] The problem with this, beyond reductionist conclusions derived from focusing on Faraday's psychology as *the* crucial factor in scientific discovery, is that acoustics was not merely a means to an end. To truly understand the place of Faraday's acoustical research and its relation to electromagnetic phenomena, we must emphasize the labor behind these experiments as well as his continuing belief that they were of philosophical and religious value. Instead of using hindsight to interpret these acoustic inquiries as a step in the "discovery" of electromagnetic induction, we should examine how they remained important to Faraday after 1831. Although Faraday focused his attention on electromagnetism during the autumn of 1831 and throughout 1832, he continued to promote his work on crispations and acoustic figures. In April 1832 he wrote to the French mathematician André-Marie Ampére (1775–1836), promising to send papers on his recent electrical research, as well as his "paper on the *arrangements of particles on vibrating surfaces*," which he had already sent to experimental authorities across Europe.[130] Again in June 1832, he attached his works on vibrations to those on electricity when writing to the German chemist Eilhard Mitscherlich (1794–1863) and later that summer dispatched both electromagnetic and acoustic papers to Ørsted, who shared in Faraday's belief that all natural phenomena were governed by similar laws and that the transmission of forces, either electric, light, magnetism, or heat, were connected.[131]

Throughout 1832, Faraday clearly wanted his earlier work on crispations and acoustical figures to accompany and be read alongside those more sensational papers outlining electromagnetic induction. This was confirmed by Faraday's earliest understanding of electromagnetic phenomena. In March he wrote to the Royal Society, requesting to deposit in its strong box a sealed note which outlined some of his early views on electromagnetism. As the phenomenon was so new, he wanted time to investigate these ideas, but was eager that his initial thoughts be recorded, especially that "magnetic action is progressive, and requires time; i.e. that when a magnet acts upon a distant magnet or piece of iron, the influencing cause . . . requires time for its transmission," and that magnetic induction was "performed in a similar progressive way." As a result, he was "inclined to compare the diffusion of magnetic forces from a magnetic pole, to the vibrations upon the surface of disturbed water, or those of air in the phenomena of sound," convinced that "the vibratory theory will apply to these phenomena, as it does to sound, and most probably to light."[132] Clearly then, the visual observation of acoustic figures was important to Faraday's understanding of how forces, such as electromagnetism, operated: these sonorous inquiries were far from trivial. It is telling that in July 1831 he conducted crispation experiments and by August he returned to the problem of making electricity from motion and magnetism, which had proved elusive for over a decade, and on the first day of recommencing these inquiries realized that while a steady movement made no effect, the continued making and breaking of a circuit produced electromagnetic induction. It was this momentary impulse, or glitch, which was so important.[133] Vibratory theory remained valuable to Faraday throughout the 1830s as he struggled to explain the new phenomena and the interaction between matter and electricity.[134]

Faraday and Wheatstone's sonorous investigations were at the heart of sound's transformation into a scientific object. And the two investigators were part of a growing network of natural philosophers who mobilized the ear within experimental practices. Indeed, Faraday and Wheatstone's scientific prominence encouraged experimentalists across Britain to take the sonorous seriously in the investigation of nature. In 1844, for example, J. P. Marrian published what he hoped would be a significant contribution to the growing interest in electromagnetic phenomena. Featured in the *Philosophical Magazine*, Marrian's "On Sonorous Phaenomena in Electro-Magnets" asserted that the key to understanding electromagnetism was acoustic observations. Repeating Faraday's famous electromagnetic induction experiment by constructing a helix out of copper wire wound round a soft iron bar and then causing this to move within an electric circuit, Marrian had set out to measure the conducting powers of different metals, but noticed that "at the instant that the magnetism is imparted *a sound is given from it*, and another sound is again perceptible when the galvanic current is broken." By varying the thickness and length of the metal bar employed, Marrian heard noises of differing intensities, observing that the sounds produced corresponded to "the tonic of each bar, and I could imitate it by striking the end of the bar with any metallic substance."[135] This experimental work constituted a sonorous investigation of electromagnetism. In October 1844, the English metallurgist John Percy (1817–89) wrote to Faraday, drawing his attention to Marrian's work, but Faraday was unimpressed.[136] He had seen the experiment before and knew of the phenomenon, "but as I heard of it suspect it is merely some conduction of the sound of the snap at the place of making or breaking contact."[137]

Faraday was more interested in sound's potential for investigating heat. In early 1829 a workman named Arthur Trevelyan (1802–78) was spreading common pitch with a red-hot plaster iron when, on placing his hot poker on a lead rest, he heard a musical note coming from the iron. In November 1830, he went to David Boswell Reid's (1805–63) practical chemistry class at the University of Edinburgh. After describing the poker phenomenon to Reid, they transformed the accident into an experiment and Reid persuaded Trevelyan to present the results to the Royal Society of Edinburgh, which he did through two papers during the winter of 1830–31. Over the next two years he continued to investigate the phenomenon, testing a selection of heated metals on various cold metal rests to replicate the incident and measuring their reactions by the musical notes produced. He found a hot brass bar on top of a cool lead rest to cause a great vibration, while applying pressure to the heated metal bar or cutting a v-notch into its surface also amplified the note. Returning to his original iron poker on a cold lead rest, he placed the lead on a pianoforte and found the poker's tones could be changed by striking notes on the instrument. Publishing his findings in 1833, Trevelyan concluded that musical pitches were contingent on the two bodies being metallic, with uneven surfaces and of a great variation in temperature, and attributed the vibration "to the usual mechanical changes which caloric occasions in passing from one substance into alterations of temperature."[138] There was no sound produced on a smooth surface because "the caloric enters into every part of the block equally." "Caloric" was understood to be the fluid substance of heat and this experiment was conceived of as a way of examining its movement.

Trevelyan's Royal Society of Edinburgh papers aroused much interest from Britain's scientific community. In early 1831 the Scottish science writer John Robison (1778–1843) sent sketches of Trevelyan's experiments to John Herschel, confident that the astronomer would be intrigued.[139] Faraday was also excited by Trevelyan's sonorous investigations, repeating these experiments at the Royal Institution in April. The use of sound to investigate the movement of heat made an obvious subject for the Royal Institution's Friday Evening audience. Instead of a hot poker, Faraday employed a heated piece of brass, four inches long and half an inch thick, to produce musical sounds.[140] He showed his viewers how the pitch could be increased by applying pressure to the brass with a small stick, entertaining his listeners by moving through an entire octave, before removing the pressure to return to the brass's fundamental note. Faraday contended that this sonorous phenomenon resulted from a rapid repeated heating and cooling which caused the brass to vibrate between its two places of contact, almost like an oscillating pendulum, producing a musical note.[141] Faraday felt Trevelyan's singing poker to be an obviously entertaining subject, but in 1831 it was also a way of investigating the movement of a force which was impossible to observe visually.[142]

Trevelyan continued to explore heat and sound between 1833 and 1835, but he remained an unknown figure.[143] He later conceived of an entire philosophy of nature based on the principle that variations in heat, conducted through metal, produced musical tones. In 1843 Trevelyan sent this work, his "Theory of Heat," to Cambridge mathematician William Whewell (1794–1866) in the hope that he would promote his writings. Based on his observations of the conversion of heat into sound during his poker experiments, Trevelyan asserted that all forces in the universe, including galvanism and magnetism, were forms of heat, and that there was "one natural law which is invariable, and which continues, and will continue to eternity, the conduction of heat."[144] Whewell's response to this communication is unknown, but Trevelyan could not find any journal to publish his work, being unknown in prominent science networks. Trevelyan complained of this to Whewell, reporting rejections from the *Chemical Gazette*, the *Phrenological Journal*, and the *Electrical Magazine*, the editor of whom warned that such a theory would "involve a complete revolution in our established views."[145] Whewell kept Trevelyan's unpublished papers but seems to have offered no encouragement to the aspiring experimentalist.

Despite these later failings, Trevelyan had initially enjoyed scientific attention, and no one had taken his singing poker experiments more seriously than James David Forbes (1809–68). Having studied at the University of Edinburgh from 1825 and, attending Thomas Chalmers's (1780–1847) lectures on theology and Reid's on practical chemistry, Forbes became a Fellow of the Edinburgh Royal Society in 1831 and professor of natural philosophy at the University of Edinburgh a year later.[146] Forbes had attended Trevelyan's Royal Society of Edinburgh paper on January 17, 1831, and subsequently witnessed Faraday's treatment of the subject at the Royal Institution in April, but he doubted both explanations for the musical phenomenon. After two years of experiments, in 1833, Forbes published his own account of Trevelyan's observations, agreeing that for a sound to occur, a groove on either the bar or block in the direction of the bar's axis was desirable. He concurred with Faraday in attributing the sound purely to vibration, rather than Trevelyan's assertion that sound also came from the movement of air through the cut groove. Forbes took the experiments much further, employing them to construct a catalog of metals which attempted to unite their physical properties and classify them by their relative vibratory power through heat conduction. As a cold substance, he found lead superior to all others in creating sound and so took lead at 65° degrees as a standard against which to measure other metals heated to 212° degrees. Forbes supposed that a resulting sound was contingent on the relation between each metal; the cold rest had to be a metal of a lower vibratory power than the heated one. Hot copper could make sound with cold brass, but hot brass could not sound with cold copper. This characteristic Forbes described as a metal's potential "vibrating energy" and he argued that this was related to each metal's potential to conduct electricity. Sound was construed as a way to observe relations between the conduction of different invisible forces. Remarkably, Forbes asserted "that when the best data are collected, the order of conducting powers of the metals for *heat* and for *electricity* is the same." He found that for heat, electricity, and vibrations, he could arrange metals into groups: namely gold, silver, and copper; zinc, brass, iron, and platinum; tin and lead; and antimony and bismuth.[147]

Given these apparent connections, Forbes was mystified that Trevelyan's discovery had not excited more interest. After all, within Britain's industrial culture of the 1830s and 1840s there was a great deal of work on this idea of a science of heat, or what eventually became "thermodynamics," especially since French engineer Sadi Carnot (1796–1832) had analyzed steam engines in terms of their efficiency in turning heat, or caloric, into work. He asserted that motive power was the product of the fall in caloric from a high to a low temperature: work from heat was contingent on temperature variations.[148] With such industrial concern over the relationship between motion and heat, Forbes urged natural philosophers to treat Trevelyan's singing poker experiments seriously. His work showed that the sound's intensity was due to "the amount of caloric which passes into the block [which] must increase with the conducting power of the material." Forbes attributed the sound to "repulsion" as much as conduction, which he suggested was an entirely "new species of mechanical agency in heat," ensuing from the cold metal's repulsing of heat from the hot metal. As he put it, sound depended on repulsion, and "repulsive energy depends on the differences of these conducting powers."[149] This view remained unchallenged until Tyndall revisited the question of singing pokers in 1853 as a Friday Evening Discourse subject. As he told Faraday, while Forbes thought it "essential that different metals must be used, I have had an instrument made . . . and obtained it with a distinct tone yesterday, which continued a quarter of an hour, from copper *on copper*." Tyndall was also able to produce musical tones from quartz samples, sufficiently loud "to be heard throughout the theatre when everything is perfectly quiet."[150]

Together, Faraday and Wheatstone had put sound at the very center of 1830s British science. Although frequently employed for the study of diverse natural phenomena, beyond those purely acoustic, such as heat, electricity, light, and magnetism, their cultivation of Chladni's acoustic plate experiments and publications in *Philosophical Transactions* had enhanced sound's scientific value. In 1830s Britain these acoustic figures took on increasing relevance, in part because of the renewed interest in invisible forces following Faraday's observation of electromagnetic induction in 1831. There was some significance to biographer John Hall Gladstone's (1827–1902) quip, when reflecting on Faraday's unifying account of nature, that no "other mortal had ever opened his ears so fully to the harmony of the universe."[151] Similarly, on Faraday's death in 1867, the journal *Punch* combined notions of musical harmony with Faraday's scientific investigations, recalling how his experiments continued,

> Till out of seeming chaos order grows,
> In ever-widening orbs of Law restrained,
> And the Creation's mighty music flows
> In perfect harmony, serene, sustained.[152]

From the Sonorous to the Telegraphic

At the 1834 meeting of the BAAS in Edinburgh, the place of sound in science was up for discussion. Before the Mathematics and Physics Section of the association, the English physician Charles James Blasius Williams (1805–89) presented his thoughts on the production and propagation of sound. Williams warned his audience that natural philosophy was far too focused on light and that this explained why "so simple and comparably easy a science as acoustics should have been so tardily developed, and that much of its recent advancement should be referrible [sic] rather to the illustrations which it affords to the sister science, optics, than to its own intrinsic value." The problem, as he saw it, was that sound was too often only of interest by way of analogy to other natural phenomena. He hoped "that some *master spirit* will take up the subject of acoustics, not only as an interesting and instructive link between the mechanical sciences and those subtler ones of light, heat, and electricity, but also for its own sake."[153] After this appeal to the BAAS's mathematicians and natural philosophers, he reviewed the state of acoustics in Britain. The great failure in recent years, he asserted, was Wheatstone's inability to conduct the sounds of the voice via wooden conductors with any power or quality. The Enchanted Lyre had not succeeded "in transmitting, *by any con-*

trivance, the sound of the voice or a flute through a solid conductor without very great loss in the intensity of the sound." To crack this problem would, Williams contended, require a true understanding of the molecular elasticity of solid bodies. He believed the answer would be found through a better understanding of the principles of the soundboards of musical instruments.[154]

Despite Williams's misgivings over Wheatstone's inability to develop the Enchanted Lyre into a system capable of communicating human speech, sound remained an important part of Wheatstone's scientific research and was to shape his subsequent efforts to develop an electrical telegraph throughout the 1830s. In 1834 Wheatstone became King's College London's professor of experimental philosophy. In May 1835 he wrote to John Herschel, reporting how he had

> just concluded a course on Sound, in which I endeavoured to include every experimental investigation which has hitherto been made in this department of physics. Among other things I copied De Kempelen's speaking machine which talks tolerably well, and cries as naturally as a living child.[155]

Wheatstone did not exaggerate; for his eight lectures he had prepared a huge catalog of experimental displays, including water tanks to simulate acoustic waves and musical instruments. He wanted to give "a complete view of the present state of our knowledge concerning the laws of sound" and hoped to "bestow more attention on a branch of science which has ministered so efficiently to our intellectual wants and pleasures, and which has of late years become so interesting to the philosopher from the strong analogies which its phenomena present with those of light and heat."[156] Indeed, during the 1830s, the King's College London lecture series on light, electricity, heat, and magnetism all included acoustic phenomena. Featuring more than 140 experiments and visual displays, Wheatstone's 1835 lectures were not just an account of acoustic knowledge but a vision of how such knowledge had shaped the very latest scientific research. Regarding sonorous impulses, there were undoubted connections "between the laws of these vibrations and those of other vibrating bodies."[157]

After explaining the theory of a vibrating string, how to measure vibrations, and the relationship between ratios of vibration and harmonics in lecture 1, Wheatstone proceeded to architectural acoustics in lecture 2, describing how "sounds are propagated with great distinctiveness along a smooth surface, as the surface of water or a wall."[158] This was something that could be experienced in the whispering gallery of St. Paul's Cathedral, where "a person speaking even in the lightest whisper at one point is heard distinctly by another person placed at another point as if the voice was close to his ear." Wheatstone reported that he had taken advantage of the recent construction of the Colosseum in Regent's Park to examine this phenomenon. There, he had an

> opportunity of trying a few experiments on the circulation of sound round a circular wall. . . . A temporary gallery was suspended from the ceiling and led from the stage whence the panorama is viewed, to the circumference of the building. . . . At the end of this gallery near the wall I observed the following effects. On pronouncing a word it was repeated a great many times, extremely loud at first, but successively weaker until it became no longer audible. A sentence being uttered the last syllable appeared only to be repeated and this was echoed as many times as before.[159]

Wheatstone's 1832 Royal Institution paper on the vibrations of columns of air in cylindrical and conical tubes reappeared as lecture 3, followed by a discussion of longitudinal vibrations in lecture 4 and of Chladni's acoustic figures in lecture 5. Outlining the history of these experiments, he quoted Chladni himself:

> "I had observed that a plate of glass or metal gave different sounds, when it was pinched and struck at different places. . . . Where I applied the bow to a round plate of brass, fixed at its middle, it gave different sounds, which compared with each other, were equal to the squares of 2, 3, 4, 5 &c. . . . The experiments upon the electric figures which are formed upon a plate of resin, discovered and published by Lichtenberg . . . made me presume that the different vibratory motions of a sonorous plate might also present different appearances, if a little sand or some other similar substance were spread on the surface."[160]

This quote not only provided listeners with a narrative of discovery but attributed Chladni's success to his ability to make connections between different phenomena.

Wheatstone's audiences could try Chladni's experiments for themselves at home. Wheatstone gave "practical directions as will enable any person to repeat the experiments," outlining how the plate must be "put in motion at a vibrating part by a violin bow, after having spread a little sand on the surface, which is repelled by the agitations of vibrating parts and accumulates upon the nodal lines." Glass was preferable, being easy to procure, while thin plates were better than thicker plates, "being more easy to put in motion, . . . to obtain more complicated figures, larger plates must be employed. In order that the figures shall be regular, the thickness must be everywhere the same." This replicable quality of Chladni's work was of great value to the dissemination of acoustic knowledge. Wheatstone specified that the plate had to be held firmly, but that those "who have not sufficient strength . . . may make use of a pair of nippers with pieces of cork or leather fastened to the nipping ends, or of a vice screwed to the table." While "simple figures, which are those accompanied by the lower sounds, will appear more easily if the bow be moved with greater pressure and less velocity," making complicated patterns, "the sounds of which are sharper," required greater velocity but less pressure. Recommending either sand or iron filings as vibrating particles, Wheatstone asserted that the "materials are easily to be obtained, and the great variety of figures, which with a little practise may be obtained from plates of different forms, will amply compensate for the time employed in repeating the experiments."[161] Acoustics was a subject which, he hoped, would encourage audiences to themselves became experimentalists. The great strength of Chladni's figures, as Wheatstone saw it, was that anyone could produce them within the comfort of their own homes.

In subsequent lectures Wheatstone dealt with the velocity of sound, its communication through solid conductors, and the movement of vibrations through musical instruments, providing experimental demonstrations with a broad selection of arrangements of solid conductors including tuning forks in connection with wood, metal, and glass rods, guitars, lutes, harps, and violins. While showing how these conductors received impulses from a perpendicular vibratory motion, he once again drew on the links between sound and light; there were "intimate analogies which subsist between the transmission of sound through solid conductions, and the propagation of light," but he briefly summarized that "sound, when its vibrations are all made in parallel planes is exactly analogous to polarized light."[162] Wheatstone continued to work on this analogy between light and sound into the 1840s, constructing devices to model the motions of the two impulses. Most celebrated of these was his wave machine which represented the motion of ether particles. Consisting of a series of wire rods, each tipped with mother-of-pearl beads, this moved in various patterns to simulate the propagation of a wave of sound or light, modeling the relationship between molecular movements and musical phenomena.[163]

Between May and June 1837, Wheatstone delivered a new lecture series uniting acoustics with light, heat, magnetism, and electricity through the theme of measurement. With these lectures, commencing at 8 p.m. on Tuesdays, made public, they were not confined to specialist audiences. On May 9, Wheatstone discussed the measurement of the pitch of sound, including tuning forks, monochords, and tonometers. The following week, he demonstrated how to measure the intensity and velocity of sound, and discussed the acoustic works of Laplace and Newton. In weeks 3 and 4 Wheatstone moved on to the measurement of light, and then in weeks 5 and 6 he outlined the measurement of temperature and heat. On June 20, he examined the measurement of magnetism and in the concluding week, electricity.[164] The structure of this syllabus is significant in that it placed sound as the fundamental phenomenon capable of measurement.

Wheatstone spoke with authority, knowing how productive acoustic knowledge could be to broader philosophical works, notably concerning telegraphy. From 1834, he had commenced investigations into the rate of transmission of electricity along copper wire. Taking the principle of observing an oscillating rod by creating a continuous luminous line of reflected light from the rod's path, as used in his kaleidophone, Wheatstone made "a series of experiments relating to the oscillatory motions of sonorous bodies, . . . to ascertain whether, by similar means, some information might not be gained respecting the direction and velocity of the electric spark." By comparing this oscillating luminous path of a moving rod with the motion of another moving body, whose velocity was known, it was possible to calculate the rod's velocity. In the same way that the reflective beads of the kaleidophone made vibratory motions visible as luminous lines, a rotating mirror made the motion of a spark observable as a continuous path. As Wheatstone explained, "Vibrating bodies afford many instances for investigation. . . . A flame of hydrogen gas burning in the open air presents a continuous circle in the mirror; but while producing a sound within a glass tube, regular intermissions of intensity are observed, which present a chain-like appearance, and indicate alternate contractions and dilations of the flame corresponding with the sonorous vibrations of the column of air."[165] From his kaleidophone, Wheatstone knew it was possible to make a vibration appear as a continuous line.[166] His experiments on the measurement of the speed of electricity eventually involved four miles of wire; in the vaults beneath the buildings at King's College London he calculated that electricity had a velocity of 250,000 miles per second.

It was for his telegraphic work that Wheatstone would become best known. As with his investigation into electricity's velocity, this involved the adoption of practices cultivated in the study of acoustics, especially those employed in his Enchanted Lyre experiments. Fifteen years of acoustic experiments had provided Wheatstone with a range of practices and techniques for constructing telegraphic communication apparatus; it was through the electric telegraph, rather than solid conductors of sound, that Wheatstone would realize the promises of virtually instantaneous communication made in his Enchanted Lyre demonstrations.[167] Wheatstone's work caught the attention of William Fothergill Cooke (1806–79), who would help transform the natural philosopher's experimental arrangements into a working telegraphic system. After discussing telegraphy with Faraday in late 1836, Cooke met Wheatstone in February 1837 to consider making a practical communications network. They subsequently took out a patent for a "five needle" electric telegraph in May, quickly securing a contract with the Great Western Railway (GWR) and, in 1844, winning a twenty-year contract with the Admiralty to connect London and Portsmouth at £1,500 per annum.[168]

Wheatstone's work on sonorous communication fed directly into his development of telegraphic apparatus. Questions over insulating deal rods were equally applicable to insulating copper wire for electrical impulses, as were problems of conduction and making a transmitted signal observable. While soundboards made sonorous vibrations audible, Wheatstone applied his acoustic experience to conveying human speech over vast distances. Indeed, the transmission of speech had always been an ambition of Wheatstone's, even with the earliest experiments of his Enchanted Lyre. While the transmission of music was amusing, it was of "far less importance than the conveyance of the articulations of speech," which Wheatstone's experiments showed to be possible; as long as the solid conductor was connected to the neck or head, contiguous to the larynx, such powers seemed close to being realized. Furthermore, "if all the articulations were once thus obtained, the construction of a machine for the arrangement of these into syllables, or words, and sentences, would demand no knowledge beyond what we already possess."[169]

In part, this is why he took so much interest in talking machines, which might be connected by telegraph to transmit speech. Wheatstone was particularly excited by the work of Robert Willis on speech-producing automata, the most celebrated eighteenth-century example of which was Wolfgang von Kempelen's (1734–1804) talking device, modeled on the human physical form and developed in 1784. Willis constructed his own version of this machine, including a reed and a system of pipes adapted in length with a plunger to produce different vowel sounds.[170] Wheatstone repeated Willis's experiments on vowel sounds, convinced that by carefully observing the similarities of sounds constituting vowels and consonants, it would be possible to reduce human speech to a few acoustic laws which could then be produced mechanically.[171] Wheatstone rebuilt Kempelen's talking machine and presented a paper on it to the BAAS in 1835. In 1840 this was still on display at King's College London, and, in 1860, the thirteen-year-old Alexander Graham Bell (1847–1922) saw the machine while in London.[172] While the communication medium changed from sound to electricity, Wheatstone's original goal of transmitting speech remained.

Not only did Wheatstone's acoustic knowledge shape his telegraphic work, but he incorporated it into his efforts to communicate using electricity. In 1878 the German-born engineer Carl Wilhelm Siemens (1823–83), reflecting on the history of telegraphy in his presidential address to the Society of Telegraphic Engineers, observed that it had been Wheatstone, in uniting telegraphy with acoustics, who had initiated the development of the telephone. Siemens explained that, in 1859, Wheatstone

> devised an arrangement by which the sounds of a reed or tuning-fork, or a combination of them, could be conveyed to a distance by means of an electric circuit.... In striking any one of the tuning-forks different currents were set up, which caused the vibration of the corresponding tuning-fork at the distant station, and thus communicated the original sound.[173]

Wheatstone's electrical communication had, in this way, finally resulted in a sound-regulated system. When, in 1876, King's College London's professor of natural philosophy William Grylls Adams (1836–1915) delivered a course of three Royal Institution lectures on Wheatstone's research, he began with a session on his acoustic investigations, before subsequent lectures covered electricity and telegraphy. Included were experimental displays of the kaleidophone and acoustic figures. Through these lectures, Adams set out a vision of scientific investigation in which Wheatstone's acoustic knowledge had been central to his electrical and telegraphic work.[174]

Conclusion

Acoustical marvels confused the senses by making what was heard incompatible with what was seen. They amused, drew popular and scientific audiences alike, and they raised philosophical questions over the nature of sound. Between 1828 and 1832, Faraday and Wheatstone oversaw the transformation of Pall Mall's premier sonorous exhibitions, including musicians and musical instruments, into objects of scientific discourse. This was achieved by refashioning the character of these performances. In showrooms and shops, the Jew's harp and Enchanted Lyre appeared mysterious, offering inexplicable manipulations of nature: the focus of these popular musical displays was on the wonder of the musician's skill or the instrument's deception. Yet repositioning these same performances within the Royal Institution involved shifting the emphasis away from the mysterious to the explanatory. Rather than trying to bewilder audiences, Wheatstone and Faraday used musical exhibitions to illustrate philosophical observations on nature. And as is clear from Faraday's lecture notes, this entailed the restyling of such amusements as experimental procedures, delivered as part of an empirical analysis of sonorous phenomena.

This radically altered sound's position within Britain's scientific culture. In many respects, the fashionable, elite, metropolitan audiences of the Royal Institution were not that different from those of Pall Mall shops. After all, both were located in London's well-to-do West End. Similarly, the same journals reported on these exhibitions, be they aimed at scientific specialists or those broadly interested in the arts and sciences. The boundaries between popular entertainment and elite scientific engagement were unclear and flexible. But through instruments like the kaleidophone, Enchanted Lyre, and concertina, the place of sound in British science undoubtedly changed during the 1820s and 1830s. Such contrivances and musical performances were ideally suited for London's showrooms and shops but, as Wheatstone and Faraday demonstrated, also provided valuable resources to mobilize at the Royal Institution. In their adaptation from showroom amusements to scientific objects, such musical performances provided useful forms of entertainment to secure new audiences for philosophical subjects. Eulenstein's career, and the journey of the Jew's harp from obscure toy to musical and scientific instrument, illustrate the growing prominence of music in London's science, as do the performances of Godbè and Mannin. Drawing on the obvious novelty of their musical skills, these were not just of aesthetic value but illustrated and imparted serious sonorous knowledge claims. At the Royal Institution, audiences were encouraged to listen, rather than just hear, and to consider the philosophical explanation for the sounds they encountered, rather than just enjoy the performance. Yet in employing music in this way, sound ended up providing new forms of experimental investigation. While music captivated London audiences in the 1820s and 1830s, the study of sound not only popularized new knowledge but helped to manufacture it. Sound's transformation from object of entertainment into scientific discourse was, by the mid-1830s, well underway.

✳ CHAPTER 2 ✳

A Harmonious Universe
Herschel, Whewell, Somerville, and the Place of Sound in British Mathematics, 1830–70

> *But that Herschel, for example, who "broke the barriers of the heavens"*
> *— did he not once play a provincial church-organ, and give music-lessons to stumbling pianists?*
>
> GEORGE ELIOT, *Middlemarch*[1]

In his *Encyclopaedia Metropolitana* treatise on sonorous phenomena, published in 1830, the celebrated British astronomer John Herschel (1792–1871) introduced a powerful analogy to explain how sound waves moved through communicating media. Comparing the particles of a transmitting body, be it solid, air, or water, to ears of corn, Herschel likened sonorous impulses to gusts of wind. Just as the sound emanating from a musical instrument would make surrounding particles of air vibrate, so wind blowing over a cornfield would make each ear oscillate upon its stalk, producing what would appear as a single wave sweeping through the corn (fig. 2.1).[2] The value of Herschel's cornfield analogy was that it allowed readers to conceive sound as a force that was conveyed but, unlike traditional sonorous analogies that taught that sound moved like water, did not risk implying that sound communicated particles over distance. Instead, Herschel's cornfield made it clear that sound was an impulse, transmitted by oscillating particles that, nevertheless, remained in a fixed location. It was to prove a persuasive analogy: over the next decade, leading science writers across Britain would replicate Herschel's windblown corn in their own accounts of sonorous phenomena as they strived to instruct increasingly broad social audiences on the science of sound. Importantly, however, this analogy would take on theological significance as sound informed grand visions of a divinely created universe, governed according to a few unifying laws of nature.

Herschel offered a scientific account of sound that was acceptable to Britain's religious audiences, but not all natural philosophy was so com-

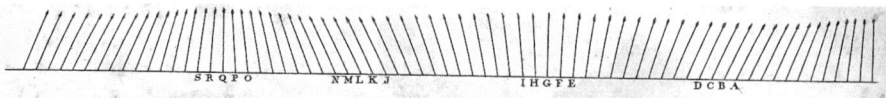

FIGURE 2.1 John Herschel's cornfield analogy, displaying the impulse of a gust of wind over ears of corn. Herschel, "Sound," plate 1. Public domain.

patible with faith. During the 1830s and 1840s, science often appeared as a politically dangerous body of knowledge which threatened to undermine the authority of traditional institutions, particularly that of the Anglican Church and Crown. New chemical understandings implied that all in the universe, physical and spiritual, might be the result of matter in motion, while geology raised concerns over the age and history of the Earth. In astronomy, the observation of nebulae and theory that these were where new stars were born had worrying ramifications for notions of how the universe was created. Collectively, new knowledge appeared to be fueling crude materialism and conceptions of a godless universe. In contrast to this, as James Secord has argued in his *Visions of Science* (2014), science writers like Herschel and Mary Somerville produced alternate accounts of nature to those socially controversial, establishing visions of a unified and ordered universe which were not materialistic but founded on theological convictions.[3] Nor was this a one-way relationship in which science writers merely wrote within a religious framework. Instead, their works shaped and guided church teachings on matters of natural philosophy. And while science writers aimed their works at Britain's reading public, clergymen offered guidance to a much broader social audience, including illiterate artisans, rural laborers, and urban workers as well as gentry and polite middle-class listeners. The spoken word of the sermon was far more accessible than the printed treatise and required no literacy. Sermons taught lessons on the sonorous which drew both on the Bible and on leading scientific works.

Herschel's work on sound provided a scientific account of sonorous phenomena which fit within a grand vision of nature as a unified system of mathematical forces and laws. Thanks to Wheatstone and Faraday's endeavors, mid-nineteenth-century Britain could boast an impressive experimental tradition surrounding acoustics, but this was not the only analytical approach to sound, as British mathematicians put sound at the center of their investigations. In particular, sound provided a way of linking natural phenomena and showing how the universe was regulated according to a few simple laws which, in turn, suggested divine creation. Of course, such romanticized conceptions of a harmonious universe were not new in the 1830s, dating back to Pythagoras's ideal that the seven planets were related to the seven notes of the musical scale, the "Music of the Spheres." Contrary to Jamie James's claim that such notions collapsed amid nineteenth-century industrialization, and that with the "revelations of modern scientific enquiry" music was understood increasingly as a series of physical properties rather than as an ideal of nature, these sentiments endured.[4] Early nineteenth-century writers continued to unite sound and astronomy, perpetuating idealistic visions of a harmonious universe, ordered by a divine Creator. In 1823, for example, the English composer Isaac Nathan's (1792–1864) *Essay on the History and Theory of Music* reflected on the beautiful parallels between harmony and astronomy. He agreed with Pythagoras's linking of heavenly bodies and musical harmony, believing that music was central to a united vision of God's Creation. "Harmony prevails throughout the works of our Creator," explained Nathan, "it is perceivable in all living things, even to the minutest fibre of the smallest field-flower, and it is their just and symmetrical proportions which delight us by throwing a pleasing harmony over the whole."[5] Reviewing Nathan's essay in 1833, the *Westminster Review* found "the analogy understood to exist between the division of the musical string, and the distances of the planets from the sun" a compelling one and asserted that, if proven, "it points to the suspicion, in the cases of the arrangement of the planets and the particles of light, of a connexion with the successively greatest possible frequencies of coincidence in the effects of different sets of periodical impulses of some unknown kind."[6] Such a connection would, the *Westminster Review* contended, have metaphysical implications. Nor was this musical-mathematical association confined to European audiences. Columbia College professor Henry James Anderson (1799–1875), for instance, expressed in 1830 how from "the minutest vibrations of a harp-string to the magnificent oscillations of a planet's axis," everything in physics was reducible to oscillatory motion.[7]

Sound was not only an invaluable didactic and analogical device for illustrating other natural phenomena but contributed to a growing understanding of a harmonious universe, ordered and stable, put in motion by a divine Creator. As planets made grand oscillations through space, particles made circular vibrations in the propagation of sound. This chapter demonstrates how, in mid-nineteenth-century Britain, there were close intellectual links between music and sound on the one hand, and mathematics and astronomy on the other, and that these connections were inseparable from broader theological accounts of nature. Focusing on three leading British science writers during the 1830s and 1840s, namely Herschel, Somerville, and William Whewell, it becomes clear how such works of natural philosophy influenced religious accounts of sound and music, as projected through sermons and religious tracts. While both Whewell and Herschel had been undergraduates at Cambridge and endured the rigorous training for the university's Mathematical Tripos Examinations, Somerville was herself a skilled mathematician. They all shared similar understandings of the universe that were highly compatible within the elite Anglican universities of Oxford and Cambridge, where discussions over science and mathematics accompanied the daily experience of musical religious practices in churches and chapels. Increasingly, as will be shown, scientific understandings of sound were interlinked with moral teachings over the place of music within worship and the arts' metaphysical value for congregations, as the Church of England underwent a musical renaissance as part of broader reform efforts surrounding its rituals and social position.

John Herschel's Sound

With state patronage, high status, and considerable organization, astronomy was the most prestigious of early nineteenth-century sciences. Yet it could also be troubling, with the popular and controversial nebular hypothesis suggesting that the universe was constantly developing. In his *Exposition du système du monde* (1796), Pierre-Simon Laplace (1749–1827) made exactly this point, alleging that nebulae, collections of gaseous matter, were the birthplace of stars and planets. This offered indications of the universe's past and had ramifications for notions of change on earth. Despite this, astronomy set a standard for the investigation of nature with valuable lessons on how to organize the physical sciences.[8] Although France had dominated late eighteenth-century astronomy, it was the Germanic states that led the discipline during the early nineteenth century; a point confirmed by Berlin Observatory's locating of the planet Neptune in 1846. Britain's astronomical tradition was growing too, especially following George Biddell Airy's (1801–92) overhaul of the Royal Observatory of Greenwich during the 1830s and 1840s. But the most celebrated figure in nineteenth-century British astronomy was unquestionably John Frederick William Herschel.[9] For Herschel, the study of the celestial accompanied that of the sonorous.

As the son of the astronomer and musician William Herschel (1738–1822), who discovered the planet Uranus in 1781, John grew up in a world of telescopes and organs. A brilliant mathematician, he matriculated at St. John's College Cambridge in October 1809, graduating as Senior Wrangler and First Smith's prizeman in 1813. Along with mathematicians

George Peacock (1791–1858) and Charles Babbage (1791–1871), he had led the Analytical Society as an undergraduate, which sought to reform Cambridge's Newtonian mathematical tradition by introducing Continental-style mathematical analysis. France was a center of mathematics during the early nineteenth century, but its analytical methods were controversial in Britain due to their revolutionary connotations. With the peace of 1815, such mathematics slowly became more acceptable at Cambridge.[10] In 1816 Herschel joined his father in conducting astronomical observations.

William Herschel had fled Hanover during the Seven Years' War (1756–63); a musician in the Hanoverian army, he took up a new post as an instructor of music to a small military band in northern England. From 1762 he worked in Leeds, endeavoring to "become known as a composer . . . always my chief ambition."[11] Organist of Bath's newly built Octagon Chapel from 1766, he later took the directorship of the Bath Orchestra. The chapel, on Milsom Street, had been designed specially to facilitate organ music and became a fashionable institution among Bath's respectable society: here its patrons could enjoy comfort and quality music away from poorer social elements. William composed several symphonies and concertos but, with his only successful composition a set of six sonatas for the harpsichord, he became increasingly interested in the study of nature, which he considered a more engaging intellectual challenge; it was, he lamented, "a pity . . . that music is not a hundred times more difficult as a science."[12] He attempted to write a treatise on harmony in 1773, but on the failure of this effort dedicated the next five years to the construction of his own seven-foot focal-length telescope with which to pursue astronomy. Music had led William to this astronomical interest; while in Halifax in March 1766, he purchased the Cambridge music theorist and mathematician Robert Smith's (1689–1768) work on harmonics. That in turn encouraged him to read the Cambridge professor's work on optics and to start making celestial observations.[13] In 1781 he reported a previously unrecorded comet to the Royal Society, which turned out to be a completely new planet. Celebrated as the discoverer of Uranus, Herschel gave up his musical career and moved to Slough in 1786, where he built a colossal forty-foot focal-length telescope with a forty-eight-inch diameter mirror.[14]

This combining of astronomy and music was far from unique. For example, William Herschel's contemporary, Charles Burney (1726–1814), approached music in the same way as astronomy, in that he treated both disciplines as objects of natural history, while the astronomer Adam Walker (1731–1821) built models both of the planets orbiting the sun and of musical instruments, and he exhibited these alongside each other in his astronomical lectures.[15] Likewise, in early nineteenth-century Britain it was mathematicians and astronomers who worked to produce greater accuracy in sonorous measurements. At Woolwich arsenal, this had musical implications, with scientific precision paralleling precision in music: the armed forces were, after all, the country's largest employer of musicians.[16]

In 1820, John Herschel and his father built a new twenty-foot focal-length reflecting telescope for the search of nebulae and double stars. William identified some 2,500 star clusters and 848 double stars (about 100 nebulae had been known in 1780), to which John added 3,346 double and multiple stars between 1820 and the publication of *A Treatise on Astronomy* in 1833. During four years at the Cape of Good Hope, he found another 1,708 nebulae and 2,103 double stars; returning to England in March 1838, he published these observations in 1847 as a vast astronomical catalog, *Results of Astronomical Observations Made at the Cape of Good Hope*. Despite this industry, the younger Herschel remained interested in mechanics, chemistry, optics, electricity, and acoustics, emphasizing the importance of empirical data in the study of nature.[17] As early as 1824 Herschel had discussed sound with the celebrated English natural philosopher Thomas Young (1773–1829). Young had been disappointed at Herschel's lack of engagement with his own works, telling him that "I think you could have found many things in them *more* worthy of being mentioned," and sending Herschel a volume of his research on the propagation of sound. Nevertheless, Herschel freely admitted to lacking any musical talent, reflecting in 1858 that his "want of power of playing on the piano however quite prevents my having any sound notions of practical harmony." Yet he did have experience of architectural acoustics, having assisted the Reverend Francis Wilson in 1831 to position sounding boards around his pulpit to increase the power of his weak voice. Herschel was also an associate of the church musician Charles Wesley (1757–1834) throughout the 1820s.[18]

Herschel made a substantial contribution to the science of acoustics, beginning with his influential entry "Sound" in the *Encyclopaedia Metropolitana*, which he had drafted by February 1830. It was probably the most cited publication on sound in English during the mid-nineteenth century. Its timing was significant, in the midst of the interest in acoustics which Faraday and Wheatstone cultivated in London. Starting with the propagation of sound, Herschel explained the relationship between velocity, air density, and temperature before outlining how sound traveled through particles. After recalling experiments on echo which he had made beneath the Menai Suspension Bridge in 1827 with Charles Babbage, he addressed architectural questions of acoustics in relation to cathedrals, churches, and theaters. Of windows, recesses, hangings, and carpeting, these were "to Sound, what black spaces in an apartment would be to light."[19] After relating this to architectural acoustics, Herschel outlined the mathematical theory of sound's transmission, which he likened to tremors along stretched chords or waves through water. However, he emphasized that "the *wave* which advances on the surface of water—the sinuosity which runs along the stretched cord—are neither of them things, but *forms*." Fearing readers would find the concept that particles moved only a little in the propagation of sound hard to comprehend, he employed the analogy of a cornfield, explaining how the

> waves in a field of standing corn, as a gust of wind passes over it, afford a familiar example of the relation between the motion of the wave, and that of the particles of the waving body comprised within its limits.... The gust in its progress depresses each ear, in its own direction, which, so soon as the pressure is removed, not only returns, by its elasticity, to its original upright situation, but by the impetus it has thus acquired, surpasses it, and bends over as much, or nearly as much, on the other side; and so on alternatively, oscillating backwards and forwards ... till it is brought to rest by the resistance of the air.[20]

The propagation of sound through oscillating particles worked in the same way. This analogy illustrated the essential difference between a wave and moving matter: sound was not a projection of particles but an impulse. Herschel's analogy was to prove influential in future explanations of this crucial distinction.

The following sections of Herschel's "Sound" subjected musical sounds and the communication of vibrations to mathematical analysis. For the former of these, he described how the audibility of sound was contingent on the speed of vibrations, while the regularity and rapidity of impulses determined if the sound was musical. Herschel then proceeded to explain the theory of how string and wind instruments vibrated and set air into simultaneous movement, relating this to Chladni's acoustic plates. In his final section, Herschel drew attention to the communication of vibrations, with particular reference to Savart's work in placing sand over musical instruments to reveal their nodal lines. Savart's memoirs, communicated to the Royal Academy of Sciences in Paris, constituted what Herschel thought were the leading writings on the subject, thanks to his use of lycopodium dust instead of sand.[21] Herschel concluded by drawing attention to Robert Willis's work on mechanically producing vowel sounds. The imitation of human speech was, he surmised, the acme of acoustic science.

Surprisingly, though, Herschel did not comment on Wheatstone and Faraday's work, especially startling given Herschel's treatment of solid conductors. He observed that wooden sticks could clearly convey sounds but also invoked Chladni's work to communicate sound through metal wires and the observation of sound's propagation through rock in Cornish mines. Using Chladni's calculations on the speed of sound through various solid conductors, Herschel observed how "we may estimate the time it requires to transmit force, whether by pulling, pushing, or by a blow, to any distance, by means of iron bars or chains. For every 11090 feet of distance the pull, push, or blow, will reach its point of action one second after the moment of its first emanation from the first mover." Such speed ensured that the communication of sound seemed almost instantaneous; however, "were the sun and earth connected by an iron bar, no less than 1074 days . . . must elapse before a force applied at the sun could reach the earth. The force actually exerted by their mutual gravity may be proved to require no appreciable time for its transmission. How wonderful is this connection!"[22] Yet while believing such sonorous communication to be a significant part of acoustics, Herschel made no mention of Wheatstone's work. More curious was Herschel's treatment of resonance, or vibrating columns of air. "The Jew's-harp is an instrument much mistaken and unjustly contemned. Nothing can exceed the softness, sweetness, and delicacy of this instrument," claimed Herschel, testifying to his own witnessing of Eulenstein's music. He then proceeded to explain exactly how the Jew's harp worked and its implications for understanding resonance:

> That the instrument itself vibrates in unison with the note it calls forth, is evident from the fact that when merely held before the open mouth, or lightly retained between the lips, its Sound is feeble and scarce audible; but acquires a great accession of force when brought in contact with and firmly held between the teeth. . . . It is more probably by a series of subordinate vibrations going on in the tongue while oscillating, that the sympathy is established.[23]

The similarity of this account of solid conductors to Wheatstone's 1828 paper "On the Resonances or Reciprocated Vibrations of Columns of Air" is so obvious that it is inconceivable that Herschel did not have it to hand when drafting "Sound." Furthermore, Herschel had witnessed

Wheatstone's experimental displays in Pall Mall in 1825. Soon after this, Wheatstone wrote to Herschel thanking him for his encouragement and inviting himself to Herschel's home at Slough, where he hoped to "exhibit a little philosophical instrument which I think I shall call the Kaleidophone." Wheatstone was confident that his kaleidophone experiments would "tend greatly to elucidate the real theory of undulatory motion, and to establish more intimate relations between sound and light." He also asked Herschel for advice on publishing his findings, as well as the input of "your superior mathematical knowledge."[24]

Later, in 1833, Herschel requested a copy of Wheatstone's Jew's harp paper and invited him to Slough to repeat his "elegant experiments on light & sound" over dinner with guests. Wheatstone accepted the invitation and "put in order a few acoustical and optical experiments," along with copies of his Jew's harp paper and his more recent *Philosophical Transactions* publication on acoustic plates. After his dinner experiments, Wheatstone presented Lady Herschel with his new "symphonion," with which she was greatly enamored. In 1835 Wheatstone personally repaired this instrument to ensure it remained in good order while the Herschels were at the Cape of Good Hope.[25] Both scientifically and socially, Wheatstone and the Herschels were on excellent terms. The question, then, is why did Herschel intentionally ignore Wheatstone's acoustic contributions? The answer was political: Herschel deliberately omitted references to recent British sonorous investigations as part of a wider controversy within British science in which he was to take a leading part through his association with Charles Babbage.

Babbage had grown up in the Devonian towns of Totnes and Teignmouth before going to Trinity College, Cambridge, in 1810, and he migrated to Peterhouse in 1812. Babbage, politically radical, had met Herschel as an undergraduate. On inheriting over £100,000 from his father in 1827, he settled in London, where he focused on building his famous calculating machine and stood unsuccessfully to represent Finsbury in Parliament in the elections of 1832 and 1834.[26] His most determined program for reform, however, unfolded during 1830 with his publication of *Reflections on the Decline of Science in England*. Here Babbage argued that English science was in decline because of its corrupt, aristocratic institutions, especially the Royal Society. For Babbage, such organizations demanded reform, with French institutions offering a model of how to reorganize science. In Herschel, he found an influential ally.[27] Babbage demanded a new, nonaristocratic president of the Royal Society in 1830 and persuaded Herschel to stand for the post. Although losing the election 119 votes to 111 later in 1830, Herschel's "Sound" played an important part in Babbage's campaign. Babbage and Herschel's correspondence regarding the decline of science accompanied their discussions of "Sound." On hearing reports that the government had agreed to finance Babbage's difference engine in February 1830, Herschel joked that he had heard "that the Chancellor of the Exchequer has at length declared the Engine 'Government property' which in the way it is mentioned seems to carry a pleasing sound—but as I have learnt by some considerable attention to the subject of Sound . . . [I have] no confidence" in such well-sounding phrases. Babbage responded by requesting a copy of Herschel's "Sound" before finalizing his Decline.[28]

Having completed the text of "Sound" in February 1830, but knowing of Babbage's coming attack on the Royal Society, Herschel added an extensive footnote to his article, emphasizing that the content came almost exclusively from foreign journals.[29] Herschel observed how he had drawn largely on

> the Annales de Chimie, and we take this *only* opportunity distinctly to acknowledge our obligations to that most admirably conducted work. Unlike the crude and undigested scientific matter which suffices . . . for the monthly and quarterly amusement of our own countrymen, whatever is admitted into *its* pages has at least been taken pains with, and, with few exceptions, has sterling merit.[30]

Herschel made it clear that French science was dominant and that the Annales de Chimie was superior to any English publication, having contributed "to the high scientific tone of the French *savons*." Furthermore, the extensive footnote lamented that good English science was appreciated in Europe before it was in Britain. Unfortunately, this respect was not mutual, as Herschel accused British philosophers of ignoring Continental science: in England, "whole branches of continental discovery are unstudied. . . . It is in vain to conceal the melancholy truth. We are fast dropping behind."[31]

This damning footnote was a late addition to support Babbage's declinist assertions, but it was also an injustice to Wheatstone's acoustical work, which built so painstakingly on Continental science. Herschel did make reference to Thomas Young, but this was not an inconsistency because both Herschel and Babbage felt that Young's death in 1829 was part of the narrative of decline in English science. However, it seems probable that Herschel deliberately omitted, even removed, all references to contemporary experimental acoustical investigations in London. Acoustics was a science that, in 1830, it was hard to claim was in decline in Britain.

Herschel later complained to Whewell that he regretted the controversial "Sound" footnote and its subsequent appropriation in declinist debates, believing that Babbage had taken his words out of context and expressing disappointment with Babbage's "violent" style.[32] The footnote effectively made Herschel a leader in Babbage's campaign.

Yet Herschel's "Sound" was not only politically significant but also influential in the broader study of nature, being one of three closely connected contributions to the *Encyclopaedia Metropolitana*—with "Physical Astronomy" and "Light" —which collectively carried theological implications for understanding natural phenomena. In his paper "Physical Astronomy," Herschel argued that as human knowledge developed, so the number of laws of nature identified could be reduced and that these laws were evidence of a divine architect. "Modern science... has given us proofs of the agency of one general and intelligent cause throughout the whole system of nature," alleged Herschel. The laws governing heavenly bodies were united; all movement was subject to the law of gravity, which commanded the oscillatory motions of the heavens in a way analogous to magnetism.[33] He expanded these connections in "Sound" and "Light," the latter published in 1827, before uniting these phenomena directly in *A Preliminary Discourse on the Study of Natural Philosophy*, published in 1831.[34] This work appeared as part of Dionysius Lardner's (1793–1859) *Cabinet Cyclopaedia* and, as Secord has shown, was as much an instruction on character and thinking as it was a scientific treatise. Selling over seven thousand copies in a few months, Lardner's work was aimed at a broader readership than Herschel's *Encyclopaedia Metropolitana* entries, being technically undemanding and, thanks to steam-press printing, low-quality paper, and pink-calico binding, cheap. While Babbage highlighted examples of poor scientific and moral conduct in his *Decline*, Herschel offered a model of ideal behavior for seekers of truth.[35] In terms of structure, he began with the phenomena of force, moved on to the communication of motion through bodies in the form of light and sound, and then considered astronomy. In this progression, sound played a useful role, laying a foundation for his analysis of light.

Herschel asserted that knowledge of nature was drawn from sensory impressions of external objects, conveyed to the human mind, which then connected these observations to draw out unifying laws. Some of these impressions were sonorous, comprised of vibratory motions passing through elastic bodies which, Herschel explained, produced musical notes if occurring in a rapid uniform manner. Sound therefore had a special place in the senses because, as Herschel continued, harmony, which depended on the "recurrence of coincident impulses to the ear," was perhaps "the only instance of a sensation for whose pleasing impression a distinct and intelligible reason can be assigned."[36] Unlike other forms of beauty, music was reducible to mathematical laws.

Yet Herschel said comparatively little about sound itself before examining light. While he spent almost fifteen pages of his treatise exploring vision, sound featured as a three-and-a-half-page chapter introduction, being little more than a precursor to, and analogy for, optical phenomena. But Herschel made constant reference to the propagation of sound when discussing light, maintaining that, beyond sound, there were few branches of physics "which promise at once so much amusing interest, and such important consequences, in its bearings on other subjects, and especially, through the medium of strong analogies, on that of light."[37] With this, he recognized that acoustics was entertaining, that it could be used to investigate nature broadly, and that it offered "strong analogies" on subjects including, but not limited to, light. If, as Herschel observed, identifying relations between natural phenomena revealed evidence of divine laws, then sound was of paramount philosophical and theological value.

Herschel's treatise thus appealed to audiences beyond those purely scientific, including those with musical concerns. For instance, William Mullinger Higgins's (1808–1882) *The Philosophy of Sound* recommended in 1838 that music follow the philosophical model which science provided. As lecturer of natural philosophy at Guy's Hospital in London, Higgins thought the natural sciences provided an attractive epistemological example for music to follow. In explaining what sound was and how it was propagated, he drew heavily on Herschel, quoting extensively from the *Encyclopaedia Metropolitana*; this was, he believed, a framework for future scientific treatises. Higgins repeated Herschel's description of Chladni's experiments before explaining the motions of vibrating air particles. As he put it, "Imagine a string to be a single row of indivisible particles, or, as they are sometimes called, molecules. . . . Within a certain limit, the close connexion between these particles may be disturbed without being destroyed." Likewise, while an external force could vibrate the molecules of this string without breaking it, this phenomenon could be witnessed with magnets, which if "suspended near to each other will soon arrange themselves by magnetic attraction." So it was with the particles of a sounding body. While such knowledge of nature was not obviously useful, Higgins contended that it was important, observing how, "by ascertaining the power of different substances in the transmission of sound . . . there may be found a connexion between this and other properties of matter," including for heat and electricity. However, Higgins lamented how undeveloped the "science of music" was, finding it peculiar that music students, while trained to perform and hear, learned little about sound itself. To understand music's emotive power over the minds of all social classes, musicians should, ideally, have both "imagination" and "scientific knowledge."[38] Such a reading of Herschel's work on sound engendered both moral and philosophical implications.

Somerville's Harmonious Universe

Herschel was not alone in employing sound to project a unified view of nature. During the 1830s there was a growing consensus that electricity, magnetism, light, heat, and sound were all connected. At the same time, parallels between these forces and the motions of the heavens made for intellectually satisfying visions of science. The latest observations on the physical sciences could in this way be connected with the most established body of knowledge of the day, which was astronomy. Herschel therefore provided something of a template for how to portray nature and its laws, and no one seized on this idea more famously than his friend, Mary Somerville (1780–1872).

Originally from Jedburgh, Somerville had a Presbyterian upbringing in Edinburgh and Burntisland but later became sympathetic to Unitarianism, which stressed the power of God without asserting that Christ was divine. On occasions, she secretly attended a Unitarian chapel, despite renting an Anglican pew for respectability. Music was important in her childhood. She rose early every day for four or five hours of piano lessons with her Italian teacher, Corri, whom she recalled "taught carelessly, and did not correct a habit I had of thumping so as to break the strings." Nevertheless, she did learn "to tune a piano and mend the strings, as there was no tuner at Burntisland," and later "got over my bad habit and played the music then in vogue: pieces by Pleyel, Clementi, Steibelt, Mozart, and Beethoven."[39] Somerville subsequently attended Alexander Nasmyth's (1758–1840) newly opened academy for ladies in Edinburgh, where she studied geometry and persevered with the piano. On marrying, she moved to London, but on the death of her first husband returned to Edinburgh. Remarrying in 1812, her new husband and cousin William Somerville (1771–1860) encouraged her scientific interest. In 1816 she returned to London and in the same year met the Herschel family, spending a day with William and Lady Herschel at their home in Slough. William showed Somerville his telescopes and musical books. She noted how well these two pursuits combined, recalling that he made us examine his celebrated telescopes, and explained their mechanism; and he showed us the manuscripts which recorded the numerous astronomical discoveries he had made. They were all arranged in the most perfect order, as was also his musical library, for that great genius was an excellent musician.[40]

Along with his telescopes, William introduced Somerville to his son John and the two quickly became friends.

Throughout the 1820s, she frequented London's parties, concerts, the opera, and grand balls, socializing with the city's elites. She was introduced to Young, witnessing "him accompany his sister-in-law with the flute, while she played the piano." On another occasion, Somerville performed on her piano while her husband performed mineralogical examinations with a blowpipe, but when he accidentally produced arsenic fumes, she collapsed mid-song. During one season, she

> subscribed to the Concerts of Ancient Music.... The music was perfect of its kind, but the whole affair was very dull. The Philharmonic Concerts were excellent for scientific musicians, and I sometimes went to them; but for my part I infinitely preferred hearing Pasta, Malibran, and Grisi, who have left the most vivid impression on my mind.[41]

Both science and music were central to Somerville's daily experiences and, in 1826, she published her first paper, reporting on the magnetizing power of sunlight. Thanks to this piece and her networking, Henry Brougham (1778–1868) invited Somerville to write an English version of Laplace's unifying treatise, *Mécanique Céleste* (1798–1825) in 1827. Brougham intended this as an addition to the publications of his Society for the Diffusion of Useful Knowledge, but while Laplace's work appeared as a foundation for a godless revolutionary state, Brougham wanted to produce an improving treatise for the laboring classes, believing education to be the best route for social improvement.[42] Somerville shared in these sentiments, observing that "Astronomy like the other sciences, enlightens the age, and brings the human mind to perfection; Love of science and literature softens the manners, and makes men better and happier ... it influences their morality." Despite these shared convictions, Somerville's ensuing treatise was too long and expensive for Brougham's purposes, but he thought her manuscript promising and suggested Herschel read it and act as an advisor.[43] The work subsequently appeared in 1831 as *The Mechanism of the Heavens*.

Sound played a valuable didactic role in *Mechanism*. Somerville explained how all planetary motion was subject to the law of gravity, combined with a first impulse which had propelled all celestial bodies into space and set them on their trajectory. All movement was, therefore, in a state of beautiful "stability and harmony," suggesting the benevolence of "the Great First Cause." As with planetary motions, Somerville described how both light and sound were propagated by elliptical movements. These were analogous, she continued, because sound was

> propagated by the undulations of the air, its theory is in a great many respects similar to that of light. The grave or low tones are produced by very slow vibrations, which increase in frequency progressively as the note becomes more acute. When the vibrations of a musical chord, for example, are less than sixteen in a second, it will not communicate a continued sound to the ear; the vibrations or pulses increase in number with the acuteness of the note, till at last all sense of pitch is lost. The whole extent of human hearing, from the lowest notes of the organ to the highest known cry of insects, as of the cricket, includes about nine octaves.[44]

Heat was also comparable to sound. The "excitation of heat and sound are not only similar, but often identical, as in friction and percussion; they are both communicated by contact and by radiation," wrote Somerville, maintaining that these undulatory laws were also similar to gravity. She asserted that "Not only the sun and planets, but the minutest particles in all the varieties of their attractions and repulsions, nay even the imponderable matter of the electric, galvanic, and magnetic fluids are obedient to permanent laws."[45] Herschel shared these sentiments over nature's order, congratulating Somerville that *Mechanism* read "like one of your own Planets, in a true orbit," holding together as a "harmonious maze . . . like the heavenly bodies themselves."[46] Herschel stated these views publicly in the *Quarterly Review*, while the *Edinburgh Review* placed her work alongside Newton's *Principia*, and the *Monthly Review* claimed the volume exceeded Laplace in its linking of natural phenomena.[47] Despite slow sales, the University of Cambridge adopted the volume as an advanced mechanics textbook in 1837.

Following the enormous success of *Mechanism*, Somerville worked toward a new volume that would demonstrate how all natural phenomena were connected, which she published in 1834 as *On the Connexion of the Physical Sciences*. This united account of nature emphasized the beauty of Creation and God's determining role in the scheme of things. Somerville combined an "Almighty Architect" with an all-encompassing universal system in which the single unifying force, gravity, was evidence of God.[48] The purpose of Somerville's *Connexion* was to set forth a vision of nature as a unified system, regulated by a very limited number of laws, "ultimately [to] be referred to the same agent."[49] Adopting a similar structure to Herschel's *Preliminary Discourse*, she began with gravity, astronomy, and planetary movements before then considering tides, atmosphere, and weather, and then sound and harmony. Acoustics thus bridged astronomy and the physical sciences of light, heat, electricity, and magnetism. Featuring only briefly, sound's role in linking celestial to terrestrial phenomena was similar to Herschel's use of acoustic science: he was certainly influential regarding this ordering of the physical sciences which emphasized the unifying principles of all nature. Somerville therefore divided her book into fourteen sections on physical astronomy, two on matter, two on sound, six on light, three on heat, eight on electricity and magnetism, two on descriptive astronomy, and one to conclude.[50]

For her account of sound, Somerville rewrote Herschel's "Sound," a copy of which he sent her on its publication, in a more accessible, less mathematical, fashion.[51] She replicated Herschel's cornfield analogy, identifying this visual image as an effective way of informing her readers' conception of how sound moved. Like Herschel, she described how each stalk, in a fixed position, oscillated "backwards and forwards in equal times like a pendulum." She then outlined definitions of music, pitch, and a brief theory of harmonics before describing how sound traveled through various media. While solids provided greater sonorous velocity than fluids or air, the single requirement for sound's propagation was that its conveying matter vibrate "in unison as a system." This was why sound traveled poorly through an effervescent glass of champagne, with bubbles creating an inconsistent structure: like light, the passage of sound was impeded when passing through a heterogeneous fluid or atmosphere. Quoting Herschel, it was obvious "that sound as well as light must be obstructed, stifled, and dissipated from its original direction by the mixture of air of different temperatures, and consequently elasticities. . . . The analogy between sound and light is perfect." Somerville's second chapter on sound covered harmony, describing how strings and columns of air vibrated, dividing into areas of rest and oscillation. The best way to observe this was, she contended, by strewing sand over a vibrating glass plate. The impulse of the bow which vibrated the sand offered a model of the astronomical movements of celestial bodies put into motion in Creation. For Somerville, then, sound revealed the beautiful order of nature: she asserted that musical instruments were systems of isochronous vibrations, outlining how "by exposing an extensive undulating surface, which transmits its undulations to a great mass of air, the sound is much reinforced ... the sounding-board and the whole instrument are agitated at once by all the superposed vibrations."[52] She advised curious readers to try this by placing a vibrating tuning fork on the sounding board of a pianoforte and to observe how all sympathetic notes would vibrate.

Somerville's subsequent study of light developed concepts explored in her two acoustic chapters. The movement of light was analogous to that of sound; it was "supposed that the particles of luminous bodies are in a state of perpetual agitation, and that they possess the property of exciting regular vibrations in the ethereal medium, corresponding to the vibrations of their own molecules." Light was, in this way, likened to the sympathetic sounding of resonant notes. As with sound, in light's propagation, particles were not projected but set in oscillation. Returning to Herschel's cornfield, she explained that in the transmission of light and sound,

> each particle vibrates perpendicularly to the direction of the ray; but these vibrations are totally different from, and independent of, the undulations which are transmitted through it, in the same manner as the vibrations of each particular ear of corn are independent of the waves that rush from end to end of a harvest-field when agitated by the wind.[53]

There were of course differences between waves of sound and light, but these did not undermine undulatory theories of these phenomena: although an acoustic analogy, Herschel's cornfield proved useful for defining light.

Sound was equally prominent in Somerville's account of heat, as she asserted that "the principal phenomena of heat may actually be illustrated by a comparison with those of sound. The excitation of heat and sound are not only similar, but often identical," especially the way radiant heat's effect of "raising the temperature of a body upon which it falls resembles the sympathetic agitation of a string, when the sound of another string, which is in unison with it, is transmitted to it through the air. Light, heat, sound, and the waves of fluids, are all subject to the same laws of reflection, and indeed, their undulatory theories are perfectly similar." Somerville conceded that, were the analogy perfect, "the interference of two hot rays ought to produce cold, since darkness results from the interference of two undulations of light, silence ensues from the interference of two undulations of sound." Likewise, sound differed from light and heat as its propagation required a denser medium, so that "its intensity diminishes as the rarity of the air increases; so that at a very small height above the surface of the earth, the noise of the tempest ceases, and the thunder is heard no more in those boundless regions where the heavenly bodies accomplish their periods in eternal and sublime silence." Here Somerville positioned the relationship between sound and heat within a broader vision of a harmonious universe. Indeed, she went further, connecting these phenomena to physiology. The tremors of elastic media, whether in the form of heat, light, or sound, proved that the human frame was "an elastic system" which received vibrations, or undulations, conveying "the mysterious influence of matter on mind."[54] In this sense, knowing how sound and light operated carried epistemological significance; only by understanding how these phenomena acted on the body could reliable observations of nature be made.

Kathryn Neeley has defined Somerville's application of earthly phenomena to those cosmic as the "scientific sublime," emphasizing the poetic blending of aesthetic, religious, and scientific qualities.[55] Somerville's intertwining of religion and science provided a vision of the universe as an ordered system which appealed to broad audiences, especially those religious, with nature's unifying laws of undulation evidence of design. As Somerville put it, gravity "must have been selected by Divine Wisdom out of an infinity of others, as being the most simple, and that which gives the greatest stability to the celestial motions."[56] She supposed that scientific investigation might one day prove an ultimate principle, greater than gravity, which would embrace "every law that regulates the material world," from the movements of planets to the "minutest particles." She concluded *Connexion* by describing how nature was "replete with ether, and traversed in all directions by light, heat, gravitation," and explainable by acoustic analogy: the universe was a system of forces and elastic bodies. Sound was central to this assertion, with Somerville observing that the

> mutual attraction of the celestial bodies disturbs the fluids at their surfaces, whence the theory of the tides and the oscillations of the atmosphere. The density and elasticity of the air, varying with every alteration of temperature, lead to the consideration of barometrical changes, the measurement of heights, and capillary attraction; and the doctrine of sound, including the theory of music, is to be referred to the small undulations of the aerial medium. A knowledge of the action of matter upon light is requisite for tracing the curved path of its rays through the atmosphere, by which the true places of distant objects are determined, whether in the heavens or on the earth. By this we learn

the nature and properties of the sunbeam, the mode of its propagation through the ethereal fluid, or in the interior of material bodies, and the origin of colour.[57]

Above all, this account of nature suggested divine foresight: there was no greater "proof of the Unity of the Deity as those purely mental conceptions of numerical and mathematical science."[58]

Somerville's first edition sold out in February 1834 and remained in print for forty years, including ten editions and over 17,500 copies.[59] The Irish journalist William Cooke Taylor (1800–1849) praised the volume and its movement from laws celestial to laws minute which, in his opinion, made it second only to Herschel's work in importance. Taylor informed readers of the *Athenaeum* that "the sections on Sound [were] the best in the book" and reflected on the advantages of the senses beyond vision in making knowledge. Quoting the French writer, Bernard Le Bovier de Fontenelle (1657–1757), he asserted that all philosophy "is founded on these two things, that we have abundance of curiosity, and very bad eyes." Taylor agreed that the senses were imperfect but felt Somerville's linking of subjects highlighted the strengths of human reasoning.[60] David Brewster provided sterner criticism in the *Edinburgh Review*. The book would "be very popular," he assured Somerville privately, but he doubted it would diffuse knowledge of the universe's material laws because of its absence of illustrations.[61] This particularly undermined her account of acoustics, a subject which he claimed she had dealt with too briefly, with "the curious and almost enchanting subject of the vibration of solid bodies . . . discussed in little more than two pages." Without diagrams, he continued, the account was insufficient, because no description of sound could "convey to the mind a just idea of the exquisite symmetry which characterises the movements of vibrating solids." Specifically, he criticized Somerville for overlooking the acoustic figures of Chladni, Savart, and Wheatstone which, for Brewster, were crucial to connecting the physical sciences. As he put it, there

> is no branch of physics which addresses itself so agreeably to the eye, or appeals with such force to our wonder, as that of acoustic figures; and, connected as it is with the theory and practice of music, we must implore Mrs Somerville to give it, in another edition, a more favourable consideration.[62]

This lack of attention toward Wheatstone's work was probably due to Somerville's collaboration with Herschel: given the links between their accounts of sound, it is little wonder she overlooked recent acoustic research.

Somerville responded to Brewster's review with a substantially revised second edition of *Connexion*, published in 1835. She stuck with Herschel's cornfield but included references to Wheatstone and Savart. She expanded her account of harmonics, noting that new knowledge of vibrating springs had led to the construction of some wonderful new instruments, among the "most perfect of these" being Wheatstone's symphonion, concertina, and aeolian organ. Somerville then delivered a detailed account of the vibrations of solid bodies, describing Chladni's observation of symmetrical forms through vibrating plate experiments.[63] Having acquired images of Wheatstone's acoustic figures from John Lubbock (1803–65) at the Royal Society, as well as Faraday as a reviewer for her revised draft, her new edition included four pages of plates with seventy-three diagrams, sixty of which were acoustic figures. Somerville linked these diagrams to details of Wheatstone's 1833 paper in *Philosophical Transactions* and emphasized the value of his work on resonance for physical science generally. That one body's vibration could cause another's sympathetic vibration was a phenomenon directly observable in astronomy, as it also was with pendulums: the motion of one moving body could determine the movements of another.[64] This effect of vibrating bodies finding unison over time, which Wheatstone explored in the form of resonance, Somerville applied to unite the physical sciences, emphasizing that these oscillations

> are to be traced in every department of physical science. . . . Such are the tides, which follow the sun and moon in all their motions and periods. The nutation of the earth's axis also corresponds with the period, and represents the motion of the nodes of the moon . . . [and] the acceleration of the moon's mean motion represents the action of the planets on the earth reflected by the sun to the moon.[65]

The unison of sounding organ pipes appeared comparable to celestial motions, but, while astronomy demanded mathematical understanding, "all the phenomena of vibrating plates may be exhibited by sand strewed on paper or parchment, stretched over a harmonica glass or large bell-shaped tumbler." Sonorous experiments made the vast motions of the heavens minute and intelligible to nonspecialist audiences. Similarly, Wheatstone's experiments on the transmission of sound through vibrating metallic rods, connecting musical instruments to sounding bodies were, Somerville asserted, of great value, allowing the transmission of the sounds of an entire orchestra over vast distances.[66] In total, while her first edition included twenty-five pages devoted to sound, her second extended to thirty-one.[67]

Significantly, illustrations and revisions on sound were not the only changes she made to her second edition: she also emphasized sound's role in connecting the physical sciences. In the first edition's introduction, Somerville's second paragraph made general remarks on astronomy and motion, but in her second edition this was completely replaced with a musical comparison. Somerville's revised paragraph claimed that "astronomy affords the most extensive example of the connection of the physical sciences" because in it

> we perceive the operation of a force which is mixed up with every thing that exists in the heavens or on earth; which pervades every atom, rules the motions of animate and inanimate beings, and is as sensible in the descent of a rain drop as in the falls of Niagara, in the weight of the air as in the periods of the moon. Gravitation not only binds satillites to their planet, and planets to the sun, but it connects sun with sun throughout the wide extent of creation, and is the cause of the disturbances, as well as of the order, of nature: since every tremor it excites in any one planet is immediately transmitted to the farthest limits of the system, in oscillations, which correspond in their periods with the cause producing them, like sympathetic notes in music, or vibrations from the deep tones of an organ.[68]

Given Somerville's substantial revision of her account of sound and that the acoustic figure illustrations were the most obvious addition to the second edition, it is clear that the value of the sonorous had increased. In 1834, sound had merely bridged various natural phenomena, but in 1835, by incorporating Chladni and Wheatstone's experimental investigations, Somerville put acoustics center stage, using sonorous vibrations to unite celestial oscillations with those experienced on earth. This opening musical analogy remained in subsequent editions.

Wheatstone, for one, thought that Somerville's second edition, with its revamped section on sound, was excellent. When, in 1835, Faraday wrote to Wheatstone requesting recommended readings on acoustics for a female friend, Wheatstone could suggest only three of worth. Claude Pouillet's (1790–1868) *Éléments de physique* (1832) was the most recent French work on the subject, but in English Wheatstone's preference was between Herschel's "Sound" and Somerville's *Connexion*. "Sir J. Herschel's article in the Encyclopaedia Metropolitana is an admirable treatise, but the subject is treated too mathematically to allow me to recommend it to a lady for perusal," observed Wheatstone, but he thought Somerville's *Connexion* eminently suitable for a female audience.[69] Wheatstone provided just one caveat: that the second edition, that of 1835, be acquired, rather than the 1834 first edition. Somerville's reworking of sound clearly made a great deal of difference for Wheatstone. Nevertheless, when suggesting literature on sound experiments for male readers, Faraday instead promoted Herschel's treatise, observing in 1836 that

> With respect to modern experiments on sound I am not aware of any work on the matter indeed I know from frequent enquiry there is no good compendium[.] Herschells [sic] treatise in the Metropolitana (Encyclopedia) is I understand good & I think that there are some good French and German works but do not know them.[70]

Evidently both Herschel's and Somerville's work on sound found approval with Faraday and Wheatstone, but which text was preferable was contingent on a reader's gender and perceived mathematical understanding.

The Sound of Whewell's Philosophy

Among Somerville's readers was the Cambridge polymath William Whewell (1794–1866), who reviewed the first edition of *Connexion* in the *Quarterly Review* in 1834. Applauding her effort to connect the sciences, he thought Somerville's greatest success lay in her use of analogies.[71] Whewell knew Somerville personally, having invited her to Cambridge in 1832, following the publication of her *Mechanism*.[72] Similarly, having read Herschel's treatise "Light" in 1828, Whewell wrote to him encouraging him to write a tract on acoustics; he would "be very glad indeed to see Sound treated in the same way."[73] Whewell conscripted Herschel and Somerville's accounts of sound within his own philosophical works, examining the sonorous as a tool for making knowledge and exploring its epistemological implications. Both religiously and mathematically, Whewell shared similar ground to Somerville and Herschel, but he invested the study of sound with increased theological and moral significance.

Born in Lancaster, the son of a carpenter, Whewell's rise had been meteoric. He went to Cambridge in 1811 and graduated as Second Wrangler and second Smith's prizeman from Trinity College in 1816. Best known for coining the term "scientist" in 1834 and later the term "physicist," he was ordained an Anglican priest in 1825, before becoming professor of moral philosophy at Cambridge in 1838. His philosophical reputation grew throughout the 1830s until, in 1841, Conservative prime minister Robert Peel (1788–1850) recommended Whewell's appointment as master of Trinity College to Queen Victoria.[74]

Seeking to unite the latest philosophical knowledge of nature within a theological framework, Whewell published his Bridgewater Treatise, *Astronomy and General Physics*, in 1833. Within this, Whewell used Herschel's account of sound as evidence of a creating intelligence. Like Herschel and Somerville, the observation of laws governing nature appeared to Whewell as proof of a "Governor of the world." As he surmised, by "observing the laws of the material universe and their operation, we may hope . . . to be able to direct our judgement concerning the government of the universe" and to learn "something of the character of the power which has legislated for the material world." Whewell reversed the ordering framework of Herschel's *Preliminary Discourse*, with heat, electricity, and magnetism preceding sound and light in book 1, then moving on to cosmic arrangements in book 2 and religious implications in book 3. Once again, sound performed a bridging role between terrestrial and astronomical phenomena. Like Somerville, Whewell used Herschel's cornfield analogy, observing that the little oscillations of fixed stalks of corn provided the best possible analogy for sound's communication. In the relationship between human speech and hearing, and the atmosphere's potential to communicate sound, Whewell saw marvelous evidence of design. "[Is] it by chance that the air and the *ear* exist together? Did the air produce the organization of the ear?" wondered Whewell, before answering that "the only intelligible account of the matter" was that "one was made for the other; that there is a mutual adaptation produced by an Intelligence which was acquainted with the properties of both." Sound appeared a God-given gift, with music its highest form. "How nicely are the organs adjusted with regard to the most minute mechanical motions of the elements!" exclaimed Whewell.[75]

Whewell developed these connections in his 1837 *History of the Inductive Sciences*, with acoustics presented at the junction between astronomy and what he termed the "Secondary Mechanical Sciences," including light and heat. After the secondary mechanical sciences, Whewell moved into what he described as the mechanic-chemical sciences of electricity, magnetism, galvanism, and electromagnetism, before concluding with the pursuit of mineralogy, botany, zoology, and geology.[76] The history of acoustics featured centrally within the secondary mechanical sciences as it relied on "the operation of a *medium* interposed between the object and the organ of sense.... The sciences of this kind which require our notice are those which treat of the sensible qualities, Sound, Light, Heat; that is, Acoustics, Optics, and Thermotics." The history of acoustics was, Whewell taught, peculiar in having formed much earlier than those of heat or light, serving to inform reasoning on these phenomena. Ancient philosophers Vitruvius and Aristotle had recognized that sound traveled through air and likened this passage to the movement of waves through water. This initial observation made acoustics unusual in the sciences because instead "of having a series of inductive Truths ... we have a series of Explanations, in which certain experimental facts and laws are reconciled ... with the general doctrine already in our possession." Rather than moving "towards a great discovery, like Universal Gravitation," knowledge of sound rested on "acknowledged truths, the production and propagation of sound by the motion of bodies.... Instead of *Epochs of Discovery*, we have *Solutions of Problems*."[77] According to Whewell, the progress of acoustics was not the sum of empirical observations but an ancient moment of intuition and was, therefore, typical of an idealistic approach to knowledge production.

Whewell was philosophically an idealist rather than a strict empiricist. For him, science was not merely the collection of sensorial evidence but a view for recognizing and establishing truths. It was the mind's creative faculties that dominated Whewell's philosophy: in making relations between immediate sensations and those previously sensed, human reasoning took all experienced sensations, connected and interpreted them, and fashioned new ideas. He argued that the first act of creating knowledge was having an idea and that studying nature was morally improving because it led to a greater realization of God's creation. Whewell maintained that knowledge was to be cultivated for its own sake and not reduced to the pursuit of physical pleasures and practical applications.[78] Yet this drew criticism from Utilitarian philosopher John Stuart Mill (1806–73), who singled out Whewell as the champion of *a priori* philosophy. Such intuitionist reasoning, based on preconceptions, alarmed Mill, who feared that such an emotional basis for science encouraged social conservatism. As Mill argued in his 1843 *A System of Logic*, an epistemological framework in which opinions comprised proofs upheld existing inequalities such as slavery and social elitism. Emphasizing a scientific approach based purely on sensory experiences, Mill picked out Whewell as the leader of intuitionist philosophy; and, it was true, Whewell's philosophy carried political connotations. Whewell believed both science and society progressed through continuous change and gradual reforms rather than revolutions.[79] Seeing himself as an intellectual guardian of state institutions, particularly the established Church of England, Whewell sought to confirm science's moral, cultural, and intellectual worth by emphasizing the human mind's creative power in interpreting sensory impressions.[80] Whewell's philosophical position developed over time, and while he maintained in 1837 that the mind's contribution to knowledge production was to organize and order facts, by 1840 he attributed a much more active role to the mind in developing concepts.[81]

The history of acoustics very much fit within Whewell's philosophical vision. It perfectly illustrated how ideas and intuitive reasoning induced gradual advances. He represented the progress of acoustics as a series of problems, such as that of vibrating strings, how pipes sounded, and the modes in which columns of air and solid bodies vibrated. Through solutions to each of these acoustic problems, Whewell showed how the science of sound progressed gradually over time. Newton, for example, had resolved much of the relationship between sound's velocity and its propagating media, while Chladni had provided the greatest advance to the problem of various vibrating bodies, having "enriched acoustics with the discovery of the vast variety of symmetrical figures of Nodal lines." Making vibration visible had helped acoustics move toward general laws and principles. In the science of sound, Whewell continued, the "number of observed laws and relations of the phenomena of sound, is already very great; and though the time may be distant, there seems to be no reason to despair of one day uniting them by clear ideas of mechanical causation." Whewell confidently asserted that acoustics would eventually become "a perfect mechanical science."[82] Whewell developed this account of acoustics in 1840 in his *Philosophy of the Inductive Sciences*. He again invoked Herschel's cornfield analogy, explaining how "if we look at a field of standing corn, when a breeze blows over it, we see waves like those of water run along its surface."[83] Whewell attributed this account of sound as an impulse to Newton, who had supposed that each particle must oscillate like a pendulum, but it was Herschel's cornfield that had made this conception most comprehensible.

The real point of Whewell's *Philosophy*, however, was to show how truths could be recognized, as he expanded on the epistemological balance between the senses and ideas. While the reception of senses could be passive, ideas were the works of an active mind. Different types of science involved varying relations between sense-based experience and imagination. "Pure" sciences like geometry and mathematics relied on "the Ideas of *Space*, *Time*, and *Number*." The mechanical sciences, such as hydrostatics and astronomy along with the secondary mechanical sciences, dealt with the motions of bodies, either fluids or solid bodies, and were subject to the principles of force. To know force was to know the cause of motion. In the universe, all celestial bodies moved from the same original cause, or force, and this causation suggested the existence of a Deity. In acoustics, causes occurred as sources of vibration, but acoustics also depended on the idea of a communicating medium.[84] For sound, this was air, but it was harder to identify the media of heat and light.

At stake here was the question of what could be known; and scientific reasoning allowed sensory impressions to be interpreted. As Whewell explained, we "see and hear and touch external things, and thus perceive them by our senses; but in perceiving them, we connect the impressions of sense according to relations of space, time, number, likeness, cause." Importantly, Whewell argued that the identification of truths was not purely visual, but aural: attaining knowledge required both sight and sound. The eye precisely perceived space, unlike sound, which was all around, yet the ear boasted two "remarkable prerogatives": it could "perceive a definite and peculiar relation between certain tones, and it can clearly perceive two tones together." This power of detecting musical concords, of fifths and octaves for example, was "extremely precise." Whewell believed that such a universal characteristic was remarkable: there was "nothing analogous in vision. Colours have certain vague relations to one another . . . but this is an indefinite, and in most cases a casual and variable feeling." Hearing's exactness was astonishing, observed Whewell, capable of detecting any discord. As all notes were specific vibrations, and concords were simple vibratory ratios, the ear was a mathematical organ unlike any other. He explained that "according to the undulatory theory of light, such ratios occur in colours, yet the eye is not affected by them in any peculiar way." The other epistemological advantage of the ear over the eye was its perception of rhythm. While vision dealt in space, Whewell explained that humans "hear sounds in *time*." The perception of time was not unique to hearing, but as sounds were always arranged in order, the ear was an especially effective measurer of time. Whewell's scientific framework therefore depended on both seeing and hearing for perceptions of time and space.[85]

Sound had another great value in the production of knowledge. Ultimately, the aim of science was the measurement of natural processes and, for this, sound had certain advantages over vision. The ear had a profound ability to measure acuteness or graveness, loudness, or number of vibrations. The harmonic division of musical notes dated back to Ptolemy's *Harmonics*; for Whewell this system constituted an ideal standard. He argued that as the ability to measure acoustic vibrations was "nearly universal among men; for all persons have musical ears sufficiently delicate to understand and to imitate the modulations corresponding to various emotions in speaking," there was something peculiarly accurate and consensual in the "science of Harmonics."[86] The almost universal detection of musical relations, especially octaves, fifths, fourths, and thirds, meant that the science of sound had a stable basis for securing consensus. The standard of eight indisputable notes covering an octave was an example for other sciences to emulate in seeking solidarity. Although the ear could be fallible, by use of a monochord, Whewell explained how absolute notes could be attained by different ratios of lengths. If a C required a string length of 180 parts, D would be 169, and E would be 144, "thus the musical relations are reduced to numerical relations, and the monochord is a complete and perfect *Tonometer*." In comparison to sound's numerical perfection, scales of color and heat were inferior.[87] Whewell thus portrayed harmonics as a standard for the measure of natural phenomena.

As much as Herschel and Somerville, Whewell offered his readers an account of nature in which sound and music were, like all natural phenomena, divinely created. Collectively, these science writers worked within a shared religious framework for Britain's literate population, traditionally an upper- and upper-middle-class readership. By the 1830s and 1840s, this was a rapidly growing social group, extending to the lower-middle and artisan classes. Aided by reduced-price printing, these science writers were reaching a broader audience than ever before. Yet they provided just one source of guidance on questions of nature, including those regarding sound and music. A far more accessible venue for acquiring knowledge of nature remained the pulpit, with the sermon a rival to the scientific tract; one that could be listened to by all, regardless of literacy and learning. However, the writings of scientific authorities like Whewell were not competing with but complementing and informing the theological interpretations of natural phenomena in Britain's churches and chapels. Among Somerville, Herschel, and Whewell's readers were clergymen, eager to mobilize leading works of natural philosophy and combine them with Biblical lessons in sermons and tracts.[88]

Pulpits and Pendulums: Sermonizing Sound

The Church of England and Britain's numerous nonestablished dissenter movements maintained their position as leading arbitrators on questions of natural philosophy for significant elements of the population throughout the mid-nineteenth century. Religious commentators offered alternative, but not necessarily competing, instruction on nature to that provided in scientific treatises. Far from being in opposition, scientific and religious accounts of sound and music shared common territory and often presented complementary views of nature. Within the Anglican tradition, this is hardly surprising, given the intellectual dominance of the universities of Oxford and Cambridge over the established Church of England. Both Oxford and Cambridge were principally Anglican seminaries, training clergy for the established Church. In 1840, for instance, 59.2 percent of students to matriculate at Oxford and 66.7 percent of those at Cambridge were sons of Anglican ministers, and a high number of these followed their fathers into the Church.[89] Mathematics featured prominently within both university curriculums, with Cambridge boasting a well-established mathematical tradition and Oxford providing separate honor schools for mathematics and *literae humaniores* from 1802, though it was possible for undergraduates to take both subjects. While the works of Whewell and Herschel, both Cambridge wranglers, were valuable for undergraduates, Somerville's writings fit neatly into Oxbridge's Anglican culture of mathematics, with Whewell introducing her *Mechanism* as a student textbook at Trinity in 1832.[90]

However, these universities were not just places where philosophical knowledge of sound could be acquired; they were also sites of musical practice. Students destined for careers in the Church of England could learn mathematical principles for harmony and melody in the works of Somerville and Whewell, and at Oxbridge such studies accompanied the musical experiences of college chapels. In the early 1840s, the only Cambridge colleges that boasted choral traditions were Trinity, King's, and St. John's. At Oxford, things were little different, with New, Magdalen, and Christ Church Colleges the only ones to offer full choral services, generally of a low musical standard.[91] At Cambridge, collegiate musical culture had made some progress thanks to the appointment of Thomas Attwood Walmisley (1814–56) as organist of Trinity and St. John's in 1833.[92] Increasingly, however, Cambridge's colleges underwent a musical renaissance, largely thanks to a network of Trinity-based clergymen with shared interests in music. These reformers had firm scientific connections and secured the patronage of none other than Whewell. This renewed interest in collegiate music was, in fact, part of a much broader reappraisal of music's religious value that Church of England clergymen cultivated throughout the 1830s and 1840s.

Although the Anglican Church had witnessed much evangelical influence during the eighteenth century, stressing the value of scripture and the Atonement, the mid-nineteenth century witnessed a revival in High Church Anglicanism, with its emphasis on the importance of the Church itself as an institution which traced its authority back to the medieval

Catholic Church.⁹³ Post–French Revolution fears that traditional social order had broken down and distrust of political radicalism had contributed to a growth of popular medievalism, especially under the guidance of Augustus Pugin and John Ruskin. Anxieties over Anglican autonomy appeared confirmed by persistent state-driven ecclesiastical reforms, including the Test Act's repeal in 1828, the 1832 Reform Act, the 1834 Poor Law Amendment Act, and the 1835 Municipal Corporations Act, which were all interpreted as attacks on the privileges of clergymen.⁹⁴ Following John Keble's (1792–1866) sermon condemning the Whig government's reduction of the number of Irish bishoprics, delivered in Oxford University Church in 1833, John Henry Newman (1801–90), Edward Pusey, and Richard Froude united to resist further state interference. Under Keble's lead, they allied traditional Anglican churchmen in the defense of the doctrine of apostolic succession through what became known as the Oxford Movement or Tractarianism, following the publication of *Tracts for the Times* (1833–41). They promoted ritualism to align Anglican worship with Roman Catholic practices, including eucharistic vestments, incense, and unleavened bread in the Eucharist, in order to renew Anglican authority by emphasizing its connections with the medieval Catholic Church.⁹⁵ Within this reforming movement, churchmen were eager to define music and determine its moral value, with musical practices important to efforts to reorganize church worship. This central role of music within ecclesiastical and theological reforming efforts was partially attributable to the preexisting choral traditions of Oxford and Cambridge. Around these, a distinct network of Oxbridge clergymen provided the most coherent nineteenth-century religious effort to define music's moral value, culminating in the publication of *Hymns Ancient and Modern* in 1860.⁹⁶ Importantly, these were Anglicans who were often familiar with the works of Herschel and Somerville and, at Cambridge, would have been well acquainted with the domineering figure that Whewell presented as the influential master of Trinity College.

The Cambridge-based Ecclesiological Movement was primarily concerned with the revival of Gothic church architecture, but the promotion of choral worship was also part of its efforts for reforming the Anglican Church and cultivating a spiritual renaissance through Catholic forms of worship. Ecclesiology, described in 1847 as "the science of Christian Aesthetics," included all decoration, clothing, and music, which originated within the musical revival of Cambridge's college chapels during the 1840s and 1850s under the leadership of John Neale and Benjamin Webb. Wary of evangelical hymn singing, these High Churchmen preferred the metrical chanting of the Psalms. From 1836 Neale lived with Cambridge's professor of astronomy, James Challis (1803–82), who prepared him for the university examinations. Neale took a scholarship at Trinity College in the same year, where Webb matriculated two years later, with both undergraduates gradually moving toward High Anglicanism. In 1839, Webb and Neale established the Cambridge Camden Society to promote the study of Gothic architecture; it grew quickly, recruiting Whewell, two archbishops, sixteen bishops, and thirty-one peers and MPs as members. Trinity's college chapel music had improved throughout the 1830s, but Neale and Webb's interest in sung worship, combined with Whewell's appointment as master in 1841 and his association with the influential music teacher and composer John Pyke Hullah (1812–1884), added fresh energy to this musical revival. In 1846, for example, Whewell had specifically called for the poor delivery of the college's chanting to be remedied, while Webb took up singing lessons with Hullah between 1842 and 1844.[97]

However, it was not just the cultivation of sacred music that Trinity fostered: it was also home to a network of churchmen keen to promote theological accounts of sonorous phenomena, emphasizing music's spiritual value alongside the study of nature. The composer of several hymns, Christopher Wordsworth (1807–85) was a fervent High Churchman. The son of Whewell's predecessor as master of Trinity, Christopher Wordsworth (1774–1846), he attended his father's college from 1826, graduating in 1830 as the Senior Classics in the Classical Tripos and fourteenth Senior Optime in the Mathematical Tripos Examinations. A regular guest of Whewell's at the Master's Lodge, Wordsworth became headmaster of Harrow school in 1836 and bishop of Lincoln in 1869.[98] Intolerant of dissenters and a promoter of the links between Anglicanism and Roman Catholicism, Wordsworth also emphasized that to know nature was to understand God's governing laws. As Wordsworth put it, "Reason will . . . convince us that the world is from God and that the Bible is from God. . . . Reason proves that nature and Scripture are like two books written by the same divine hand."[99] This theological interpretation extended to Wordsworth's views on sonorous phenomena, reflecting his familiarity with leading works of natural philosophy. As archdeacon of Westminster, he gave a sermon in 1868 entitled "Sacred Music" at the Cathedral Church of Llandaff. Wordsworth informed his congregation that the "Divine Creator of the Universe appears to have dictated the use of Music in the Public Worship," having

> placed us in an element of a fluid and elastic nature, capable of receiving sounds by impulsion, and conveying them by vibration in a regular measure like that of the waves of the sea. He has adapted the human

ear to the reception of these sounds, and has attempered the human soul to the influences of sound transmitted by the air, and penetrating by the avenue of the ear, so that the in-most feelings of the heart are swayed, and the most secret thoughts of the mind are controlled, by sounds communicated from without.[100]

Echoing Whewell's sentiments, Wordsworth asserted that sonorous communication constituted a "complex mechanism" which proved divine design. Both nature and the Bible delivered examples of the sonorous celebration of God. As Wordsworth put it, through the whistling of wind, rustling of trees, and the "roaring of the mighty ocean in the gloom of night, the floods clapping their hands in joy beneath the summer sun, these celebrate the praise of God." Therefore, the production of beautiful church music evoked nature and inspired devotion. The Bible confirmed this relationship, with Wordsworth observing how music accompanied great Biblical moments of trial, from the Israelites' singing on crossing the Red Sea to God's fear-inducing "awful music" at Sinai and Revelation's prophecy that "an overture of that still more dreadful sound which will one day break upon the world, and echo through the vault of heaven, and pierce the lowest depths of earth; when the Archangel shall blow the last trump, and summon the world to Judgement." Such "sacred music" had "genuine power," capable of lowering the walls of Jericho and, indeed, the pride of all Jerichos to come. Musical worship was moral but, Wordsworth warned, only if done in faith, there being "one instrument above all others which they must endeavour to keep in tune, —*that* instrument is the *heart*. The *heart* is the chief musical instrument in every Christian choir."[101]

Wordsworth was not alone in these sentiments. Fellow Trinity graduate Edward Miller (1799–1877) shared in this scientifically informed theological account of sound. Having matriculated at Trinity in 1817, collecting his MA in 1825, Miller became a deacon in 1823 and curate of Bognor in 1838. For Miller, there was "a mystery" in music that science could not explain. In his 1848 sermon celebrating Smethwick Old Chapel's new organ, Miller explained that the

> elements of Music are sleeping around us in the viewless air, and we can awaken them when we will; perhaps, when the grossness of our mortal nature is purged away, we shall know they *never* slept; but rose and fell as God's breath did stir them, evermore! The spirit of music dwells in the elements of Air, — that wondrous fluid, with its thousand pulsations; and which therefore may be called, in a sense more

extended than that of the text, the "musical instrument of God." Every breath we draw contains within itself the soul of melody; and how wonderful are the organs of a man, which enable him, by a mere act of volition, at once to inhale the vital air, and while one part is bounding with life-blood through his heart and veins, to embody the rest in a concord of sweet sounds![102]

This theologically informed account of sound's propagation through the "fluid" of air displayed Miller's familiarity with natural philosophy. Similarly, human-built church organs could not compete with the wonder of organs divinely manufactured. "How small is the diameter of the human throat, and how small its measure," mused Miller, "yet it will distinctly give the same note with the pipe of an organ eight feet in length! And the valve which covers it, and plays with electric swiftness . . . is, as you know, a very little thing; yet with the contraction and expansion of the throat, it will utter a scale of seventeen degrees." While applauding church organs and the sounds they produced in vast cathedrals, Miller cautioned against forgetting music's divine origins, as it was in faith, rather than knowledge, that man should trust for euphonic beauty. As he put it, "if abstruse science had sufficed, to resolve and combine the chromatic scales of harmony," then "mathematical sages . . . where Vishnoo is worshipped, and the Crescent has dethroned the Cross" would have equaled the marvels of Christian church music. However, Miller believed this had not been the case as true music "was the gift of God; a gift which human genius alone was powerless alike to conserve or to create." In contrast to Christian music, he informed his congregation that they would "find that the Music of infidels is rude and barbarous, fit only for the ears of children or savages." Here, religious accounts of sound took on highly racialized implications. According to Miller, perfection in the "noble science" of "harmony, and above all, Sacred Harmony" demanded divine inspiration.[103]

There was a similar religious interest in the nature of sound and the moral value of music at Oxford. Although both Magdalen and New Colleges boasted choral traditions, it was Christ Church College that was very much central to this musical concern. Between the 1810s and 1840s this college enjoyed a period of high academic attainment, with large numbers of first-class degrees achieved and excellent undergraduate performances in the classics and mathematics.[104] Famously, future prime ministers Robert Peel and William Gladstone both achieved double firsts in mathematics and *literae humaniores*, uniting the scientific study with that of the classics. Just as at Cambridge, Christ Church was home to a network of Anglican ministers who combined scientific knowledge of sound with religious accounts of music. For instance, Walter Kerr Hamilton (1808–69), Christ Church graduate and bishop of Salisbury from 1854 until 1869, stressed music's metaphysical value in a sermon on choir singing in 1860. There was, he felt, "no sensible means of influencing the soul so powerfully as the regular, orderly disposition of a few high, and low sounds." He distinguished between melody, which was sound "so arranged in order according to the laws of science as to make what is called a tune," and harmony, which was a far more complex arrangement of sounds. Both were acceptable for worship, but he preferred melody's potential to unify a congregation.[105] Likewise, Henry Liddon, an influential member of the Oxford Movement who studied at Christ Church from 1846, believed music to be central to Anglican authority. Extremely High Church, Liddon was instrumental in establishing the Tractarian bastions of Keble College and Pusey House in Oxford and became increasingly concerned with church rituals, including music, believing these essential to Church authority.[106] His sermon "The Sights and Sounds of Christendom," preached at St. Paul's Cathedral on September 11, 1870, was a defense of Anglican ritualism which emphasized the importance not only of believing by seeing but also by hearing God's word. Four years later, as canon of St. Paul's Cathedral, he was incensed at the Conservative government's proposed Public Worship Regulation Act to restrict the use of incense, music, and colored vestments. This legislation was in response to growing fears that such Catholic ritualism was being used to reunite the Church of England with Rome, but, for Liddon, it represented an unacceptable encroachment of the state into Church affairs.[107] In this context as a ritualistic practice, church music carried political connotations.

Similarly, having matriculated at Christ Church College in 1819, William Gresley (1801–76), the prebendary of Lichfield, promoted a highly Catholic vision of the Anglican Church in which music was central. On September 12, 1852, at St. Paul's Church in Brighton, he delivered "A Sermon on Church Music," highlighting Biblical examples of music's use. Psalm 150 gave authority to the use of instruments in worship, and it was the organ that Gresley felt most appropriate for use in Gothic architecture.[108] Though alarmed at the advance of atheism, Gresley had much praise for the "wonderful discoveries" that "men of science have made" but warned that within this group was a small number who used scientific knowledge to deny the existence of God. Yet he was happy to draw on the positive vision of nature which philosophers like Somerville evoked, which was one of mystery, not just matter. Almost three decades after its publication, Gresley cited Somerville's account of harmonic laws as evidence of Creation, observing that "sound might have been all one monotonous note; but God has imparted the power of harmony to sounds, that men may sing His praises, and experience the exquisite delight of music."[109] Gresley, then, was not rejecting scientific knowledge outright but providing a religiously informed account of music and sound, grounded in earlier works of natural philosophy which he considered morally acceptable.

Clearly, at both Oxford and Cambridge, religious and scientific accounts of sound were highly compatible. Herschel, Somerville, and Whewell complemented and informed spiritual authorities throughout the Church of England, far beyond the theological bastions of Oxbridge. However, few commanded so much influence over Church musical practices as John Jebb (1805–86), whose sermons and tracts promoted the spiritual value of music while drawing on the latest philosophical works on the nature of sound. Born in Dublin and educated at Winchester College and Trinity College Dublin, he brought together Anglican evangelicalism with the revival of ecclesiastical antiquity through what he termed "Catholic Evangelicalism," this being the unification of Catholic practices with an emphasis on Biblical readings.[110] Jebb argued that scientific knowledge was essential to religious music, and he took a leading role in the cultivation of English choral music. In early Victorian Britain, Leeds Parish Church was the only parochial church in England to have an organized choir, established around 1815. Determined to improve his choir's standard, the vicar of Leeds, Walter Hook (1798–1875) consulted Jebb, a family associate, who recommended harmonic Anglican chants to emphasize the ancient gravity of sung worship. Hook followed Jebb's recommendations, appointing Samuel Sebastian Wesley (1810–76), the son of composer Samuel Wesley (1766–1837) and grandson of Methodist hymnodist Charles Wesley (1707–88), as organist in 1842. Once Exeter Cathedral's organist, Wesley brought much skill to Leeds.[111] John Hullah was particularly impressed with Wesley and Jebb's efforts in Leeds, approving of the great moral value they ascribed to musical experiences: all agreed on music's transcendental capacity and promotion of devout sentiments.[112] Jebb subsequently published lectures on these musical reforms, including guidance to separate the singing of the choir from that of the congregation in order to preserve the quality of the music, and Leeds's church became a model for how other parochial churches could introduce cathedral choir music into daily worship. In the 1830s, choral worship had been limited to cathedrals and a few Oxbridge colleges, but by the mid-nineteenth century such practices had become widespread.

After the absence of choral services from Anglican Church worship for some two centuries, music was now at the very center of the Oxford Movement's cultivation of medievalism, and it accompanied the more extensive Gothic Revival.[113] It was from Hook and Jebb's work at Leeds and the improvement of collegiate music that the choral revival of the 1840s expanded.[114]

At the third Annual Festival of Church Choirs, held in 1862 at Norwich Cathedral, Jebb outlined precisely the relationship between sung church worship and knowledge of nature. In his sermon to the festival congregation, he claimed that sacred music resulted from divine inspiration and that the Apostles had sanctioned its liturgical use, but he also stressed that vocal and instrumental church music should be of a high standard.[115] Sacred music had to be practiced in reference to the laws of sound, unlike medieval music, which had been largely defective

> because its principles were untrue, and inconsistent with the physical laws of sound; that the ignorance of the octave (which exists in nature, and is therefore part of God's natural laws) resulted in the imperfect expedients of the tetrachord and hexachord, consequently in a defective and inaccurate melody; and that harmony was all but unknown, and when first attempted, was handled in such a manner as to be at variance with the real principles which advanced science has recovered.[116]

Imperfect music was, therefore, not sacred as it was inconsistent with God's nature. Jebb identified Roman Catholic forms of sung worship as especially guilty of such imperfection, having adopted "the strange and really unscientific methods of the Grecians; who were confessedly ignorant not only of all harmony, but of the real scale and relation of progressive sounds which exist in nature." Musical knowledge's progress as a science, he asserted, facilitated superior forms of worship, and Man had a moral duty to understand acoustics for the improvement of church music. "We know that the capabilities of the science itself are inexhaustible, because that science is made up of the immutable laws of sound created by God himself," observed Jebb. "It has combinations as numerous as the sentiments and aspirations which Holy Scripture suggests." Jebb informed his audience that before the corruption of Gregorian chant, the music of the Old Testament had been composed in reference to nature, being "scientific; that is, that it employed and used . . . the principles and combinations of physical science; in other words, obeyed the laws of

God." The advance of modern science and its rediscovery of the divine laws of harmony would enhance worship. Jebb advocated that "this sublime science" of music be cultivated only by true Christians, rather than some professional but faithless musicians. Indeed, he warned his congregation that the rising professionalization of music was typical of an increasing "spirit of secularity" which was threatening the church.[117] Jebb's sermon did not go unnoticed. Later, in October 1862 at Dresden in Staffordshire, John William Hewett responded with a sermon of his own, rejecting Jebb's requisite of perfect music for choral worship, contending that imperfect music was equally sacred, if performed with true religious feeling.[118] According to Hewett, sung worship was about creating unity and arousing godly devotion rather than aesthetics, and he warned congregations not to get so "carried away by the beauty of the outward expression of worship" that they forget the spiritual nature of a service.[119]

Church of England efforts to define music and its place in nature were, clearly, far from consistent and not confined to the Oxford Movement or High Church Anglicans. Edward Copleston (1776–1849), bishop of Wandaff from 1827, enjoyed church music but rejected the Oxford Movement's veneration of mystery and ritual. After a scholarship from Corpus Christi College Oxford and a fellowship at Oriel College during the 1790s, Copleston was by the 1820s a leading liberal Tory churchman, but by the 1840s was alarmed at Tractarianism's apparent hostility to science and promotion of "mysticism."[120] Indeed, Copleston took a keen interest in the sciences and met privately with Wheatstone in 1840 at King's College London, where he was astounded by an exhibition of his "mechanical apparatus for an electric telegraph" that boasted a velocity of 160,000 miles a second. This was a power far superior to Tractarian claims of mysticism, argued Copleston, which demonstrated how the "subjugation of nature" provided man with more "powers" than any "pretended magic."[121] The telegraph was thus placed in stark contrast to spiritual mysteries. What then of music's emotive power and mystery? Copleston was equally cautious over music's role in faith. In his sermon to the first annual meeting of the Society for the Improvement of Church Music, held at Usk in July 1848, Copleston reflected on the abuses of sacred music. The Old Testament taught that outward forms of devotion were worthless if done without conviction, but if congregations were careful not to undermine sincerity with aesthetics, music was "the most powerful of those agents which awaken past sensations, with a kind of magical force," suitable for worship.[122] While Copleston found High Church mysticism offensive, the magical powers of science and music were to be celebrated.

Conclusion

The most common form of authority in nineteenth-century Britain did not come from the laboratory or scientific treatise but from the pulpit. And this very much extended to the nature of sound and, in particular, the moral value of music. Indeed, this philosophical understanding of sound's physical properties, its ordering laws, and its place within God's Creation was inseparable from the spiritual worth of music, its place in worship and society, and its metaphysical character. Such an understanding drew directly both on the Bible and on the works of leading natural philosophers, and few were so prominent as Herschel, Whewell, and Somerville. Each of these science writers worked within a religious framework that employed sound as an example of divine creation, stressing how the laws governing natural phenomena throughout the universe were united. While sound provided analogies and visual reasoning for linking heat, gravity, electricity, magnetism, and light within an essentially harmonious system of divine order, music constituted a spiritually powerful natural phenomenon that obeyed mathematical principles. Anglican clergymen eagerly invoked such natural philosophy within their sermons, providing lessons on the order and beauty of nature. Faith was to be built on God's two volumes, nature and scripture, and in the 1830s and 1840s, sound and music featured prominently in both.

Invested with theological significance, sound served as an intellectual link between earthly and astronomical phenomena. As planets made oscillations, so did particles conveying sound, and both were comparable to oscillating ears of corn. This is why Herschel's cornfield analogy was so often repeated: once the idea of a sonorous impulse without the projection of particles was established, explaining light and heat became easier, as did their connections to stellar movements. Among natural philosophers and clergymen, there was a shared belief that a universe ordered by a few constant laws suggested intelligent design, and there was no better example of how natural phenomena were united than through the study of sound, which offered an invaluable analogy for heat and light. Waves of sound could be more directly experienced and understood than those of either heat or light, in part because they could be visualized more fully following Chladni's acoustic investigations. It is no coincidence that sound became increasingly prominent in works seeking to unify nature following Wheatstone and Faraday's experimental investigations between 1828 and 1832, but through Herschel, Somerville, and Whewell's writings, these acoustic investigations took on increased religious significance. Such science was not a rival source of knowledge to that provided through the pulpit and sermon. Alternative authorities within the Church of England drew on this philosophical knowledge of sound, incorporating it within their own spiritual insights on the metaphysical value of music.

PART II

Contesting Knowledge

Mathematicians, Musicians,
and Sound Measurements

CHAPTER 3

The Problem of Pitch

Mathematical Authority and the Mid-Victorian Search for a Musical Standard

> *Of all the fine arts, music is the one which has the most influence on the emotions, and it is this influence which the legislator must encourage.*
> NAPOLÉON BONAPARTE, LETTER, JULY 26, 1797

On hearing in 1847 that Charles Barry's newly constructed Houses of Parliament had been built with particular regard to "the science of Acoustics" for aiding political discussions, the satirical journal *Punch* proposed a musical system of parliamentary discourse. Speaking "in conformity with the principles of harmony," the Lords on "one side of the House will be compelled to speak in a different key from that adopted by the members on the side opposite." Hoping for debates "set to music," *Punch* recommended front-bench ministers "talk in four flats" while opposition members could "begin with a B sharp when calling on any member of the Government."[1] Such notions of a musical parliament were clearly humorous, but this in fact parodied a series of acoustic experiments within the new legislative building. At the invitation of a politically diverse group of MPs who were concerned with the acoustics of the Commons and Lords chambers, Scottish chemist David Boswell Reid had employed the buildings to examine how a room's shape, materials, and atmosphere affected the transmission of speech. Although Reid's experimental investigations resulted in acoustically poor debating chambers, his work embodied a genuine political desire to mobilize scientific knowledge in the construction of a new parliament building.[2] Yet for all that MPs sought to control sound within their own working environment, it was difficult for Victorian British politicians to exert influence within the wider musical world. Despite William Whewell's romanticizing of the absolute measurement of pitch and the ear's mathematical accuracy, determining what frequency constituted a musical note was challenging. This regulation of musical sounds was especially difficult in Britain due to its liberal political consensus, which emphasized the government's minimal role in ordering society and the economy. With the exception of a few radical MPs, Parliament was committed to this *laissez-faire* tradition in which individuals were, as far as possible, left to regulate themselves. Within this political framework, the impetus for regulating musical sounds was not to come from Parliament or the government but from independent musicians, natural philosophers, and politicians. They would work, not through state apparatus, but through private musical and artistic organizations. And they would have to face the challenge of exerting authority over musical practice without wielding legislative power. At stake here was the question of who ordered British musical life.

Throughout the early nineteenth century there was a broad consensus among European musical communities that the pitch at which orchestras were playing music was rising. In the mid-1830s, an A-above-middle C in Paris was generally sounded between 435 and 443 vibrations per second, but by the 1850s most institutions played at a higher frequency. The Conservatoire de Paris's A had increased from 436 to 446 between 1847 and 1856, while the Lille Opéra was at 451 by 1854.[3] In 1834 a congress of musicians at Stuttgart attempted to resolve this escalation, setting C-above-middle C at 528, but this "Stuttgart pitch" secured little support. This was not just a European problem; the New York *Musical Review* published an article calling for pitch regulation in 1839. The anonymous author, "Regulator," asserted that a standard pitch was essential if music was to be cultivated as a science and recommended that American musicians "get some standard of pitch from Europe—perhaps the Philharmonic Society's, of London—and by it let all forks be tuned, and by the forks let all *instruments* be tuned; and let us have, at least in this country, a *standard*."[4] While most agreed that there should indeed be an international musical standard, the question of how many vibrations per second should define a specified note, usually either A- or C-above-middle C, was immensely difficult to resolve. Should the measure be one agreed on by musicians and instrument makers, or mathematicians and natural philosophers? Generally, musicians favored slightly higher frequencies than mathematicians, who preferred a pitch based on the observation that a string sounding the lowest audible C vibrated sixteen times per second, with each subsequent C in the scale increasing by the power of two, so that the frequency of a C-above-middle C would be 512. French mathematician Joseph Sauveur (1653–1716) made similar calculations based on the sounding of an organ pipe which, when increased or reduced in internal volume, produced proportionally different pitches. This mathematical ideal of C512 was unpopular with musicians who, accustomed to playing at higher frequencies, found it flat.

Little wonder, then, that attempts to standardize musical pitch quickly became controversial. Initially, it was France that seized the initiative, with Emperor Napoléon III sanctioning a commission to introduce musical uniformity throughout the country, which in 1859 established a national unit of pitch of A435. As Fanny Gribenski has shown, until the early twentieth century, the French state was at the forefront of pitch reform, dominating negotiations over musical standardization.[5] Following France's example, but lacking the centralized authority of the Napoleonic state, British musical and scientific audiences responded with a similar scheme for eradicating pitch disparities. Several months after the announcement of the French measure, the Society of Arts established a committee of musicians, mathematicians, experimentalists, instrument makers, composers, and self-proclaimed musical connoisseurs to agree on a British standard; however, this enterprise rapidly became controversial. John Herschel advocated C512 while others favored following France's standard of A435, which would equate to a C of 522. Others promoted a higher pitch altogether, preferring a C in the high 520s or low 530s. In these discussions, the limit of scientific authority in musical matters became very clear. For all of Faraday and Wheatstone's experiments and Herschel, Whewell, and Somerville's writings, the question of pitch was something musicians were eager to resolve for themselves. As much as scientific individuals tried to define sound, they struggled to exert influence over musical culture. Scientific knowledge was not inherently authoritative and it did not equip its promoters with irrevocable determining agency. Between 1859 and 1860, in Britain at least, the disciplines of mathematics and music became increasingly defined through the search for an absolute standard of pitch. The distinctions between different groups, such as natural philosophers and instrument makers, or mathematicians and musicians, were ambiguous in the early nineteenth century, but they became more evident through this controversy, with musicians eager to seize authority over their art.

Both Myles Jackson and Bruce Haynes have demonstrated how the standardization of musical pitch has historically had little relation to any natural reference but resulted from negotiations between musicians, instrument makers, and physicists: musical measures are socially, culturally, and economically contingent. Focusing on sound in nineteenth-century Germany, Jackson showed that these exchanges were far from harmonious, constituting a power struggle between different interest groups.[6] In nineteenth-century Britain, however, these groups were only loosely defined. The boundaries between the sciences and the arts were unclearly marked, and participants often shared common values, with individuals such as Herschel eminent both musically and mathematically, and musicians like Frederick Ouseley keen to appear scientific.[7] In this context, no one was inherently authoritative in the process of unifying pitch, as protagonists of contrasting skills and knowledge had to fashion themselves as trustworthy guides over what constituted a suitable standard. It was not always self-evident who was strictly a "musician" and who was predominantly a "mathematician": such epithets were self-styled. What was more important was who was deemed a credible judge of what a musical standard should be. While Herschel invoked nature as the source of his authority, others mobilized arguments grounded in the aesthetics of different frequencies or practical experience as musical performers or instrument makers. This chapter therefore delivers a detailed prosopographical account of the actors involved in the 1859 pitch committee, not only to show how the Society of Arts attempted to establish an inquiry of great authority in the wider musical community but to illustrate how varied the interests and experiences of these personnel were.

Efforts to make musical pitch uniform cannot be seen as an isolated question of standardization but were part of broader discussions over measurement. From James Joule's paddlewheel experiments to calculate a mechanical value for heat during the 1840s, to Britain's legalization of a new imperial system of weights and measures in 1855, to debates over adopting the metric system throughout the 1850s and 1860s, questions of measurement were prominent in industrial Britain.[8] Similarly, as Britain's submarine telegraphy network expanded during the 1850s, the problem of electrical measures became increasingly urgent. By calculating the total resistance of a broken cable and then comparing this to a known resistance standard, a telegraph engineer could know how far out to sea the break had occurred. In 1861 the BAAS established a Standards Committee which campaigned for the establishment of national electrical units. Physics laboratories, such as Cambridge's Cavendish Laboratory, opened in 1874, were very much part of this imperial communications project in which measurement was so crucial.[9] While there have been several rich histories dealing with these questions of measurement, British attempts to standardize musical pitch in the late 1850s have received little attention.[10]

This chapter explores the makeup of the 1859 pitch committee, both in terms of the credentials of its participants, their various backgrounds and experience, and their areas of knowledge and skills. This committee was intended to invest the agreed-on pitch with authority in broader society and to secure the standard a consensus. The chapter then analyzes the reception of the committee's decision, with particular emphasis on Herschel's campaign for C512, before examining the standard's subsequent implementation. As Graeme Gooday has shown in his study of Victorian electrical standards, the practice of constructing new measurement standards is far from objective: the key concern in processes of measurement involves trust. To understand how a measurement standard is accepted or rejected, we have to ask who trusts who, what do they trust, and why do they trust one standard over alternatives.[11] The pursuit of standards is an effort to make measures of natural phenomena appear objective and self-evident, but such work is always subjective and culturally contingent.[12] This was equally true of musical units: the question of pitch was not simply about mathematics or aesthetics, but about who could be trusted to determine a standard that was musically acceptable.

However, the process of musical pitch unification, arguably, concerned more diverse social audiences than any other nineteenth-century standard. Reflecting the disparate communities that were invested in matters of music, pitch standardization straddled boundaries between the arts and sciences, industry and manufacturing, the military and the church. With bands a prominent part of the armed forces, a revival in church music, theological understandings of the human voice and ear, and substantial popular and elite musical cultures, individuals of scientific learning had much competition in determining a unifying measure. Here was a standard of unrivaled social complexity, and it was this complexity that raised troubling political questions. With so many interested parties, it rapidly became clear that scientific knowledge alone commanded little authority to govern within the confines of a politically liberal state.

The Pitch Committee of 1859–60

Appointed on July 17, 1858, in response to growing complaints over pitch escalation from vocalists, a French commission to investigate pitch gathered information from across Europe and America. A portion of the commission's report examined the state of pitch in England, with London pianoforte makers Broadwood and Sons sending information and three tuning forks to Paris. The first fork, used at the London Philharmonic Society during the 1830s, was a quarter-tone lower than the average Parisian pitch. The second, used to tune Broadwood pianos, was much higher, and the third, belonging to the Philharmonic Society, was higher still.[13] The 1858 French commission agreed on A435 in February 1859, and its decision became law from July as the standard known as the *diapason normal*. This lowering of pitch came despite French composer Hector Berlioz's warnings that such a reduction would be expensive.[14] France's standardization of pitch did not go unnoticed in Britain, with the *Spectator* lauding Napoléon's commission, which included Italian composer Gioachino Rossini, as an eminent scheme to enhance "national unity." The journal lamented how high pitch had risen over the previous fifty years, with many of George Frideric Handel's works now impossible to sing, the higher notes having become too sharp, as well as disrupting the intervals between notes. As the *Spectator* surmised for its reform-leaning readership, this "confusion of the scale . . . is one amongst the millions of examples of that eccentricity in the mechanism of nature which forbids human systematizing. We cannot reduce nature to our narrow idea of 'perfection.'"[15] Nevertheless, it attributed the musical error

> to a moral cause. The composer desires to make his work "brilliant"; he throws it rather high in the scale. The performer desires to produce a "brilliant" execution, and he tunes his instrument rather sharp . . . in the ambition of brilliancy, the singer, the instrument performer, and the composer are constantly working upwards.[16]

This "mania" for "brilliancy" appeared unstoppable without an accurate standard to regulate musical performances and prevent composers from producing works which were so high that they damaged the voices of performers. The *Spectator* noted that the "song of birds" would be an excellent "natural standard," were it not that the "delicate chirp of the bird . . . eludes systematized reductions to our larger and more precisely divided gamut." It was France, however, that seemed to have the solution. What was wanted was a piece of metal fixed to a certain vibratory frequency, in a way similar to the French construction of the meter, which was determined by "systematically based measurements upon a natural standard": that standard being a quadrant of the earth from the equator to pole. The French were, the journal continued, "a systematic people."[17]

The *Spectator*, with its customary liberal-radical agenda, saw political implications in this musical question. If musicians could not behave responsibly with pitch without governance, then this principle extended to the economy, given that in "civilized countries commerce is the handmaid of music," and both, the *Spectator* argued, required regulation. Furthermore, the *diapason normal* highlighted political differences between Britain and France. While "in a country like France a gracious Emperor, strengthened by a sufficient reverence for music," could, the journal asserted, restrain musicians, this was difficult to do in a liberal, *laissez-*

faire country. For the *Spectator*, rising pitch offered a powerful lesson on the perils of an unregulated art, the limits of liberal government, and the desirability for state intervention when individuals could not be trusted to behave responsibly.[18] This was of course a moment in which, regardless of subsequent historical revisions, Britain prided itself on its liberal system of government, in which the economy was thought better regulated by natural laws and *laissez-faire* thinking than by parliamentary legislation.[19]

Britain might not have had an emperor, but it did have the Society of Arts, founded in 1754 to promote the nation's arts, manufactures, and commerce. After receiving a Royal Charter in 1847 (it became the Royal Society of Arts in 1908), this was an institution of growing influence, located just off the Strand in Westminster on John Adam Street. With Prince Albert as president between 1843 and 1861 and a largely aristocratic governing body which included leading scientific and commercial individuals as well as celebrated engineers like John Scott Russell and Robert Stephenson, the society exerted growing influence. On learning of the recent French standardization of pitch, Harry Chester (1806–68), a vice-president of the society, was convinced of the urgency of such musical regulation. In the spring of 1859, he wrote to the chairman of the society's council, Charles Wentworth Dilke (1810–69), explaining that as there could not be government legislation for such a measure in Britain, the society should establish "a voluntary law" for the nation's musical communities to adopt.[20] An opera enthusiast and future Liberal MP, Dilke was a keen promoter of the arts and manufacturing, having helped organize the Great Exhibition of 1851 and reported on subsequent exhibitions in Dublin, New York, and Paris.[21] He eagerly embraced Chester's suggestion and recommended to the Society of Arts council that it appoint a committee to introduce a British musical standard. Lacking legislative power for enforcing any measure it agreed on, the council recognized the importance of ensuring that those conducting the inquiry were recognized as authorities in their disciplines, professions, and crafts.

As a result, the appointed committee boasted diverse scientific, ecclesiastical, engineering, musical, and industrial credentials. For the large part, they were already a closely connected network, with Dilke and Chester joining fellow society council members Thomas Sopwith (1803–97) and William Hawes (1805–85). While Sopwith provided engineering experience, Hawes boasted strong musical connections, with his associates forming the majority of the committee. Hawes had taken an interest in cathedral music during the 1840s, suggesting reforms to St. Paul's Cathedral's choir.[22] As the younger brother of Benjamin Hawes (1797–

1862), an MP and husband to Isambard Kingdom Brunel's sister, Sophia Macnamara (1802–78), William Hawes had influential musical and scientific associates. Brunel himself had wedded into the musically gifted Horsley family, marrying Mary Elizabeth Horsley (b. 1813), the daughter of composer William Horsley (1774–1858) and sister of Charles Edward Horsley (1821–76), also a member of the pitch committee. The Horsleys and Brunels enjoyed musical evenings together in South Kensington during the 1830s, with Felix Mendelssohn and the Haweses frequent guests.[23] Another of Hawes's associates, the London-born conductor and organist George Thomas Smart (1776–1867), consented to join too.[24] Smart regularly corresponded with Hawes on musical subjects, including reviewing Hawes's sister's compositions, and was a friend of William Horsley, often attending musical performances with him.[25]

Also appointed was an associate of both Smart and Prince Albert, the organist and composer George Job Elvey (1816–1893), who had been the organist in Oxford at New, Christ Church, and Magdalen Colleges before appointment as organist to St. George's Chapel at Windsor in 1835, where he twice rebuilt the chapel's organ and gave Albert lessons in harmony.[26] In 1856, Elvey applied unsuccessfully for Cambridge's chair of music on the death of its incumbent, Thomas Walmisley. The duties of this post consisted of performing services each Sunday at Trinity, King's, and St. John's Colleges and the University Church of St. Mary's. As vice chancellor, Whewell oversaw the competition which was decided by an open poll of the University Senate.[27] Elvey's successful opponent in this campaign, the violinist and composer William Sterndale Bennett (1816–75), would join him on the pitch committee. Although successful at Cambridge, Bennett had failed to win George Thomas Smart's support (Smart favored Charles Edward Horsley), but Bennett did secure Herschel's patronage, who asserted that if Bennett was "really disposed to raise that very low nonentity the Musical Professorship into a worthy and efficient position—by giving lectures in which the principles of the physical science of sound shall be made (as at a *scientific* University they ought to be) an integral feature... then in that case all I *can* do to forward his election, I will." Evidently Herschel's promotion of Bennett's scientific credentials carried weight, as Whewell allowed Bennett to canvass in Trinity College. He subsequently took the post by 174 votes to Elvey's twenty-four and Horsley's twenty-one.[28] The pitch committee united rivals Elvey, Bennett, and Horsley, as well as Bennett's eventual successor, George Alexander Macfarren (1813–87), and Edinburgh University's Reid Professor of Music, John Donaldson (1789/90–1865). Having reformed Edinburgh's musical syllabus in 1858, Donaldson constructed a new "School of the

Theory of Music," complete with a museum of instruments, concert hall with organ, and introduced Chladni, Savart, and Wheatstone to the curriculum.[29] Together, this network of academics had been influential in the rationalization of Britain's musical education.

Along with this academically impressive network, the Society of Arts recruited some of the nation's leading composers and conductors, such as Julius Benedict (1804–85), Alfred Mellon (1820–67), director of both the Haymarket and Adephi Theaters, and Henry David Leslie (1822–96), whose short-lived National College of Music in Piccadilly in 1864 and Royal College of Music between 1878 and 1883 were both attempts to establish an English rival to Continental conservatories. Having edited the works of Bach, Beethoven, Mozart, and Handel, composer Cipriani Hambley Potter (1792–1871) contributed considerable historical knowledge to proceedings.[30] Musical performers were of course prominent, with Henry Charles Lunn (1817–94) and John Goss (1800–1880), the organist of St. Paul's Cathedral, and celebrated violinists Henry Griesbach (1798–1875) and Henry Gamble Blagrove (1811–1872) joining the inquiry. Organists Edward John Hopkins (1818–1901) and Robert Kanzow Bowley (1813–70) were present, as was the composer and pianist Otto Moritz David Goldschmidt (1829–1907), husband to Swedish soprano Jenny Lind (1820–87). Most famous of all was the Westphalian pianist Charles Hallé (1819–95), who had moved to Paris in 1836 before then fleeing to England during the 1848 revolution. With London swamped with Parisian musicians, Hallé moved to Manchester and established his own concerts but was often frustrated in his new adopted home, lamenting in 1860 that he had to "work again at that wretched English music."[31]

Adding a military voice to the proceedings was James Waddell, bandmaster of the First Life Guards from 1832 until 1863, indicating the importance of the army to pitch standardization: this disciplined organization could help to diffuse any musical measure agreed on throughout the empire. Beyond the army, however, it was the inclusion of leading cathedral musicians that reflected the considerable interest in pitch reform of those from religious musical traditions, especially from within the Church of England. Most prominent of these was Frederick Gore Ouseley (1825–89), who informed the pitch committee on matters of church music. As a child, Ouseley displayed an astonishing sense of pitch; aged three, he apparently "once told his sisters the exact name and pitch of a clap of thunder," and later "frightened an unmusical nurse by telling her that a clock had just struck in the key of B♭ minor. This sense of absolute pitch never deserted him."[32] At Christ Church College Oxford, between 1843 and 1846, he joined a network of Anglican churchmen that agreed with John Jebb's view that the Book of Common Prayer should be sung. He took a doctorate in music from Oxford in 1854 and, a year later, became the university's Heather Professor of Music, introducing student examinations to supplement the existing requirement of an original composition. In 1858, Ouseley established his own parish church and college at Tenbury, Worcester, where he worked to establish "a model for the choral service of the Church" within an Anglican framework.[33] Few from within the Church of England delivered so much instruction over the place of music within worship. When Ouseley put his own views of music and faith into a sermon in December 1860, he lamented that so many thought the use of music in churches to be frivolous. On the contrary, he asserted that music was inherently spiritual.[34] Ouseley offered solid religious guidance to the pitch committee. Indeed, his input was part of a broader Anglican influence within the committee which drew directly on Britain's recent revival of church music, as well as theological understandings over the divine creation of the voice.

Equally concerned with religious music was John Hullah (1812–84), the Worcester-born music teacher and composer, and student of William Horsley. Professor of vocal music at King's College London since 1844 and prominent within Charles Kingsley's Christian Socialist movement, he believed that "music contained within it a moral force which could refine and cultivate individuals and encourage a sense of value and worth within the community."[35] After working to improve Trinity College Chapel's music, he was well known to Whewell, but more important for the pitch commission was that Hullah owned a tuning fork of $C512$.[36] He was not alone in bringing forks to the inquiry. Perhaps the two most important forks in existence at the time were Handel's $A423$ and an earlier version belonging to John Shore (1662–1752), an associate of Handel and also Sergeant Trumpeter to the Court, who had first conceived of this device in 1711. In 1859, both forks belonged to Reverend G. T. Driffield, rector of Bow and author of several hymns.[37] Between Driffield and Hullah, the committee possessed three valuable indicators of pitch, both historical and mathematical.

Given their stake in questions of tuning, several eminent instrument makers joined the committee. Frederick William Collard (1772–1860) was, in 1859, one of Britain's leading piano manufacturers, having established Collard & Collard in 1831 with his brother William Frederick, before going into partnership with his two nephews, Frederick William Collard Jr. and Charles Collard, in 1842.[38] The extensive piano-making firm Broadwood also sent a representation to the committee, most likely Walter Stewart Broadwood (1811–93), who was an experienced tuning-fork manufacturer. Kensington-based instrument maker Erard, who produced over a thousand pianos and harps annually, sent M. Bruzard to the committee. Organ builders Gray & Davison despatched Frederick Davison, but it was the two powerful firms of Willis and Hill who were most prominent on the inquiry. The organ builder Henry Willis (1821–1901) was highly eminent in his trade, having rebuilt the organs of Tewkesbury Abbey and Gloucester Cathedral between 1846 and 1847. He then established his own workshop at Foundling Terrace on Gray's Inn Road in 1848 and won fame for his Great Exhibition organ, which was later installed in Winchester Cathedral.[39] This success earned Willis contracts for the new organs for St. George's Hall in Liverpool in 1855 and the cathedrals of Carlisle (1856), Wells (1857 and 1891), Canterbury (1866 and 1886), Durham (1876), Salisbury (1876–77), St. Mary's, Edinburgh (1879), Glasgow (1879), Truro (1887), and Lincoln (1898). His appointment to the pitch committee provided unprecedented knowledge of organ building.[40] Also appointed was Willis's rival, William Hill (1789–1870), who had likewise impressed at the Great Exhibition.[41]

The committee thus included authoritative representatives from Britain's leading musical interest groups, including instrument makers, performers, composers, critics, soldiers, and clergymen. It was especially representative of Britain's tradition of Handelian choral performances, which was such a central part of the nation's musical culture. Yet this committee also boasted impressive scientific credentials. The Scottish physician and public health reformer Neil Arnott (1788–1874) offered medical knowledge to the investigation. Along with the appointment of Charles Wheatstone and Robert Willis, mathematics was equally well represented, including a strong network of Cambridge wranglers. Whewell was a domineering figure on the committee, finding support from Trinity-trained mathematician Augustus de Morgan and St. John's College's Benjamin Morgan Cowie (1816–1900), who had passed Senior Wrangler in the 1839 Mathematical Tripos Examinations. As well as being Gresham College's professor of geometry, Cowie was theologically active, giving a series of five sermons on sin and the Atonement at Cambridge in 1856.[42] Another mathematician, Thomas Minchin Goodeve (1821–1902), would eventually become the first professor in Imperial College London's new department of mathematics in 1869. A friend of Hullah, Bennett, John Stainer, and Herschel, the musician and mathematician William Pole (1814–1900) straddled Britain's musical, scientific, industrial, and engineering communities.[43] In addition to earning testimonials of his skill from Jebb and Samuel Wesley, Pole's most prominent supporter was none other than Ouseley, whom he studied under at Oxford between 1859 and 1860 while simultaneously serving on the pitch committee.[44] When Pole took his bachelor of music, Ouseley wrote to the president of St. John's College, requesting that he be allowed to graduate from that college, given that Pole had "passed the most brilliant musical examination I have yet witnessed," in which Elvey had been his examiner.[45] Similarly, Henry Wylde (1822–90) was respected both in musical and scientific circles, having studied at the Royal Academy of Music and acting as a juror on the musical instrument section of the Great Exhibition of 1851. Wylde emphasized music's moral value, explaining how "Music alone, of all the fine arts, may be said to derive its first impulse from the Divine Teacher," while musical instruments mobilized laws observable within "the great laboratory of nature."[46] He was particularly keen to adopt a scientific framework to music composition, outlining this in his 1872 student textbook, *Harmony and the Science of Music*.[47] Wylde was an especially apt appointment to the pitch committee, having recently assisted with Sydney Ringer's experiments with tuning forks on the passage of cardiac murmurs through blood vessels, examining how the pitch of forks changed as they traveled through bone, muscle, skin, and cellular tissue. The challenge for Ringer was to ensure accurate readings of pitch, and for this Ringer employed Wylde, whom he judged to have an "acute ear."[48]

Although the London barrister Thomas Phillips (1801–67) acted as the committee's overall chair, he delegated the direction of individual meetings to different committee members.[49] The *Times* reported on June 4, 1859, that Whewell had chaired a "meeting of a number of leading scientific men and those interested in music" at the Society of Arts to discuss a uniform musical pitch.[50] At the pitch committee's meetings of July 1, 1859, and January 11, 1860, Potter occupied the chair, while Hullah directed discussions on December 23, 1859. Pole subsequently chaired the meeting of February 2, 1960.[51] The committee's leadership was in this way shared between those possessing scientific knowledge and those of musical experience. This was indicative of how the Society of Arts hoped to mobilize the musical authority of the nation's leading instrument makers, performers, composers, conductors, organists, mathematicians, soldiers, and clergymen. It was a framework intended to build credibility with audiences involved in the practice of musical performance.

The diversity certainly satisfied the *Spectator*, which observed how the committee included "representations of mathematical science, concert conductors, pianoforte makers, composers, the teacher of the rising public, and the prince of opera-conductors." While it was confident that this British inquiry was "superior in its representation of mathematics" to France's recent effort, the journal thought France's A435 well supported by the great musical authorities of Halévy, Rossini, and Berlioz. Indeed, unless the French commission had made some sort of technical error, the *Spectator* was in favor of adopting the French standard and saw the Society of Arts' inquiry more as a review of the *diapason normal*. Although the journal feared the French pitch would be rejected as it was foreign, it was sure that the "names on the committee . . . are a guarantee that the subject will be regarded in a practical light, having in view not any merely national or corporate jealousy, but the enduring interests of art." The urgency for not just a national but an international standard had, the journal claimed, heightened because of the growth of increasingly fast forms of transport; trains and steamships made international variations in pitch more apparent, with musicians traveling more than ever before.[52]

Dilke and Chester's acknowledged liberal political values shaped the investigation's proceedings. Rather than look to establish a pitch and then seek to enforce it on the country, the committee, emphasizing the consensual nature of any standard selected, sought the opinions of the nation's music societies and performers. On August 28, 1859, it launched a national survey, circulating a letter containing three questions to Britain's leading musical organizations. The committee asked if a national standard of pitch was desirable, would such a measure be difficult to introduce, and at what frequency should it be set? The committee's secretary, the Cambridge-trained mathematician and barrister Peter le Neve Foster (1809–79), conducted the inquiry throughout the autumn, compiling substantial material. Among those who replied, a clear consensus emerged that the proposed unified pitch would be valuable and that the tone should be reduced from those in present use.[53] Along with historicist considerations regarding the integrity of celebrated musical works, this support reflected physiological concerns over the preservation of the voice. W. Mason, the conductor of Lincoln Cathedral's Choir, called for pitch to be lowered by three-quarters of a tone, complaining that during his twelve years at the cathedral, three out of four voices were "ruined before they are developed" because of escalating pitch.[54] Charles Saldman agreed, warning that rising pitch was giving "a different character to musical compositions": he proposed that a "congress of eminent musicians & instrument makers from the principal cities in Europe might meet, either at Paris or London & the question be at once settled."[55] E. Shepherd, honorary secretary of Abingdon Musical Association, thought a standard pitch would "curb the tendency of the present age for its elevation," but could not "see how this can be met except by an act of parliament."[56] Likewise, J. Dyson felt the measure a hard one to enact but, as secretary of the Windsor and Eton Royal Glee and Madrigal Society, reported that all the members of the choirs at St. George's Chapel and Eton College supported a uniform pitch.[57] Several respondents expressed confidence that the committee could resolve the matter. While one correspondent thought that the C_{512} fork, belonging to Hullah, embodied the finest pitch possible, M. Burges of Macclesfield Sacred Harmonic Society was sure "that the question may safely be left to the decision of the talented *articles* named in the circular."[58]

However, Foster's consultation did not produce unanimity. Of some forty responses to the circular, not one suggested an elevation of pitch, while twenty-nine wanted it depressed by between one- and three-quarters of a tone. Of these twenty-nine, eleven favored C_{512}.[59] Despite agreement over the value of a national standard of pitch, there was opposition to its introduction. As secretary of the Philharmonic Society, George Hogarth had put the letter before a general meeting on November 21, 1859, but reported that they resolved that introducing a uniform pitch was "attended with so many difficulties as to be impracticable in this country."[60] W. Grice reported that the Society of British Musicians had considered such musical standardization at their general meeting on November 30, and also judged "it impossible at present to carry the same into effect."[61]

Respondents also raised aesthetic objections. J. Surman of the London Sacred Harmonium Society argued that if orchestral pitch was lowered, music "would lose much of its brilliancy." He cast doubts over the seriousness of pitch escalation, observing how in oratorio music it was usually the tenors who complained of high pitch, rather than the altos or basses. As for the chorus singers, there was dissatisfaction with the works of modern composers, like Beethoven and Mendelssohn, but this was limited to male altos and the problem could be remedied by having "a certain number of ladies to sing with them." Surman thought the fault lay with the singers themselves, speculating that as their voices "have got up as well as the pitch of the instruments, how few Basses there are now that can sing many of the Bass solos in the anthems."[62] Often, he continued, tenors complained that their solos were too low for them. Similarly, Macclesfield violinist W. Holland did not find contemporary pitches too high. While he acknowledged that old organs now sounded flat, higher tones did not affect violinists and wind instruments were easily adjusted.[63]

Instrument makers echoed these doubts over a proposed standard. Despite the inclusion of "some of the principal instrument makers" on the pitch committee, this group raised objections.[64] In particular, it was those involved in instrument manufacturing for Britain's armed forces who expressed the greatest concern, reflecting the immense impact a na-

FIGURE 3.1 The 1859–60 pitch inquiry consulted a diverse range of musical authorities in its investigation, including instrument makers like Henry Distin of Leicester Square. Royal Society of Arts, PR/GE/121/10/5, Musical Pitch, H. Distin to Foster (January 5, 1860).

tional measure would have on the bands of the army and navy: in theory, these would be the most rigorous enforcers of such unification. The military musical instrument maker Henry Distin, for example, felt the pitch of C_{546}, used by the Italian Opera in Covent Garden, to be the best possible (fig. 3.1).[65] In the same fashion, Rendell Rose, of the Military Musical Instrument Manufactory, opposed any immediate pitch reduction, warning Foster that should such a measure be enforced, many instruments would have to be destroyed, necessitating "a sacrifice of the old Insts. with the Regiments and other of our customers."[66] However, manufacturers' attitudes to lowering pitch very much depended on the instruments they produced. Committee member and organ builder Henry Willis informed Foster, on missing a meeting late in 1859, "that I throw the whole weight of my influence" behind the adoption of C_{512}. As it was hard to adjust an organ's pitch, most being of a lower pitch than violins and pianos, Willis favored this reduced standard.[67]

From the national survey, the committee agreed unanimously that a standard pitch was desirable, but found the question of frequency harder to resolve. Instruments changed pitch during performances, subject to temperature, with instruments in theaters tuned sharper on stage than in the orchestra to account for such variations. A uniform pitch would therefore be difficult to achieve, especially as no one seemed able to resolve what measure this should take. Tuning forks from the late eighteenth century provided evidence that during Handel's time (ca. 1720–59), pitch had been about a tone lower than contemporary levels. The Philharmonic Society's tuning fork, used between 1813 and 1842, was examined and compared with Handel's fork, which sounded a semitone lower. However, the 1813–42 fork gave C518.4 compared to C546 for the contemporary average pitch. In short, the C and A of 1859 were almost equivalent to the D♭ and B♭ of the 1840s. This rise of a semitone in less than twenty years was attributed to a general desire for increased "brilliancy" of timbre among performers, along with an influx of "foreign vocalists, gifted with voices of exceptionally high register." Another problem was that manufacturers were, the committee found, producing instruments of higher frequencies, but while "the pitch of artificial instruments admits of alteration to almost any extent . . . the pitch of the voice, like the voice itself, admits of no alteration but at the will and by the hands of Him who made it."[68] This reference to God's Creation revealed much of the religious influences within the committee. The committee subsequently agreed that it was the human voice which should be prioritized in setting a standard, invoking both physiological and theological authority.

Although difficult to enforce, the committee was confident it could at least define what the musical standard should be. A pitch was "capable of exact measurement, and that measurement once recorded, it may be reproduced at any distance of time, without reference to any other sound whatever." "Physical science is," the committee determined, "enabled to afford this, and to bring to the aid of musical art more than one process by which such a standard may be adjusted."[69] Traditionally the number of vibrations constituting a tone could be determined either through the division of a vibrating string or the phenomenon of beats, but from 1819 Cagniard La Tour's siren, which blew pressurized air against a rotating disc with holes in it to produce sound, could be employed to quantify pitch. After 1855 it was also possible to measure vibrations using Jules Lissajous's optical arrangement in which a beam of light, reflected through a series of mirrors attached to vibrating tuning forks, produced an image from which a tuning fork's frequency could be calculated.[70]

The question, then, was at what frequency should the standard be set? For this the committee initially looked to "acoustical science," which asserted that all notes corresponded to a basic standard of one vibration per second. Therefore, a series of notes escalating through scales vibrated to increasing powers of two (2, 4, 8, 16, 32, etc.), so that a length of string that gave the lowest C with one vibration per second would give the next C up (2 vibrations per second) were its length halved, and the next C in succession when it was halved again (4 vibrations), and so forth. The committee resolved that this "*theoretical* note is found to agree so nearly with the musician's idea of the note C . . . that writers on acoustics . . . have

Handel's Tuning Fork (c. 1740)	A at 416	—	C at 499¼
Theoretical Pitch	A at 426⅔	—	C at 512
Philharmonic Society (1812-42)	A at 433	—	C at 518⅖
Diapason Normal (1859)	A at 435	—	C at 522
Stuttgard Congress (1834)	A at 440	—	C at 528
Italian Opera, London (1859)	A at 455	—	C at 546

FIGURE 3.2 The various pitches, given for both C and A, considered during the inquiries of the 1859–60 pitch committee. "Uniform Musical Pitch," *Journal of the Society of Arts*, 6. Public domain.

thus established what may be called a *theoretical pitch*, or definition of the note C." Theoretically speaking, the scale of a pianoforte should rise from 256 vibrations for middle C to 512 vibrations for C above middle, but this was not a frequency musicians employed. The problem with using acoustic theory was that if such a mathematical standard was found "injurious to musical effect," musicians would reject it without question. The committee examined the proceedings of the French commissioners, who had "been governed by considerations of a purely practical kind (their report ignoring mathematical convenience entirely)," and selected A435 (giving a C of 522), which was "a pitch, certainly not identical with the pitch of 512 vibrations, but differing from it only to the extent of ten vibrations per second" (fig. 3.2).[71] Similarly, London's Philharmonic pitch, used between 1813 and 1842, was fixed

> without reference to any mathematical or scientific text whatever. A few eminent practical musicians consulted together, and came to agreement among themselves that a certain pitch was a convenient mean, neither too high for voices nor low for instruments. . . . The authority, therefore, of practice as of theory—of art as of science—belongs alike to the pitch of C 512.[72]

Yet the committee could not promote C512 when London's average C was 546, approximately a semitone's difference.

Early on, there were clear tensions between those identifying as musicians and those deemed to be mathematicians. Oxford's Savilian Professor of Astronomy, William Fishburn Donkin (1814–69), explained that "unless there be some practical objection unknown to me, every mathematician would wish to define the pitch of the scale by assigning the series of powers of 2 . . . to the C's. This would nearly agree with Hullah's proposal, for taking 1024 vibrations . . . we should have strictly 853 1/3 for the nearest A; and I suppose Hullah's 850 is meant to represent this in round numbers." Donkin believed A850 to be a suitable standard but resented the interference of mathematicians.[73] As he put it, "I do not at present see that a mathematician, as such, has any interest in the question beyond that of preferring the number 2 as the most simple and natural base for a definition. Speaking as a musician I should say that the pitch I refer to . . . answers all the required conditions, on which musicians seem to be pretty well agreed."[74]

Instrument makers would also resist C_{512}, as such a dramatic reduction would render "the greater portion of their existing stock valueless." A subcommittee appointed to assess the progress of the new French standard reported to the full committee that there had been trouble "in enforcing the new musical diapason in France, and that authority such as would never be sought for, or obtained, in this country, has found a powerful antagonism in 'the inexorable logic of facts.'" If the French imperial state could not enforce a standard on its people by law, what could a committee hope to achieve? A shift from C_{546} to C_{512} would certainly provoke hostility. The committee therefore felt the French standard of 522 too extreme and instead thought the 1834 Stuttgart Congress's recommendation of C_{528}, half way between C_{546} and C_{512}, had "the best chance of attaining the general assent of contemporary musicians." This was a pitch the committee was sure would be acceptable for Britain's musical communities.[75]

John Herschel's Mathematical Standard

On June 1, 1860, the committee published its report; it then held a public meeting on June 5 at the Society of Arts on John Adam Street to discuss its implementation. Driffield commenced the discussion by proposing the report be adopted, observing that it was not an attempt to "dictate to the meeting, or to the musical public generally," but a carefully produced document of reliable evidence. Chester seconded this motion, stressing the urgency of reaching a moderate standard and a unanimous conclusion, but warned that, unlike in France, "in this free country" there could be no "law for a compulsory uniform pitch." The question before the meeting was not just of a musical unity but of how to manage music without legislation.[76] Driffield's motion subsequently passed; however, within the audience was John Herschel, who disagreed completely with the committee's recommendation of C_{528}.

By the late 1850s Herschel was probably the most eminent scientific authority in Britain, especially influential in questions of measurement.

As a royal commissioner on standards between 1838 and 1845, a member of the Board of Visitors for the Royal Observatory at Greenwich from 1843, and Master of the Mint from 1850 until 1855, Herschel worked on national questions of metrology. After the nation's standard weights and measures were lost in the fire which destroyed the Houses of Parliament in 1834, a new imperial system passed into law in 1855. Herschel had helped determine the measure of the new yard, but he was unsatisfied with the unit finally agreed on.[77] In 1863 he campaigned against a Parliamentary bill for the adoption of the French metric system, instead promoting weights and measures based on his own revised yard. The arguments he made at this time echoed those made four years earlier concerning pitch. He argued that any object chosen as a standard unit had either to be "imperishable in its nature," or "some ideal or resultant length," such as a fraction of the Earth's quadrant. Herschel emphasized the importance of a standard taking a physical form, observing that societies required "some rod, bar, ruler, or other standard." Above all, a standard had to have a claim to universal acceptance; as he put it, "Nature presents to us but one material *object* which combines all the requisites enumerated—viz: the globe itself." Rather than the meter or imperial yard, as a unit of distance he favored the length of a pendulum, measured to beat a second, while "placed at the extremity of the earth's polar axis." As the imperial yard was 36 inches and Herschel's proposed pendulum yard would be 39.13929 of these same inches, this would be a small change. This length was "equal to the length of a pendulum vibrating seconds *in vacuo* and at the sea-level, in the latitude of London." Herschel felt the French meter lacked a universal claim, as the Earth's quadrant was variable at different locations; it was based on the meridian passing from Dunkirk to Formentera. The French meter had been the result of revolutionary "political passions," which demanded something based on decimalization: the 100,000th part of the globe was preferable to a second, of which there were 86,400 in the day.[78] A pendulum would oscillate at different rates around the globe, subject to the Earth's variable density, but Herschel argued that if its position of measurement was fixed, this problem would be solved.

Herschel had similar demands of any musical standard: that it be referenced to nature, materially embodied in the form of a fork, and mathematically defined. Well connected with members of the pitch committee, including Wheatstone, Whewell, De Morgan, Pole, and Willis, Herschel had hosted Ouseley and his wife Harriot for lunches and dinners at his home in 1839, including a musical evening. Indeed, Harriot thanked Herschel for taking "flattering notice" of her son, providing the boy with many amusements.[79] Herschel missed the pitch committee's initial meeting but wrote to its chairman, Phillips, on June 14, 1859, advocating C_{512}. The recent tendency for pitch to become inconveniently high arose, he asserted, "from a distinct natural cause inherent in the nature of harmony—viz: the excess (amounting to about eleven vibrations in ten thousand) of a perfect fifth one-seven-twelfth's of an octave; which has to be constantly contended against . . . wherever violins or voices are not kept in check by fixed instruments."[80] The question, then, was how much should pitch be lowered? Herschel felt that if another committee would be held in twenty years to further lower pitch, then a temporary C_{522} could be acceptable, equivalent to the French standard, but, without the guarantee of a future committee to reduce the standard to C_{512}, he would not agree to any compromise. Herschel argued that C_{512} was a scientific, natural standard which,

> being the ninth Octave of a fundamental note corresponding to one vibration per second, has a claim to universal reception on the score of intrinsic simplicity, convenience of memory, and reference to a natural unit, so strong that I am amazed at the French not having been the foremost to recognize and adopt it; when it is remembered that their boasted unit length, the metre, is based on the subdivision of a natural unit of space, just as a second (a universally used aliquot of the day) is of time: the one on the linear dimensions, the other on the time of rotation of the Earth.[81]

What was required, Herschel contended, was a musical pitch which took nature as a standard, and this involved basing any measure on the vibration of a string. He confessed himself "to be more French than the French themselves" in his enthusiasm for C_{512}, which would make musical pitch "part and parcel of a complete natural metrical system which would recommend itself to all nations on its own merits."[82]

Herschel not only recommended C_{512} for its mathematical credentials but was careful to draw out its musical advantages. This standard would satisfy "the wishes of the singer and the requirements of that most perfect and charming (because most naturally affecting the feelings) of all instruments, the female voice." Above all other instruments, made from "wood, brass, wire, and catgut," this had to be protected. While a vocalist could maintain a higher pitch during her youth, C_{512} would allow her to perform "up to an age when the voice though still perfect, and in fact improved and mellowed by time and practice, is yet unable, without painful effort, to reach the extreme elevations it could accomplish without difficulty at an earlier period."[83] Herschel's proposal did not just carry claims of scientific accuracy but was in fact also a way of prolonging the careers of female singers. The pitch committee read Herschel's letter at its meeting on July 1, 1859, before it appeared in August in the *Leeds Mercury*, cuttings of which Herschel circulated to his scientific associates, including Somerville.[84]

Herschel's calls for C_{512} found some musical support, even from within the British army, whose representatives were typically apprehensive of any proposed pitch reform. In early Victorian Britain, the military was the main employer of musicians; music was important to the ordering of the army, with instruments disciplining troops on the battlefield and regulating marching. Standardized instruments, such as metronomes and pendulums, were not just valuable in scientific work but for the movements of the armed forces.[85] In February 1860, the radical MP Major General Thomas Thompson (1783–1869) wrote to Herschel requesting advice on where to purchase an accurate C_{512} tuning fork, hoping to follow the astronomer's "enlightened proposal" and employ the standard.[86] Herschel suggested Wheatstone, who subsequently denied "any connection with the tuning-fork" and directed Thompson to John Parker, a bookseller on the Strand, who stocked tuning forks made by Hullah. He produced these especially for singing lessons, maintaining that to teach vocal music, all a teacher required was a wand to beat time and a fork to check pitch.[87] At Parker's, Thompson procured a C_{512} fork, which came with "assurances on the part of Mr Hullah" as to its precision. Thompson concluded from this that Hullah had "undertaken the patronage of Sir J. Herschel's standard," but Hullah did not recommend C_{512} as a musical standard, preferring C_{528} for quality of sound. Thompson, however, doubted the fork's accuracy: it was all very well to aspire to a mathematical standard, but securing this materially in the form of a tuning fork was difficult. Although Hullah boasted of the care which went into the manufacture of each fork, Thompson wanted Herschel's views on its precision. Hullah claimed each fork was made by "some scientific process. But no light was obtained on its nature. A hazarded question whether the process was connected with a pendulum, did not appear to be understood."[88] Even when Herschel did secure musical support for his standard, it was troublesome to procure reliable instruments and there were varying conceptions over the methods and skills thought best to manufacture trustworthy tuning forks. While the pitch committee had confidence in scientific techniques for determining pitch and securing frequency in tuning forks, Thompson's comments suggest there was a problem over instrumentation. Indeed, from the 1860s, tuning forks appeared to embody perfection and precision, but this production of reliable instruments required much skill and labor. Rudolph Koenig (1832–1901), for example, a Prussian instrument maker in Paris, became famous for the high standard of his tuning forks, but these were the product of highly refined practices and the careful selection of steel.[89]

Though not on the committee, Herschel attended the public meeting which followed it, where he made a last appeal for a mathematical standard. After Phillips proposed the report be carried, Herschel queried if the precise number of vibrations was still under contention and demanded a reconsideration of C_{512}. When Potter asserted C_{528} to be a more practical musical measure, Herschel conceded that he had no right to attend the meeting "as a musician, but he attended in the interest of general science and it was his desire that some general and correct principle, easy of application to this subject, should be recognized, which he thought would place them in a position superior to their French neighbours, under compulsory legislation."[90] The only way to do this, Herschel reiterated, was by establishing a British standard of C_{512}. At the very least he wanted the committee to agree to leave the question of frequency open to future revisions, conceding that an immediate shift to C_{512} might be too sudden and that 528 was an acceptable temporary resolution, if reconsidered within twenty years.

Herschel's proposed amendment received a mixed reception. Organ builder Davison seconded Potter's calls for C_{528}, citing the protection of orchestras as his priority. H. F. Chorley seconded Herschel's proposal, but Hullah feared that a temporary solution would undermine the hard work of the committee; the previous year had been spent in an effort to find a permanent standard, but to choose a temporary pitch would undermine the report's credibility. Hullah alleged that his was not a mathematical but a musical problem, and the meeting's task was "not so much to decide what was the best pitch that could possibly be imagined—but to find out a pitch which would meet the various difficulties of the case." Herschel's amendment would, he speculated, weaken the committee's authority, as "they would appear before the public as an undecided body." Chester supported Hullah's observations and recommended the report be reworded to "admit of the excellence of the theory of 512, at the same time that it accepted practically the 528."[91] Dilke likewise agreed that C_{512} was "the proper mathematical number" and believed that while 528 was practical, the report should state 512's theoretical credentials. Leslie seconded this proposal, but Potter refused to consent to any acknowledgment that 512 was mathematically superior to 528. Reverend W. Cazdet concurred, expressing how he "did not understand why 512 should be taken as more mathematically correct than 528," especially as this was a "practical" musical question.

The meeting subsequently rejected the proposal to recognize 512 and passed Potter's original specification of C528. Hullah summed up the problem of acknowledging 512's scientific credentials as one that was dependent on the committee's limited power. What was agreed on "amounted only to a recommendation and not to the passing of a law. If that meeting were a parliament, and could force the country to adopt this view, the case would be different." As it was only a resolution, what was chosen had to be something Britain's musical community would voluntarily agree to, and Herschel's standard was too low to achieve this. Yet again, Britain's *laissez-faire* traditions informed the committee's reforming conduct, only this time, to limit the influence of mathematical theory and restrain the authority of one of the country's leading gentlemen of science. Recognizing that the adoption of his idealized standard was politically unenforceable, Herschel admitted defeat. Whewell concluded proceedings by proposing that the "Society of Arts be requested to undertake the preparation of a standard tuning fork" of C528. Hullah then advised that the committee inform the directors of the nation's orchestras, military bands, and instrument makers of this decision.[92]

The consensual nature of the committee's choice of C528 was something conveyed in the periodical press. The *Liverpool Mercury* explained how it was "an intermediate tone between the high opera pitch at present in use and that recommended as a basis in mathematical treatise." The newspaper emphasized that to secure "a general concurrence" the committee had "obtained the consent of composers, performers, musical instrument makers, and patrons of music, to a declaration that they will adopt the pitch which has been selected."[93] Similar reports appeared in the *Times* and the *Journal of the Society of Arts*.[94] In August 1860, *Chamber's Journal* reviewed the pitch committee's deliberations, which it described as an inquiry "not exactly scientific, artistic, or industrial, and yet occupying a sort of borderland between those three domains." It was an urgent concern, with the average Parisian C rising from 489 to 538 between 1699 and the 1850s. Despite Hullah's efforts to introduce uniform tuning forks for his students in London from 1842, and George Smart's attempt to set C520 as a standard thirty years earlier, little had been done to prevent this escalation. So, according to *Chamber's Journal*, the committee had been badly wanted and had collected valuable evidence on the matter, but the journal also identified a crucial difference from the French efforts to unify pitch. While "the French like to have things done for them by their government, even to the tuning of fiddles," the British approach had been to establish a commission of so much authority and influence, that people would follow its recommendation without government order. At the same time, while in France it had been doctors and musical men, such as Meyerbeer, Rossini, Berlioz, Auber, and Halévy, who had taken the lead, the British committee appeared much more diverse, including "composers, conductors, vocalists, instrumentalists, professors, amateurs, connoisseurs, and critics . . . while the scientific phase of the subject was to be watched over by Dr Arnott, Professors Goodeve, Lunn, De Morgan, Wheatstone, and Willis, and Dr Whewell."[95] What this was, *Chamber's Journal* surmised, was a demonstration of how to build a consensus in a liberal society or, rather, of how to govern without state intervention. Favoring a measure that could attain a consensus, the press expressed little regret for the meeting's rejection of C_{512}. As committee member Lunn put it, "Herschel, as a practical mathematician, fondly clung to the 512," but this could never be more than a "theoretical" dream and, as Lunn later reflected, "it should have been sufficiently obvious that this is not a mathematical, but a musical, subject." For him, it was the requirements of "voices and instruments" alone that should inform the frequency of a musical standard, and that "although mathematical science may be brought in to count these vibrations, there can be no possible practical reason why 512 for the note C should be assumed to be more correct than 528."[96] Lunn's sentiments were typical of musical attitudes to pitch standardization.

Nevertheless, Herschel's reforming zeal did not end with the question of pitch: he envisaged a new mathematically based system of harmony. If pitch should be reorganized on scientific principles then why, reasoned Herschel, should not the division of musical notes within a scale? Combined with his promotion of C_{512}, this idealized conception of a mathematically grounded harmonic system revealed the extent of Herschel's standardizing ambitions for music. It was Robert Brown's work on harmonics that really piqued Herschel's interest in this subject. In 1858, having read Brown's proposals to develop a system of musical arithmetic for the division of an octave, Herschel wrote to congratulate the author, expressing sympathy with his aims.[97] Brown published his *Introduction to Musical Arithmetic* in 1865, which expanded on his 1860 treatise, *Elements of Musical Science*. Throughout this earlier book, Brown drew heavily on Herschel's 1830 treatise "Sound," including an appendix covering "Harmonics" taken largely from it and the astronomer's own essay "Concert Pitch." Dedicating it to Herschel, Brown claimed to share his scientific approach to the study of nature and hoped to provide an authoritative account of the practice and theory of harmony, which required "no skill in mathematics" beyond simple arithmetic. While his tabled logarithms were more complicated, Brown advised his readers to "take on trust the numbers that are given as the Logarithms" of octaves, fifths, major thirds, and "grave sevenths." His work combined the "known facts" of natural philosophy and mathematics with "the ear of the musician."[98]

Herschel was not content to be cited in Brown's work but developed a system of harmonic division of his own. In 1868, he published an article in the *Quarterly Journal of Science*, "On Musical Scales," responding to recent attempts to explain the relationship between the major and minor scales in music "on the principle of the maximum of simplicity in the ratios of the vibrations of the several notes employed." Herschel was especially impressed with the "employment of the fractions expressing the ratios of the vibrations, or of the lengths of the vibrating strings, and their multiplication or division one by another, to explain the relations of musical intervals." This was a noble aim, thought Herschel, but the challenge was in producing a system of fractions which gave a clear sense of change: it was hard, for example, to get a sense of difference between 7/9 and 10/13. Herschel identified that this was a very different approach to the traditional musical study, as the "great majorities of those who study music do it as a subject *sui generis*, and not as a branch of mechanics. Their thoughts are directed to musical intervals and not to the vibrations which give rise to the sounds." While it was easy to appreciate an interval between C and E, this was not so of relations in terms of vibrations. As with the question of pitch, there were tensions over harmonic systems between musician and mathematician. Herschel, however, recommended a solution: "The use of logarithms, by the intervention of which musical intervals are treated as magnitudes susceptible of addition and subtraction, and, like feet and inches, measurable on a scale ... gets over this difficulty." The challenge was finding a natural unit for such a system: as Herschel explained, every "subject of mensuration has its natural unit: in angular measure, the circumference of the circle; in time, the length of the day; and in music, the octave." However, the representation of a musical unit was numerically awkward. Ideally, Herschel thought it would be mathematically satisfying "to regard the octave as divided into 1,000 equal intervals, each singly denoting a difference in tone so small as to be indistinguishable by the nicest ear, in the same way that a thousandth part of an inch was indistinguishable to the eye."[99]

Accordingly, Herschel designed a harmonic system with an octave of 1,000 parts, a fifth at 585, a third at 322, and a minor third at 263. This uneven division was, he claimed, acceptable in the same way that the year was divided into unequal months. For the nonmathematician, Her-

(C.)	C	D♭	D	E♭	E	F	G♭	G	A♭	A	B♭	B
Do	0	0	0	0	0	0	0	0	0	0	0	0
Re	170	170	152	168	170	170	170	152	169	170	169	170
Mi	322	338	322	338	340	322	339	322	338	340	339	340
Fa	415	415	415	415	415	416	415	415	415	433	416	415
Sol	585	585	567	584	585	585	585	585	585	585	584	585
La	737	754	737	753	755	755	755	737	753	755	754	755
Si	907	923	907	923	925	907	923	907	923	925	906	924
do	1000	1000	1000	1000	1000	1000	1000	1000	1000	1000	1000	1000

FIGURE 3.3 Herschel's intervals for various keys, each divided into a thousand parts, as a proposed mathematically musical scale. Herschel, *On Musical Scales*, 13. Public domain.

schel asserted that to "those who study music simply *as* music, without troubling themselves with ratios, logarithms, or vibrations, it will save some trouble and bewilderment to accept these numbers as they stand." Although a natural scale, he admitted that it was hard to reconcile all eight notes harmonically in such a mathematical system. For example, if the key of C was taken for a pianoforte, Herschel explained that the intervals would be C = 0, D = 170, E = 322, F = 415, G = 585, A = 737, B = 907, and C = 1,000 (fig. 3.3). Try as he might, it was impossible to provide a harmonic system in which a full octave was 1,000 parts, while the fifths and thirds were round numbers. But after experimenting with several scales, Herschel reiterated that to be able to divide the octave by 1,000 would idealistically reform music along mathematical lines, allowing for the addition or subtraction of equations to calculate intervals. If a third was 322 parts, for instance, then two-thirds would be 644 parts of an octave.[100]

Herschel's "On Musical Scales" attracted much interest in the musical community. In 1870, the Euing Lecturer of Music at Anderson's College in Glasgow, Colin Brown, reviewed the work, explaining how Herschel had asserted the chromatic scale could be divided into a thousand increments. Brown declared this a brilliant mathematical discovery but regretted that musicians had abandoned the reform of their art to mathematicians like Herschel "without either affirming or denying the truth of their calculations."[101] Another reader, W. H. Callcott, having previously read Herschel's "Concert Pitch" in Brown's treatise, went to the British Museum in October 1868 to consult Herschel's new paper on scales, confident that any remarks from the astronomer "must interest every musician."[102] However, not all of Herschel's readers approached his work out of mathematical or musical concern: some reflected on its theological implications.

Soon after publishing "On Musical Scales," Herschel received a mystifying letter from a Francis J. Hughes, the wife of a vicar living in Penally vicarage at Tenby. Having examined Herschel's harmonic proposals, she was sure the mathematician shared her evangelical views of nature, describing how her Biblical readings had taken her "into depths connected with the laws that govern the natural sciences." Hughes was convinced that "the deepest scientific truths might be traced" throughout scripture. "The type of light and seeing its echo in Creation, first caused this belief" she explained, which had led her to record

> the laws by which it appeared to me light was gradually developed through colors from Black or Darkness: but I felt, unless I could gain some proof, I should only be considered fanciful. About three years since, it occurred to me to place the twelve semitones of keyed instruments, and see what would be the result of the laws I had written down. To my surprise, without one impediment, I gained the harmony of each note of the major keys with sharps; and also the major keys with flats—each semitone sounding its own harmony and governing all the rest, the 8th being the first of a higher octave.[103]

Hughes believed that she had discovered a divine connection between color and sound which would be of interest to Herschel, especially in light of his recent treatment of musical intervals. She claimed to be "quite ignorant of science" and "not capable of understanding scientific works, and only attempted to gleam a few leading points from your article," but felt her theory worthy of attention since a local doctor who "thoroughly understands harmony as regards the rules of composition" had been "so much interested" that he had visited the British Museum in the hope of finding some authority to corroborate her connection. Hughes's somewhat eccentric letter is not historically frivolous but represents an evangelical, albeit extreme, Anglican reading of Herschel's work which sought to unite harmony and optics in reference to scripture. Herschel kept the letter, marking it "To be kept as a curiosity in its way," but it is unclear if he replied.[104] Perhaps only a curiosity, Hughes's letter is, nonetheless, immensely illuminating. Her comments reveal the social diversity of the audiences concerned by any proposed system of musical reform, be it harmonic or pitch. When dealing with questions of music, Herschel's mathematical works were subject to a broad readership with theological, as well as scientific and artistic, priorities. In moving into a question of musical reorganization, Herschel's mathematics was exposed to unfamiliar audiences.

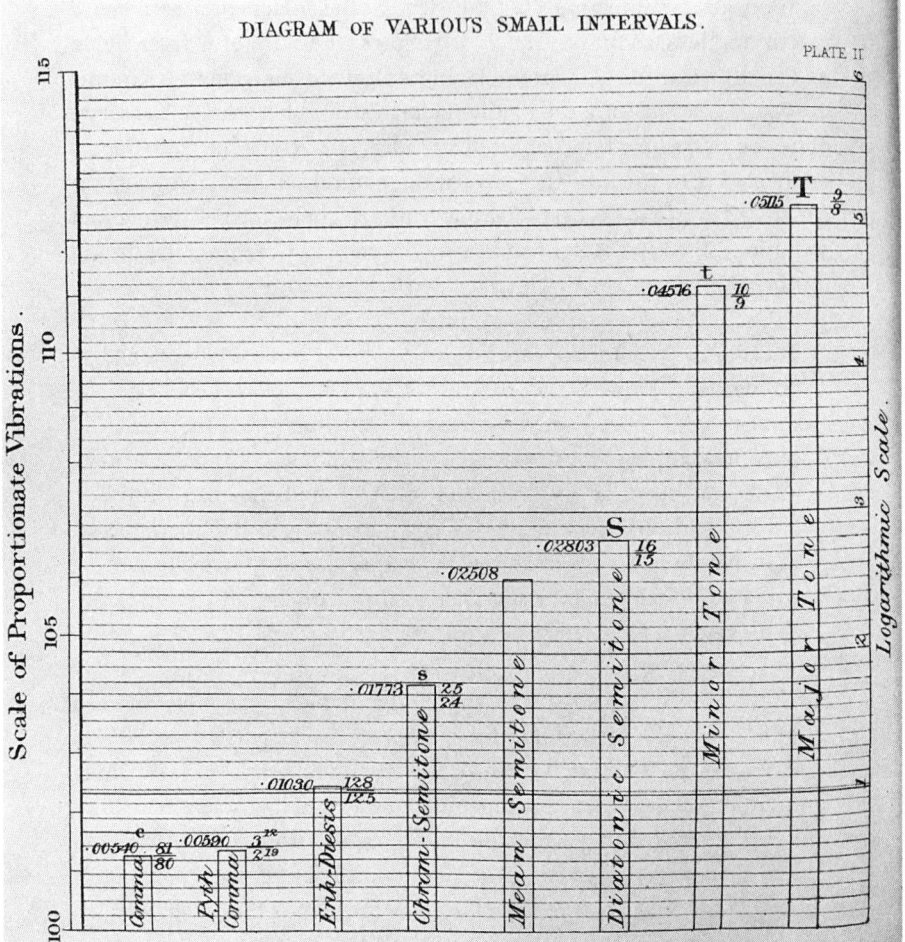

FIGURE 3.4 Pole's diagram to show the differences between small intervals. Pole, "Explanation of the Musical Scale and Its Component Intervals," plate 2. Public domain.

Nevertheless, Herschel was also contributing to a wider scientific discourse over harmony, with his work on intervals appearing in the same year as several key works on sound and, specifically, harmonic systems. The year 1868 also saw the publication of Airy's *On Sound and Atmospheric Vibrations* and Ouseley's *Treatise on Harmony*, which included Pole's "Explanations of the Musical Scale and Its Component Intervals." Pole intended to "present to the *eye* a representation of the relative magnitudes of the various musical intervals, analogous to the impression which they make upon the *ear*" (fig. 3.4). He observed that Hullah's "ladder" repre-

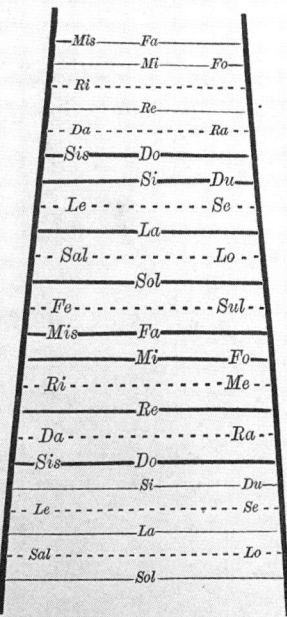

FIGURE 3.5 Hullah's "ladder" diagram of the chromatic scale, for the aid of teaching vocal music. Hullah, *Chromatic Scale*, plate. Public domain.

senting a diatonic scale had already achieved this: Herschel's old adversary on pitch was equally interested in the question of intervals, having produced his own diagram to assist singing teachers (fig. 3.5). Pole, however, claimed his own graphic demonstration of intervals was the first "with such accuracy as to render appreciable to the eye the minute differences on which the more delicate appreciation of the pitch of sounds must depend."[105] Included within Pole's entry in Ouseley's treatise was a 30-inch diagram for an octave, giving 2½ inches per semitone (fig. 3.6a, 3.6b).

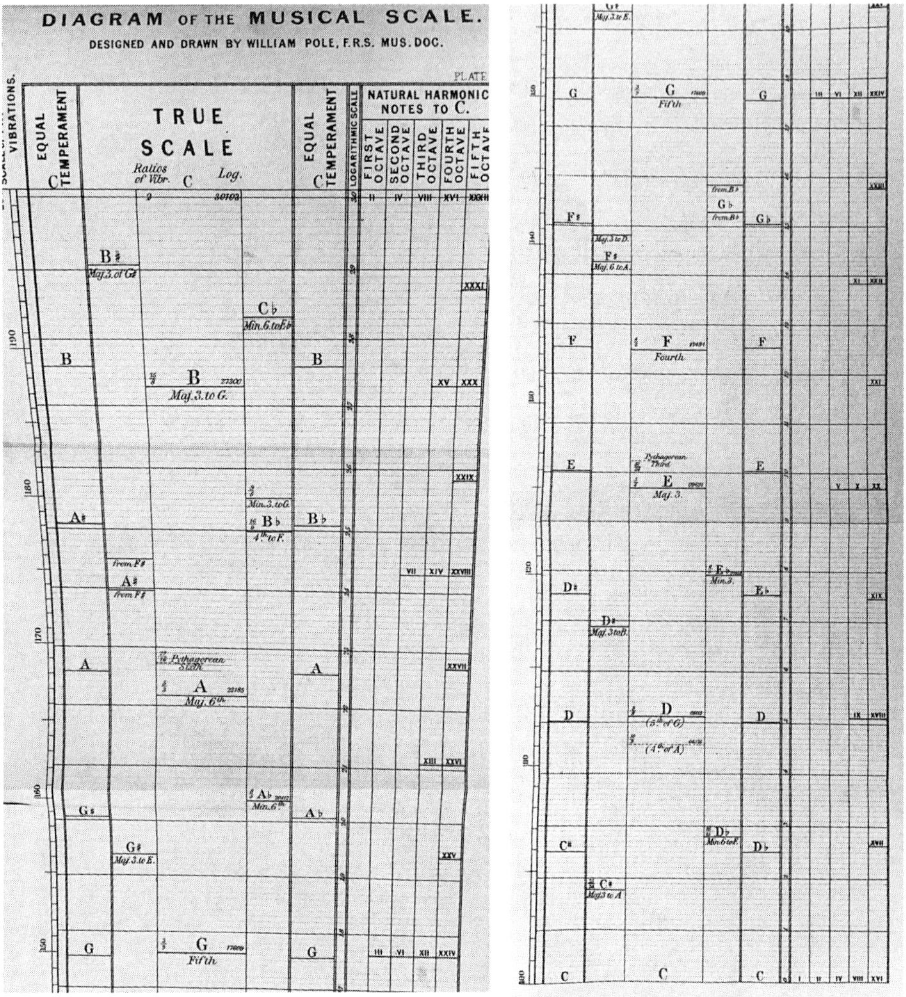

FIGURE 3.6A AND 3.6B Pole's diagram of the musical scale, intended to make visible the distances between the intervals of notes. Pole, "Explanation of the Musical Scale and Its Component Intervals," plate 1. Public domain.

Conceding that it was too difficult to show distance in terms of vibrations per second, Pole instead used logarithms to express musical intervals in terms of ratio of vibrations.[106]

Airy's *On Sound* also dealt extensively with intervals, explaining how these were not contingent on the numerical difference of vibrations between notes but between the proportion of vibrations. A scale was a matter not of any fixed vibrations per second but of fractions covering an octave: a fundamental of C = 1 would give E = 5/4 (major third),

F = 4/3 (fourth), G = 3/2 (fifth), A = 5/3 (major sixth), and 2 = C (octave). Like Herschel, Airy felt the sense of change between these fractions was unclear and outlined alternate harmonic systems employing logarithms to express intervals. He reviewed Pole's proposals, made in Ouseley's *Treatise*, but found them insufficient. Pole's logarithms for intervals ran C = .00000, D = .05115, E = .09691, F = .12494, G = .17609, A = .22185, B = .27300, and C = .30103, which Airy thought had "the convenience of being connected with the ordinary system of logarithms, but they do not offer facility for extension." Instead, he recommended Herschel's system, in which the fundamental was 0 and the octave 1,000, based on "a modulus that the logarithm of 2 is 1000." Such a scale was "convenient" for mathematicians and musicians to understand intervals, but it had limitations. As much as Herschel had wanted "to make all the semitones equal; the logarithm of the proportion for each semitone being [(logarithm of 2)/12], giving a scale of C = 0, D = 167, E = 333, F = 417, G = 583, A = 750, B = 917, and C = 1000, D = 1167, E = 1333," Airy regretted that "this system would fail at every critical point of harmony."[107]

Shortly after reading *On Sound*, in late 1868, Herschel discussed his harmonic system with Airy, who confessed to having abandoned "the logarithms, because (whatever we senior wranglers and senior musicians may think), absolute numbers referring immediately to nature are best understood." Airy explained how, regarding "F♯[,] I was at my wits' end. . . . But I do not give up my number as entirely reprobate. Though I attach little to the connexion of 6/5, 7/5, 8/5, 9/5, yet I attach a good deal to the connexion 7/4, 7/5, 7/6, in which 7 is a common base."[108] As with mathematical ideals of pitch, Herschel's and Airy's hopes for a mathematical temperament provoked hostility from musical authorities. Airy wrote to Herschel, informing him of "the indignant protest of Dr Pole against the system of equal temperament," appearing in his article "Diagrams and Tables" within Ouseley's *Treatise on Harmony*. Pole had claimed that there were

> those who contend that this system (that of twelve equal semitones) ought to be made general; that the ear ought to be forced to endure its imperfections; and that even in cases where true harmonious relations *may* be obtained . . . they ought to be abolished, and the harsh ones substituted. And unfortunately, the superficial condition of musical education in the present day tends very much to bring their results about, not only to the distress of those who, from better training, are able to appreciate the beauty of pure harmony, but to the permanent determination of the art . . . in a theoretical point of view the knowl-

edge of the true harmonic relations on which the whole structure of music is based is fast becoming obsolete.[109]

This was, Airy felt, directed at himself and Herschel, and their attempts to mathematically reform musical harmonic division. However, Airy thought Pole was confused and attributed his hostility to harmonic reform to a misreading of his, and Herschel's, proposed systems of intervals.

Late-Victorian Pitch

Following its resolution of C528 in 1860, the Society of Arts kept a close eye on international pitch standards. In 1870, it sent questionnaires on pitch to Europe's leading orchestras, including Berlin, Brussels, Cologne, Copenhagen, Dresden, Leipzig, Moscow, Stockholm, Vienna, and Naples, but this merely revealed that while some employed the *diapason normal*, few cities had an enforced standard.[110] In Britain, pitch continued to rise after the 1860 meeting, with most London orchestras playing A at 456 by the 1870s and the London Philharmonic reaching A455 in 1874. Instrument makers were having to stock tuning forks of three different pitches, including France's A435, the concert pitch of A455, and A446 for private performances.[111] In this chaotic context, nine years after the pitch committee, Lunn wrote in the *Musical Times* pressing for musical reform. Music was, he proclaimed, "an universal language," and its masters produced "immortal works" which made their emotions and thoughts comprehensible to all. This universal character gave music a somewhat magical quality, but for all this idealism, Lunn regretted that the most important measure of sound, pitch, remained irregular.[112] The only long-term solution, as he saw it, was to adopt the French C522. The 1860 report had, in his opinion, missed a trick, it being

> a remarkable instance of the independence of the English character (which, however commendable in politics, is often most reprehensible in art) that, although the fixing of the *diapason normal* in Paris was actually the cause of this meeting at the Society of Arts, scarcely any notice should have been taken of the pitch there established.[113]

While France had an enforced standard, English music remained in chaos, as demonstrated by a recent letter published in the *Athenaeum* in which the English tenor Sims Reeves (1821–1900) refused to sing at the Sacred Harmonic Society until its pitch was reduced. Lunn hoped

Reeves's letter would initiate a renewed campaign for reform. As things were, he reported that most English conductors looked on eighteenth-century choral music and many of Mozart and Beethoven's compositions with "terror": variations in pitch among instrumentalists within the same orchestras made such performances unpredictable. Ideally, Lunn wanted a conductor's pitch to "reign with despotic sway." Invoking a Biblical comparison, he surmised that "although the confusion of tongues may have placed a barrier between man and man, music—which has ever been, and ever will be, the gentle medium between man and his Creator—should at least be an universal language."[114]

Lunn had a point. At the Wagner Festival of May 1877, held in the Royal Albert Hall, Pole reported that Wagner himself had complained that singers were obliged to sing at A455. Likewise the celebrated Italian-French soprano Adelina Patti (1843–1919) refused to sing at Covent Garden in 1879, complaining that the orchestra's A455 was too high.[115] Even pitch committee members ignored the standard of C528. Most notably, when Willis completed the Royal Albert Hall's new organ in 1871, the largest instrument in the world at the time, although he had a free hand with its design, the instrument's pitch was to be determined by the committee responsible for the hall's construction. But Willis decided to do as he pleased and ignored both the committee and C528, fixing the organ to what he considered a more aesthetically pleasing pitch, this being C540. Much amused by himself, Willis recalled how the

> authorities arranged that Sir Michael Costa, Mr R. K. Bowley, then the general manager of the Crystal Palace, and some of the leading wind instrument players of the day, including Lazarus, should attend at the factory to settle the question of the pitch of the organ. They also brought a violinist but I couldn't see what a fiddler, who is a very useful man in his way, had to do with settling the pitch. However, we duly proceeded,—Costa presiding over the conclave. When they began to blow into their different instruments each man had a different pitch! It was a regular pandemonium. By and by, we settled upon something which was considered satisfactory, and we bade each other good morning.[116]

As much as Willis had worked toward standardization during the pitch meetings of 1859 and 1860, this anecdote demonstrates the extent to which manufacturers valued their autonomy to determine the pitch of their instruments. Efforts to dictate pitch were met with something close to contempt, however amusingly recalled.

The question, then, was how could this chaos be resolved? For William Henry Stone, physician to St. Thomas' Hospital in Westminster, astronomy offered the solution. In 1875, he identified the problem of pitch to be an instrumental one, with the absence of a standard producing an overwhelmingly discordant effect on British audiences. In Europe, on the contrary, every orchestra had a standard tuning fork to which each performer had to conform. Part of the trouble, Stone continued, was the human ear itself, there being "a considerable difference in even cultivated ears to the appreciation of minute shades of pitch . . . many possessing a personal peculiarity similar to what is termed 'personal error' in astronomical observation." Although a medical practitioner, Stone's veneration of astronomy's disciplining practices as something to which musicians should aspire, invoked a high standard of precision. He explained that in institutions such as the Royal Observatory at Greenwich, the individual personal error of each astronomer was calculated so as to enhance the accuracy of transit star observations. Specific practices and apparatus, notably Airy's use of the galvanic barrel chronograph from 1854, were designed for this, and, Stone was confident, "from extended experiments, that there exists a similar phenomenon in the ear as in the eye."[117] Astronomy thus provided a model for reforming music by introducing sonorous standards, reducing human error, and regulating musicians' ears. While it was the existing practice to tune an orchestra taking the oboe as the standard, this was, Stone alleged, far from ideal because temperature caused variations in this instrument. He proposed that the best standard would be a free-moving nonmetallic reed, which was little changed by temperature fluctuations. It was up to conductors to enforce this standard, with Stone arguing that they should "consider it their duty to run through the principal instruments one by one against a trustworthy standard." This "trustworthy standard" should not be a tuning fork, the note of which was "so feeble and evanescent that it is not fitted for the noise and bustle of the concert-room, and moreover it is greatly under the influence of temperature."[118] Even in the 1870s, then, it was hard to secure a consensus over how best to physically embody a disciplining standard pitch.

After two decades of discontent and discussion, the 1880s witnessed renewed standardization efforts, largely under the direction of the mathematician and phonetician Alexander Ellis (1814–90). Ellis had studied mathematics at Trinity College Cambridge from 1833, passing Sixth Wrangler in the Mathematical Tripos Examinations of 1837. Most famous for his efforts to reform the system of English spelling by basing it on the sounds of the language, Ellis's musical interests were markedly scientific. Following the publication of Hermann Helmholtz's (1821–94) *Die Lehre von den Tonempfindungen als physiologische Grundlage für die Theorie der Musik* in 1862, Ellis produced the first English translation of this seminal treatise, *On the Sensations of Tone as a Physiological Basis for the Theory of Music*, in 1875. Beyond this, he worked to catalog the various pitches of several hundred musical instruments throughout Britain and Europe, with the help of the piano tuner Alfred Hipkins (1826–1903). These surveys involved the calculating and comparing of how pitches, especially those of organs, had changed throughout time and over space, with Ellis publishing his results, "On the History of Musical Pitch," in 1880. In this, he emphasized just how inconsistent musical pitch had become. Ellis's fascination with pitch contributed to his generally eccentric appearance: along with rigorous dieting and teetotalism, he was renowned for always carrying a set of tuning forks about him in a large greatcoat he called "Dreadnought," which included twenty-eight pockets full of what he believed to be practical objects.[119]

Five years after the publication of Ellis's work on pitch, in 1885, the British ambassador to Belgium reported to the Foreign Office that the king of the Belgians, Leopold II, had issued a decree to standardize musical pitch. The year before, Giuseppe Verdi (1813–1901) had persuaded the Italian government to pass legislation fixing A at 432. In response to these measures, the principal of the Royal Academy of Music and 1859 pitch committee veteran, George Macfarren, demanded a meeting be held in St. James's Hall on June 20, 1885, to discuss the question. The result was an agreement that the *diapason normal* should be introduced throughout Britain.[120] Nevertheless, this meeting was subject to derision, with the comic journal *Moonshine* mocking that, although "not musical enthusiasts ourselves, . . . we can see plainly enough that musicians ought to select one sort of pitch and stick to it."[121] Despite such nonchalance from the periodical press, the Society of Arts launched a new committee to investigate the question of a national standard in 1886, with Macfarren as chairman and Ellis as secretary. This committee adopted the same approach as that of 1859–60, conducting a national survey to obtain a consensus over what course to take. This time, however, printed questionnaires were sent out on a vast scale, asking respondents if they wanted a standard pitch and, if so, would they accept the French *diapason normal*? In May 1886, Ellis, who provided much of the committee's direction, reported on the extent of this consultation. After nine-and-a-half long hours of sorting, Ellis counted 710 responses, having despatched between 1,800 and 1,900 circulars (fig. 3.7). Of these 710, he recorded that all favored a uniform pitch and that 630 would accept the French standard, with eighty preferring some alternative. He alleged that "Though the number who replied

[*This Sheet to be torn off and sent in the enclosed envelope to the Secretary of the Society of Arts.*]

QUESTIONS.

Questions 1 and 2 should be answered by Yes or No only, without any addition, and question 3 by the number of double vibrations in a second for either A or C (*not* for both notes, to prevent error), also without any additions. The Council of the Society of Arts will, however, be greatly pleased to receive further remarks, on separate pieces of paper, written on one side only, and duly signed.

⁂ It is particularly requested that the answers may be sent to the Secretary of the Society of Arts, Adelphi, London, W.C., not later than 30th April, 1886.

QUESTION.	ANSWER.
1. Do you consider it advisable that a Uniform Musical Pitch should prevail in England?	*Yes.*
2. If you answer Question 1 in the affirmative, are you willing to accept the French Pitch A 435, giving equally tempered C 517⅓, in England, in order that the Uniform English Pitch may be the same as the officially Uniform International Continental Pitch?	*Can only say, at present, that a lower pitch than the one in use in England is desirable*
3. If you answer Question 2 in the negative, and Question 1 in the affirmative, what is the Uniform Pitch that you would prefer for use in England?	

Name *William Rea*

Profession *Music*

Address *7 Summerhill Grove N'Castle on Tyne Organist to the Corporation &c: Conductor of Rea's Choir, Amateur Vocal Society &c:*

Date *May 6. 1886.*

⁂ The signer is requested to state if he or she is a Professor of Music or Physics, adding at what University or College, a Mus.D. or Mus.B., the Principal of any College, Academy, School, or other Institution for Music, a Teacher of Singing, a Professional Singer (adding the kind of voice), a Band Conductor, a Choir Conductor, an Instrumentalist (adding the name of the instrument), an Amateur, &c.

It is requested that names may be given, as it is desired they should appear in the list (should one be published). Ladies are, therefore, requested to prefix Mme., Mlle., Mrs., or Miss, to their names to prevent mistakes, and Gentlemen to give any title or any initials they may wish added to their names.

FIGURE 3.7 The 1886 Society of Arts pitch inquiry's circular, with the English pianist and organist William Rea's response. Royal Society of Arts, PR/GE/121/10/6, Miscellaneous Letters Relating to Pitch, William Rea to Alex J. Ellis (May 6, 1886).

were by no means the whole body of English musicians . . . they may be considered as fairly representative." Ellis broke these figures down for the committee. Of the eighty who would reject the French pitch, twelve preferred C512, thirty-one suggested C528, six wanted between C529 and C535, and seventeen were for the higher pitch of C540. In terms of the social makeup of these respondents, Ellis counted that of the 630 who were pro-French, 125 were private music teachers, twenty-eight were school teachers or institutional professors, seventy-four were organists, 108 conducted choirs or orchestras, eight were military bandmasters, twenty-five were at the Royal Academy of Music, nine were composers, twenty-four were vocalists, fifteen were instrumentalists, and thirty-nine were instrument makers or dealers. There were also thirty-nine doctors of music, of which Ellis counted twenty-five Oxbridge and five from Dublin, while there were forty-eight respondents whose primary claim to authority was holding a bachelor's degree in music, with thirty from Oxford and fifteen from Cambridge.[122]

One respondent in particular, however, caught Ellis's attention. The Belgian wind-instrument maker Charles Mahillon (1813–87) not only approved of the committee's object but claimed that "England cannot isolate herself by holding aloof from the general movement which is an accomplished fact not only in France, Spain, and Belgium . . . but likewise in Russia, where the A of 435 vibrations has been officially adopted for several years." From his perspective, this question of pitch engendered broader questions over Britain's relationship with Continental Europe. Importantly though, Mahillon alleged the problem of cost no longer remained, as "experience has recently proved in Belgium the possibility of transforming all wind-instruments—the wood included—to the normal pitch."[123] This claim intrigued Ellis because the Brussels-based C. Mahillon & Company was a highly reputable military musical instrument manufacturer. Ellis wrote back for further details. Charles's son, Fernand Mahillon, who managed the company's London branch on Leicester Square, claimed to "have altered the Instruments (*brass and wood*) of the entire Belgian army." He advised Ellis to contact the Belgian musicologist and composer François-Auguste Gevaert (1828–1908) at the Conservatory of Music in Brussels and Mr. Staps, the inspector general of the Belgian Military Bands, for testimonies as to how effectively Mahillon & Company had unified the pitch of their instruments. Mahillon estimated the cost of transforming a cavalry bugle to be five shillings, while brass instruments would be between twelve and fifteen shillings to change. Clarinets and oboes would be from fifteen to twenty shillings. If the committee doubted Mahillon's skill at making these adjustments, he asked them to "trust me with an instrument or two; I shall alter them to the low (French) pitch of 435 double vibrations, and they will then be able to judge by themselves." Although warning Ellis to expect opposition from "all the professional gentlemen," who would object to having to update their instruments, Mahillon was convinced of the imperative of introducing the French pitch to align Britain with the Continent.[124] Emboldened by Mahillon's claims, Ellis reported to the committee that the objection of the high cost of transforming instruments was no longer valid.[125]

Predictably the French pitch attracted much opposition. In early 1886, the English composer John Stainer (1840–1901) had been invited to sit on the Society of Arts' new pitch committee, but he was skeptical of the appointment, believing that all efforts to standardize pitch would fail if the *diapason normal* was the chosen frequency. He warned that previous efforts at pitch reform had been unsuccessful because of "the fact that the 'French Pitch' was proposed. I feel sure that the English nation will never adopt the French pitch."[126] Were he to be on the committee, he was sure to "be in a glorious minority. . . . I shall not in any way attempt to throw obstacles in the way of the adoption of the French pitch; but I am absolutely certain the attempt to establish it will fail; we *cannot* ask the government to throw away as useless £100,000 worth of military instruments." However, he was confident an English National Pitch of C528 could be acceptable, observing that all of Britain's colonies would adopt it, as would the Continent, eventually.[127] This imperial context was a common reference point for those opposed to the French pitch. In May 1886, in response to the pitch committee's growing preference for the French standard, the Westminster instrument-maker Augustus Charles Köhler (1841–90) printed and circulated an appeal against the *diapason normal*. He implored readers to support the adoption of the British Army's pitch which was, he declared, the only one with "any authority" in the country and used throughout the empire. Known as "Kneller Hall Pitch" as the Royal Military School of Music occupied this venue, this standard could easily become the national uniform pitch, given its employment in British and colonial regiments across the British Isles, India, Africa, Canada, and Australasia. While a shift to French pitch would be expensive and difficult, Köhler boasted that Kneller Hall Pitch was already "well established throughout the whole universe where English is spoken and British interests predominate."[128] In the context of the rampant British imperialism of the 1880s, musical pitch took on renewed national and political significance, more than it had in 1860. At stake here were not just questions over standardization and liberalism, but Britain's place in the world: was it primarily an imperial power, content to exert influence throughout its global empire, or was it equally committed to its position in Europe, as a collaborative partner to its Continental neighbours? The debates surrounding pitch reform in 1886 highlighted concerns over Britain's global role that would extend well into the twentieth century.

In the end, the 1886 committee's efforts came to nothing. The Duke of Cambridge's (1819–1904) declaration that it would be impossible to alter the pitch of the military bands effectively terminated any hopes that musical unification could be achieved. Cousin to the Queen and commander-in-chief of the British Army between 1856 and 1895, the duke was well known for his opposition to almost all proposed reforms of the armed forces. The army's ability to enforce a standard had been the committee's best hope of achieving national regulation, so Cambridge's opposition was a severe setback. All in all, Stainer's prediction had been accurate, while Mahillon's promises convinced few beyond Ellis. By the 1890s, pitch standardization seemed even further out of reach due to a growing awareness of the problem of temperature. In January 1896, the acoustician and wind-instrument maker David James Blaikley (1846–1936) read a paper at the Royal Academy of Music where he described the variations in sound possible from temperature changes. He cited recent experiments in which a bugle and a clarinet were placed in hot water and two identical instruments were placed in iced water. On removing them, it was found that each instrument differed in pitch by nearly a semitone. While the French *diapason normal* was a fine example to follow, it took no account of temperature, leading Blaikley to conclude that "a more scientific definition of pitch" was required.[129] It would seem that the midcentury hopes of standardizing pitch had been wildly optimistic.

Similar pitch unifying schemes proved equally troublesome beyond Britain. In 1885, the Austrian Ministry of Culture called for a conference to establish an international diapason, with delegations from Italy, Austria, Hungary, Prussia, Russia, Saxony, Sweden, and Württemberg all attending. This meeting recommended that A be set at 435, and this standard was later included within article 282 of the Treaty of Versailles in 1919. The work of a later conference in London in May 1939, at which France, Germany, Britain, Holland, Italy, Switzerland, and the United States largely agreed to revise A up to 440, was interrupted by the Second World War, so that by the early 1950s pitch remained an elusive problem.[130] When the Royal Society of Arts established an inquiry to standardize pitch in 1949, the papers of the 1859 and 1886 committees were read to commence the proceedings. Little progress toward a single standard had been made and the question of authority remained as difficult to resolve as it had been ninety years previous. Indeed, the chairman of the 1949 committee, Llewelyn S. Lloyd (1876–1956), was appointed for his "balance between the mathematician, the experimental scientist, and the musician." As the *Journal of the Royal Society of Arts* put it, "whether we are scientists, technicians or musicians, we can all of us trust what he says." Evidently, concerns over trust and credibility were still pertinent to who was an authority to determine the standard. Yet by 1939 it had been generally agreed that any standard pitch should be chosen by "the musician" and maintained by "the technician." The hope for a mathematically accurate, absolute frequency, so much cherished in the nineteenth century, was abandoned. Triumphantly, the Royal Society of Arts announced that "technical science" had been "kept in its place" and that it was acknowledging "the true claims of music" to determine its own standard.[131]

Conclusion

Both in 1860 and 1886, Britain's musical communities rejected a mathematical standard pitch. While in 1860 it was the country's musicians, instrument makers, and composers who refused to adopt the measure, in 1886 it was the British army and its instrument suppliers who objected to a unifying standard of any frequency. Throughout these standardizing efforts, mathematicians continually demonstrated the most idealistic interpretation of what a standard musical pitch should be. Contrary to musicians, composers, and instrument makers desperately seeking to keep the frequency of a standard pitch higher and more euphonically pleasing, Herschel and his supporters promoted a system of music built on mathematical principles. C_{512} and a harmonic scale divided into 1,000 parts were at the very center of this vision. Herschel's failure to secure a consensus for C_{512} illustrates the very limited authority of science within the arts and specifically music. Despite the fact that Whewell, Wheatstone, Robert Willis, De Morgan, and Pole were all committee members in 1859 and 1860, the theoretical arguments in favor of a mathematical standard pitch did not convince Britain's music elites of its value. What was wanted, rather, was something perceived to be aesthetically superior. In placing his schemes for the reorganization of musical practice before a broader public, with conflicting religious, artistic, and economic interests, Herschel exposed his philosophical credentials to audiences quite different from his usual scientific readership.

The Society of Arts' attempt at pitch standardization in 1860 revealed, more than ever before, what a socially diverse subject music was. This was a moment where tensions, perhaps previously unfelt, were starkly realized between mathematicians, soldiers, clergymen, musicians, composers, instrument makers, critics, and performers over who had the authority to define and discipline musical practices. Music's socially diverse character meant that, as a body of knowledge, science lacked inherent authority in the face of a popular consensus within Britain's musical, religious, and military networks. And without France's state apparatus and willingness to enforce legislation on its population, it became clear that only a compromise pitch would be acceptable. In many respects, the question of pitch revealed the difficulties of using scientific knowledge to regulate society within the framework of a liberal state. While notions of the extent to which Britain actually was *laissez-faire* tend to be overstated or romanticized, they nevertheless made it hard for the pitch committee to assert any authority. If we see this matter of standard pitch in the context of broader efforts to unify measurement standards, such as electrical resistance, then it becomes clear just how difficult it was to secure these regimes' support. The care taken in appointing such an extensive committee demonstrated an acknowledgment that, without an autocratic government, it was credibility that was so important to the introduction of a musical standard. Enforcing a measurement regime on the nation's musical community within a liberal state or, at the very least, in a state without a dominating figure such as Napoléon III, required a committee that was credible to its intended audiences. It had to combine scientific and musical knowledge with experience, musical taste, and practical skill. The question of pitch revealed the importance of credibility for governing the arts in Victorian Britain.

CHAPTER 4

Accuracy and Audibility
Mathematics, Musical Consensus, and the Unreliability of Sound, 1835–81

> *Music when healthy, is the teacher of perfect order, and when depraved, the teacher of perfect disorder.*
>
> ATTRIBUTED TO JOHN RUSKIN

Accompanying the rapid expansion of nineteenth-century Britain's railway network was the gradual spread of national standard time, taken at and projected from the Royal Observatory at Greenwich. The advance of this unifying measure at the expense of locally recorded time appeared at once to be ordering the social and business life of the nation and, after the 1884 Washington Conference, the world. Greenwich time carried with it industrial connotations and has historically been construed as inseparable from a wider cultivation of capitalistic discipline: here was a system of regulation for Britain's booming industrial economy and increasingly coal-powered transport infrastructure.[1] Electric telegraph cables conveyed time taken at Greenwich across Britain, which was then displayed from time-balls, clocks, bells, and cannons. However, these methods of dissemination engendered concerns over visual and aural communication that were scientific, political, and social. Attitudes toward the accuracy of sound and vision shaped one of the most industrial features of the nineteenth-century soundscape: that of keeping time.

This chapter examines these tensions between the eye and the ear as organs of communication through the work of the Astronomer Royal, George Biddell Airy (1801–92), the domineering figure in the Royal Observatory's construction of a national time system between 1835 and 1881. Crucially, Airy had a very low estimation of the human ear's accuracy, preferring to rely instead on visual practices both for recording and sharing Greenwich time. As will be seen, a series of encounters with musical authorities shaped this distrust of the ear and convinced him of the difficulties of securing aural reliability. Although he was keen to apply mathematics to the investigation of sonorous phenomena and encouraged mathematical students to do this, his involvement with the construction and measurement of the new bells for the Houses of Parliament at Westminster revealed that mathematical theory was quite detached from musical practice. While he could calculate an instrument's frequency of vibration, tuners, instrument-makers, and professional musicians struggled to agree over the notes sounded. Building a musical consensus was such a problem that Airy came to distrust the reliability of the ear altogether. To enhance the accuracy of the observatory's astronomical work, Airy introduced new instruments and practices to reduce human error. When he attempted applying similar solutions to musical problems, however, encouraging the use of tuning forks rather than ears alone, such instrumentation did little to resolve disagreements over pitch. Airy was, subsequently, skeptical of sound's practical value for disseminating time.

By exploring Airy's experiences of how difficult sound was to measure and regulate, it becomes clear why Britain's time system took on the visual character that it did. Jonathan Sterne has been right to emphasis the role of sound in the history of science, but Airy's work suggests the prioritizing of vision is no mere historical bias but something contemporaries actively cultivated.[2] There was nothing inherently inferior about the use of either sound or vision in timekeeping practices; alternative authorities worked to promote varying methods of time dissemination, which relied on both hearing and seeing. However, by placing questions of time regulation within their broader musical contexts, this chapter argues that visual commitments in the production and management of scientific knowledge are not historiographical constructs but were choices historical actors promoted.

Sound Theory

Between 1835 and 1881, the Royal Observatory at Greenwich expanded under the direction of the Cambridge-trained mathematician, George Biddell Airy. Born in Alnwick, Northumberland, Airy attended Colchester Grammar School before going to Trinity College Cambridge in 1819. There he excelled in mathematics under the guidance of George Peacock (1791–1858), becoming Senior Wrangler in the Mathematical Tripos Examinations of 1823. During his time as an undergraduate he studied the problem of sound in 1821 and, in 1823, noted that there "is a good deal of investigation of a mathematical nature not connected with College studies" including problems concerning musical chords and organ pipes.³ When introduced to Richarda Smith, his future wife, Airy recalled how he had fallen in love "at first sight" and found her voice particularly enchanting; her "beauty and accomplishments, her skill and fidelity in sketching, and above all her exquisite singing of ballads, made a great sensation in Cambridge." In the words of his son, Airy was "very fond of music and knew a great number of songs; and he was well acquainted with the theory of music: but he was no performer."[4]

Intermittently throughout his career, Airy showed interest in sonorous problems. In 1848, for example, he was embroiled in a controversy with James Challis (1803–82), Cambridge's Plumian Professor of Astronomy and Experimental Philosophy, over his theoretical determination of the velocity of sound in the *Philosophical Magazine*. Airy provided mathematical evidence that Challis's conclusion "that soniferous vibrations may be communicated along a limited cylinder or 'filament' of air without affecting the air which surrounds it; and that the theoretical velocity of sound is not that in which all mathematicians . . . have agreed," was incorrect.[5] Challis had mistakenly introduced a constant into his calculations which, if of a value different from zero, Airy found invalidated the mathematician's conclusion.[6] Airy personally informed Challis of his response, to which Challis confessed that his results had been startling as they were "so counter to what appeared to be established on the subject" and that he was prepared for a critical response.[7] Airy continued to work on the problem into 1849, publishing further theoretical analysis on the progress of sonorous waves in the *Philosophical Magazine*.[8]

Beyond these mathematical interests, Airy discussed more practical musical problems with Augustus De Morgan (1806–71). De Morgan, a mathematician and historian, had been at Trinity College Cambridge from 1823, where he had studied mathematics under Airy, Peacock, and Whewell, graduating Fourth Wrangler in 1827. A fine flautist, he had also been in the Cambridge Amateur Musical Union Society. When the newly established London University elected its first professor of mathematics, De Morgan secured the appointment with testimonials from Peacock and Airy.[9] In 1857 De Morgan presented Airy with a difficulty which could not be resolved by an equation. While tuning an organ, De Morgan had to collaborate with professional organ tuners, which proved challenging. He explained how the "question is about how the tuners can best count the beats—which I am instigating them to do, and which they have dropped doing, if ever they did it." The organ in question had twelve intervals, with the tuners claiming they could "distribute .234 of a mean semitone among 12 different intervals, equally, by the ear alone." De Morgan thought this claim to accuracy untrustworthy; he wanted the tuners to count beats by sounding two notes simultaneously so that when the intervals were perfect, the beats would not occur. However, the tuners preferred to rely on their ears alone, hearing each note in turn. De Morgan consulted an organ builder, who unhelpfully explained that tuners of different instruments rarely agreed over an instrument's pitch.[10] The problem De Morgan found with trying to tune the organ using beats was in timing and coordinating the eye and the ear. He told Airy that using

> a watch is a difficult thing—because the tuner cannot start at a given moment. He must, after the concord begins to sound, get his ear into gear; and he is not well up to his counting for a few seconds. Also, the watching with the eye as the half minute comes towards its end is apt to lead him into beat with the watch and not with the pipes.[11]

To solve this, he had the tuners employ sandglass timers, which could "be held in the hand until the tuner's ear goes with the beats."[12] Airy was intrigued, responding that he had "never made experiments with a sandglass and should be very glad if you or any trustworthy person would do so," but advised De Morgan to persevere with pocket watches to count beats to guarantee accuracy. De Morgan followed Airy's suggestion of persuading the tuners to work with accurate tuning forks, rather than by ear alone, but they also continued to use sandglasses.[13]

Likewise, De Morgan complained to John Herschel, reporting that "the organ tuners use no *beats*—and that their methods are as rough as possible. They attempt equal temperament by the ear alone—and do it perfectly well, considering." However, the difficulty was that the tuners did not always agree with each other and refused to use De Morgan's system of counting beats, "which is of course the true one," to resolve these differences. It seemed that "the different parts of the organ tell different stories."[14] De Morgan had already discussed the laws of beats with Herschel in 1856, concluding that it was a highly incomplete subject.[15] In 1864, De Morgan published his experiences of organ tuning and his concerns over the growing adoption of equal temperament in the prestigious *Transactions of the Cambridge Philosophical Society*, raising this musical problem with scientific audiences. Originally delivering this paper to the eminent society in November 1857, De Morgan made evident the difficulties of reconciling musical and mathematical knowledge.

In "On the Beats of Imperfect Consonances," De Morgan lamented that while there was some mathematical theory available on harmonic phenomena, notably in Dr. Robert Smith's 1749 *Harmonics*, this was unknown beyond "the hands of some of the organ-tuners."[16] After defining a "beat" as "any acoustical cycle derived from composition of ordinary vibrations," De Morgan provided mathematical formulae for both unison and imperfect beats. This "theory of beats" had practical ramifications for organ tuners, De Morgan claimed. The problem, as he saw it, was that organ-tuners preferred to use their ears rather than beats to tune organs:

> In equal temperament, for example, the tuner gets one octave into tune, with its adjacent parts so far as successions of fifths up and octaves down require him to go out of it; and the notes thus tuned are called *bearings*: all the rest is then tuned by octaves from the bearings. The method of tuning bearings, after taking a standard note from the tuning-fork, consists merely in tuning the successive fifths a little flat, by the estimation of the ear.[17]

If the fundamental note was flat due to a tuner's defective ear, then all the notes would be flat, despite being tuned correctly in terms of ratio of vibrations. Such tuning practices were "without numerical trial": there was no mathematical theory, just a reliance on the human ear. This involved a tremendous "loss of accuracy," argued De Morgan, as different tuners, "however excellent their ears, do not agree in their results." Depending on the ear meant there could be no musical consensus: two tuners would almost always tune the same organ differently. They would "produce two systems which do not agree: they take care that their tuning-forks shall give them the same standard-note; but this is all they can get."[18]

The difficulty was in converting the sound of a tuning fork to an organ through the measurement of the ear. An "old professional tuner" and associate of De Morgan assured the mathematician "that he did not believe either that any tuner gained *equal* temperament, or that any one tuner agreed with himself or with any other. He summed up by saying that 'equal temperament was equal nonsense.'"[19] Further potential for inaccuracies drew an astronomical comparison:

> In tuning the organ, I feel certain that the ear of the tuner must be much injured, for the moment, by the hideous squalling slides which the pipe sounds while the tuning-instrument is inserted and turned about at the top. He might still be a judge of a perfect unison; but I should no more imagine him able to know the fiftieth part of a mean semitone from the twenty-fifth, when his ear is just out of this abominable clamour, than I should rely on the tenth part of a second from

the wire of an astronomer who had the instant before been tossed in a blanket.[20]

Believing the state of Britain's organ tuning to be chaotic, De Morgan's solution was a system based on a standard of beats. Tuning would be done through the counting of these in relation to time, between different notes, rather than by the ear in reference to a tuning fork. As he put it,

> When beats are employed at the instrument itself, a watch is in several respects a difficult standard. The counting should begin when the ear is well *in gear* with the beats, which will not happen just at the five seconds or the quarter minute. And the employment of the eye at the very commencement of counting is confusing to the ear. A regulated metronome might be used, but I suspect it would be a troublesome instrument. A half-minute sand-glass ... would probably be found the best time-piece: this could be turned over when the ear is in repose on the beats; and the counting would begin from the tuner's own perception of his own act, with that composure which would arise from the act being in his own power.[21]

This was an alternate tuning system to that of equal temperament, based on De Morgan's own mathematical formula of beats. Here was a mathematician's solution to musical practices aimed at regulating tuners and securing standardization. This question of musical consensus and accuracy was one that would plague Airy's own inquiries into sonorous phenomena.

Despite Airy's sporadic attention to acoustic phenomena, sound was the theme of one of his most substantial works.[22] He published *On Sound and Atmospheric Vibrations, with the Mathematical Elements of Music* in 1868, an advanced mathematical textbook aimed at university students, especially those at Cambridge aspiring to achieve wrangler status in the Mathematical Tripos Examinations. Airy dedicated this work to his former tutor, Peacock, recalling how "more than thirty years ago by him I was urged to write a tract on sound for use in the University." Evidently the Astronomer Royal had wanted to write this extensive treatise for some time. *On Sound* mathematically investigated the two principal laws of communicating sound: that it diminished with distance and that its propagation occupied a time proportional to the distance traversed. Students could find formulae in this textbook for calculating the effects of different air qualities, such as its elasticity, on the velocity of sound. He explained differences between transverse light waves consisting of states

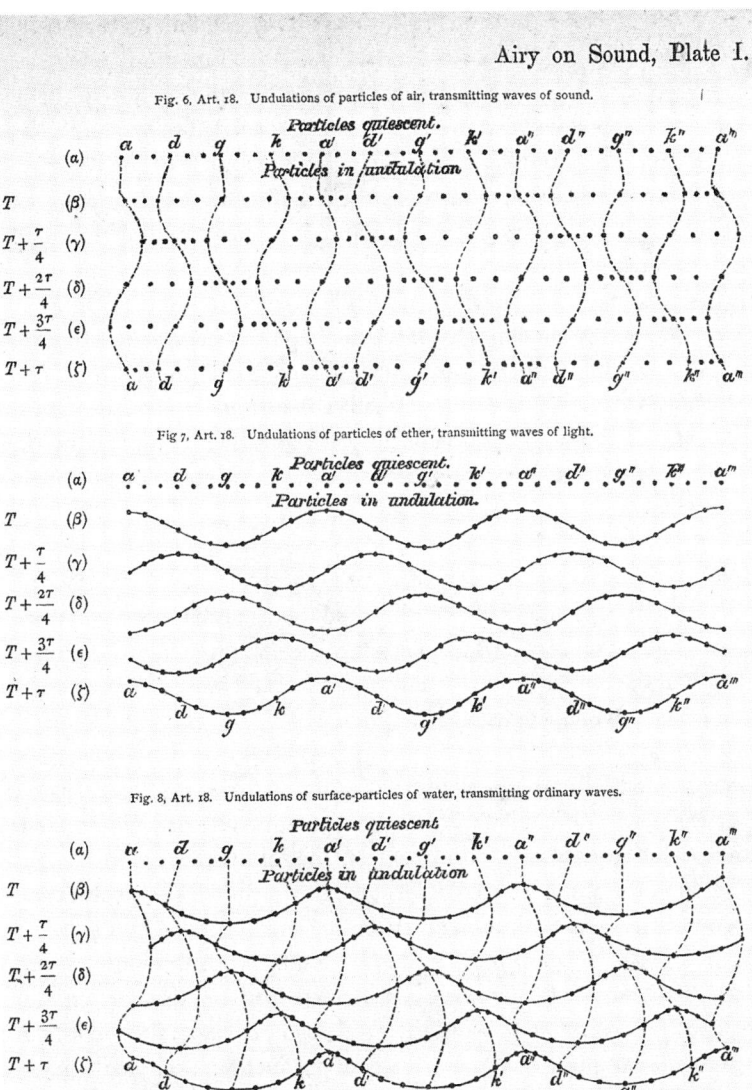

FIGURE 4.1 Airy's comparison of sound, light, and water waves. Airy, *On Sound*, plate 1. Public domain.

of elevation and depression, and sound waves which were longitudinal and consisted of states of condensation and rarefaction (fig. 4.1). Airy claimed Isaac Newton had first developed an explanation for the transmission of sound as a wave through air in the third book of his *Principia* (1687), but he also asserted that it was important to show the mechanical pressures which a wave's condensations and rarefactions produced on every separate particle. For this, students needed a calculus newer than that of Newton: the Theory of Partial Differential Equations, which took the investigation of sonorous waves beyond Newton's synthetical equations to an analytical interrogation.[23] Airy proceeded to lay out the analytical methods required for an algebraical treatment of the velocity of sound in air of varying states. He then demonstrated the application of mathematical theory to the vibration of air produced by the oscillations of musical strings.

One problem that Airy was particularly keen to impress on his student readers was the difficulty of reconciling the mathematical theory of sound's velocity with experimental evidence. Experimental techniques were at risk of error, such as the calculation of velocity in relation to the flash of a cannon. By observing the flash and then counting the time until the sound was heard, it was possible to calculate velocity, but Airy warned that

> there is a physiological circumstance, the effects of which have hitherto escaped notice, but which probably produces a sensible error; it is, that two different senses (sight and hearing) are employed in the observation of the two phenomena, and we are not certain that impressions are received on them with equal speed.[24]

He reckoned "that the perception of sound is slower . . . perhaps $0^s.2$, than the perception of light."[25] When calculating velocity, this difference could cause an error of hundreds of feet. The solution he proposed was to have two or more observers and to take the mean of their observations. Such practices were common in astronomy and, indeed, Airy alleged that the best observers of sound were those "familiar with astronomical practices." Later in his textbook, Airy provided mathematical explanations for harmonics, resonance, and the production of musical concords. He remarked that music was a very material product, built on mathematical laws, and noted how the "expression of 'sound continuing to ring in our ears' may not be so purely poetical as is usually thought."[26]

In many respects *On Sound* represented the continuation of a much bigger debate that Airy had been involved in, concerning the social and national value of mathematics itself. In 1866, Airy proposed to the vice chancellor of the University of Cambridge that the Mathematical Tripos Examinations be reformed with greater attention paid to physical mathematics, including mechanics, tides, and sound, and less focus on what he termed "Useless Algebra." As his son Wilfrid Airy (1836–1925) put it, Airy's view of mathematics was that it was "a useful machine for the solution of practical problems."²⁷ These proposed tripos reforms would make mathematics of greater social worth, according to Airy, who informed the vice chancellor that a "careful selection of physical subjects would enable the University to communicate to its students a vast amount of information . . . [that] would be felt by every student to possess a real value . . . and that it would enable him to correct a great amount of flimsy education in the country, and, so far, to raise the national character."²⁸ This call for the application of mathematics to physical subjects incurred much resistance, namely from Cambridge's Sadleirian Professor of Pure Mathematics, Arthur Cayley (1821–95), who wanted students to learn theory before applying it to natural phenomena. Throughout 1867 Airy and Cayley argued over the place of mathematics in society. While Cayley felt a university was not just about practical education but the cultivation of all science, valuing the study of "pure" mathematics "for its own sake," Airy was adamant that the university had a public obligation to produce mathematically trained national servants. The key question for him was, "In the sense in which mathematical education is desired by the best authorities in the nation, is the course taken by this national institution satisfactory to the nation?"²⁹ The Astronomer Royal thought not. In the context of this controversy over tripos reform in 1866 and 1867, it is revealing that *On Sound* appeared in 1868, demonstrating the application of mathematics to a physical subject and obviously aimed at Cambridge students. Airy had clearly been working on this around the same time as he was conducting discussions with the vice chancellor and Cayley, so it is not unreasonable to see it as part of a campaign to make mathematics more useful. At the same time, this text represented Airy's most thorough published account focused entirely on a single natural phenomenon: he believed the study of sound to be a valuable part of a useful mathematical education. But Airy knew all too well that writing for undergraduate mathematicians was quite a different challenge from working alongside practical musicians.

Building a Musical Consensus

Following the destruction of the largely medieval Palace of Westminster, home to Britain's Houses of Parliament, in the dramatic fire of 1834, architect Charles Barry was selected to build a new legislature for the nation. Absent from his original plans was the now famous clock tower, home to the Westminster Clock and great bell, Big Ben. This enormous construction was to be the premier timepiece of the age, providing accurate Greenwich time to the nation. The government immediately sought the advice of Airy, who, in collaboration with horologist Edmund Beckett

FIGURE 4.2 The recasting of the Great Bell for the Palace of Westminster's clock tower attracted quite a crowd in Whitechapel. Anonymous, "Recasting of the Clock Bell for the New Houses of Parliament," *Illustrated London News* 32, issue 913, April 17, 1858, 401. Public domain.

Denison (1816–1905), took responsibility for the project.[30] Airy's charge was to ensure the clock's accuracy, but between 1859 and 1861 it was the musical tune of the clock's bells which presented him with the greatest trouble. Indeed, during the summer of 1860, Airy experienced what a troublesome thing sound could be to measure as well as the difficulties of applying mathematics to a physical acoustic problem.

Airy's first task came in late 1859 when Charles May, Parliament's consulting engineer concerning the clock tower's construction, wrote to inform him that the great bell had cracked due to the brittleness of its alloy.[31] Warner and Sons of Cripplegate had originally cast the bells in 1856, but the great bell cracked before installation and Mears of Whitechapel recast it in 1858 (fig. 4.2). Denison had determined on an alloy ratio of 22 parts copper to 7 parts tin to obtain a sweeter sound than the traditional 22:5.5 ratio, but this high tin content made the bell fragile, with new cracks appearing in 1859.[32] Airy was confident that he could resolve the error; calculating the "impact momentum" of the falling hammer by finding the square root of multiplying the hammer's weight in pounds by its fall in inches, he claimed that the problem was the degree of resistance the bell provided against this impact. Big Ben's hammer did not seem to exert excessive force, falling nine inches and weighing 766 pounds, giving an impact momentum of 83.0301150186 (the square root of 766 multiplied by 9). Airy felt it was, therefore, the bell that was at fault.[33] He conducted experiments to confirm his doubts over the alloy, comparing samples of 22 parts copper to 5.5 tin (as was traditional) with 22:7 (Big Ben) and 22:10.3 for reference. He found, by dropping a 2-pound weight on each of these, that the traditional alloy did not break until the weight dropped from a height of twenty-four inches. In comparison, Big Ben's alloy broke at just eight inches, while the sample with the highest tin ratio broke at two.[34] Believing that the alloy was inconsistent, Airy advised the bell be turned so that the cracks were positioned at its point of least vibration.[35] Although it had cracked, Airy found that when hit, "producing an almost stunning sound, I have placed my finger . . . upon each of the chinks, and have been unable to discover any unfair opening or change." He judged that the bell was structurally secure, outlining "his views on the theory of a bell's sound vibrations. Let ACBD represent the circle of the massive sound bows upon which the sound almost entirely depends" (fig. 4.3). Due to the elasticity of its metal, the bell vibrated 160 times a second. Cracks were most likely at the parts "in which its curvature is affected in the greatest degree," which were parts A, B, C, and D. While here the movement was fast, at I, J, K, and L the vibration was minimal and the curvature less exaggerated. Airy recommended the bell be moved so that the cracks were at these places of least movement.[36]

While adjusting for cracks was easily resolved, a much harder challenge was ensuring that the great bell and the four accompanying quarter-chimes sounded accurate musical notes. Early in 1860 the government requested Airy collect information on the clock and bells. Doubting he had the musical knowledge to provide this, he wrote to De Morgan for advice on how "to discover the notes of the Quarter Chime Bells of Big Ben" and asked if he could purchase for him a handful of tuning forks or alternative instruments to accurately measure musical sound.[37] De Morgan responded that, because of its reliance on the ear, musical knowledge was unlike all other sciences and advised him that he would "have to go to those who do bells professionally. It takes a very practical ear to compare a bell and a fork with accuracy." Along with raising doubts over the reliability of tuning forks, De Morgan predicted that Airy would find it hard to reach a consensus among "bell people," who would likely disagree over the notes sounded at Westminster. De Morgan suggested he "go to a piano forte" authority who would be able to "settle" any dispute "by a set of experiments" conducted with his own practices and tuning forks.

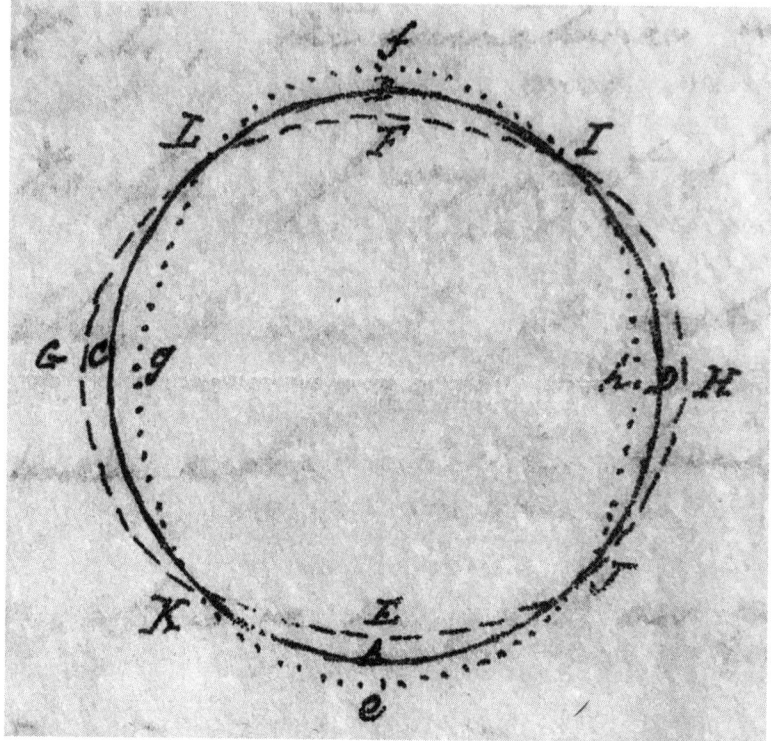

FIGURE 4.3 Airy's drawing of the vibrating circumference of Big Ben. RGO/6/609, "The Astronomer Royal Reports on the Clock and Bells of the New Westminster Palace" (April 21, 1860), 30–48, 43.

Piano tuners were more reliable, he claimed. He also recommended hearing the bells at a distance to ascertain their true notes, remarking that "I would believe the ear which took the sound at a distance": if too near to the source of the sound, the ear would be confused.[38]

Airy subsequently requested details of the intended quarter-bell notes from the clock builder Frederick Dent, who alleged that the great bell should sound an E, with the first, second, third, and fourth quarter bells giving G♯, F♯, E, and B respectively.[39] Airy ordered a set of five tuning forks corresponding to Dent's given notes from Robert Werner, a musician, in late March and on receiving them asked Werner to help him compare the forks to the bells. Werner and Airy met at Dent's shop on the Strand in April and then went to Parliament to make comparisons.[40] Werner provided the practical musical experience which Airy lacked, producing two reports for the Astronomer Royal and calculating the actual notes the bells produced in comparison to tuning forks (fig. 4.4). Using the

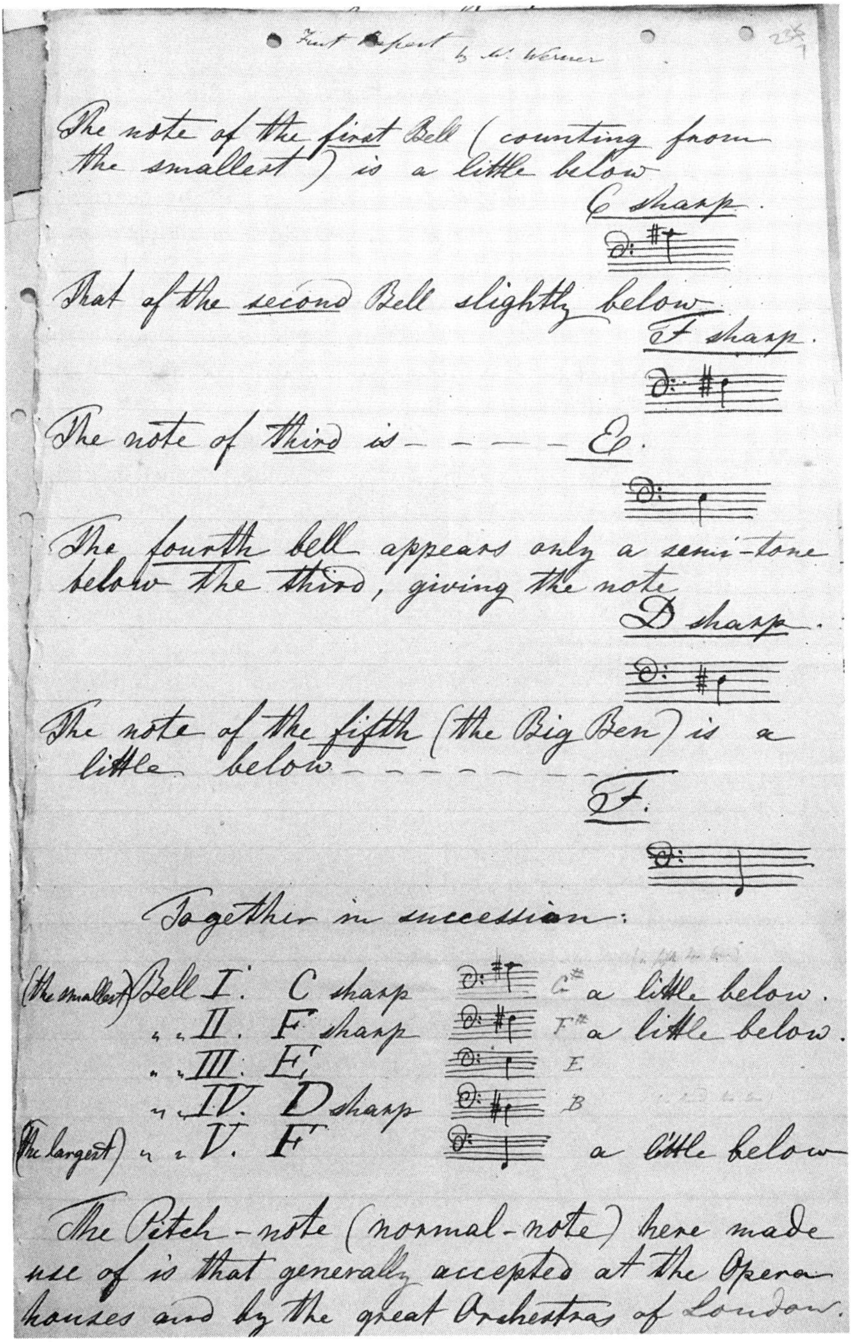

First Report by M. Werner

The note of the first Bell (counting from the smallest) is a little below C sharp.

That of the second Bell slightly below F sharp.

The note of third is — — — — E.

The fourth bell appears only a semi-tone below the third, giving the note D sharp.

The note of the fifth (the Big Ben) is a little below — — — — F.

Together in succession:

(the smallest) Bell I. C sharp — C♯ a little below.
" II. F sharp — F♯ a little below.
" III. E — E
" IV. D sharp — B
(the largest) " V. F — a little below

The Pitch-note (normal-note) here made use of is that generally accepted at the Opera houses and by the great Orchestras of London.

FIGURE 4.4 Robert Werner's first report on the notes of the Westminster bells. RGO/6/609, "First Report by Robert Werner" (April 1860), 237.

pitch of notes accepted at London's opera houses and orchestras, Werner found that the quarter bells struck C♯, F♯, E, and D♯, and the great bell an F, but he revised this in a second report, claiming that the first bell fundamentally vibrated a B, but with an A also audible.[41] He thought the E quarter too sharp but Big Ben's note to be perfect, taking the pitch of the Italian Opera at Covent Garden as his standard. The overall combination, Werner continued, "must prove contrary to the simple rules of harmony, or to the natural feeling of the inner relationship of certain chords": it would be hard to harmonize the bells as their vibrations interfered with each other. Werner concluded by providing calculations of the number of vibrations per second for each bell, based on its sound and dimensions, and in reference to a monochord.[42]

Skeptical of Werner's expertise, Airy sought a second opinion.[43] In March 1860, Airy had written to Whewell complaining that he was "never free from Government matters" and requested his old friend to send Trinity's organist to Great St. Mary's Church, Cambridge, to "accurately record the tones of the Quarter chime bells." He hoped that the organist could send him "tuning-forks or other apparatus accurately presenting to our senses in London the tones which you hear in Cambridge." Airy originally wanted the Westminster bells to sound the same notes as those of St. Mary's and now looked to them for reference.[44] Securing an accurate replication of notes from Cambridge to Westminster was hard, but tuning forks offered a means of making pitch observations mobile. Whewell followed Airy's instruction and reported that Trinity's organist, John Larkin Hopkins (1819–73), thought it "difficult to procure tuning forks to the exact pitch of Great St. Mary's chimes, the pitch having been raised since these bells were cast and tuning forks are, nearly all, now made to the present pitch." Hopkins believed that the bells had once been in the key of D but were now in D♭ and suggested that the use of a present-day D key would provide a "sharper and brighter" tone.[45] As D's pitch had changed over time, it was hard to purchase tuning forks to match Great St. Mary's chimes precisely, but both Whewell and Airy felt Hopkins a trustworthy authority on the matter.[46] In particular, Airy valued Hopkins's writing out the notes of St. Mary's bells for reference to the Westminster chimes; if they could not be the exact same pitch, then at least they might have the same intervals (fig. 4.5). He took Hopkins's report to Westminster where he resolved that the new bells were "not well harmonized."[47]

When Airy certified the finished clock on April 21, 1860, he noted a generally high degree of workmanship but had reservations over the notes of the great bell and quarter chimes. With the help of "two able musicians" he examined the bells again, this time in comparison to a vio-

FIGURE 4.5 Hopkins's report on the chimes of the bells of Great St. Mary's Church, Cambridge. RGO/6/609, "Chimes of Great St Mary's, Cambridge" (April 1860), 258.

lin, and found that while St. Mary's gave A, D, E, F♯, Westminster's produced "D (a little sharp), E (very sharp), F♯ (a little sharp), F and the Great Bell sounded an E." While Airy had intended the Westminster bells to chime B, E, F♯, G♯, with the great bell giving an E, the chimes of St. Mary's were actually A, D, E, F♯, with the great bell a lower D. He cited Hopkins as a credible authority on this matter and attributed the variation to "a difference in the standard pitch of reference."[48] Airy provided a theoretical explanation for the problem:

> When expressed mathematically, by the corresponding number of vibrations of air in a given time, the proportions are as follows, for the four quarter bells, 6, 8, 9, 10. This is by far the best series of four harmonized sounds that can be formed within the compass of an octave.

If these were in perfect turn, the ratio would give E, D, E, F♯, G♯, but Airy found that the "simplest numbers that I can form, for giving the relation of the vibrations from the actual bells (conceived in perfect time) are 9, 16, 18, 20, 27. It will be seen at once how ill related are these numbers."

Worse still, after being rung, the bells caused a horrid beating which filled the air about them, conveying "the most certain evidence of the want of harmony."[49] Airy speculated that one bell might have an especially strong ring that caused the other bells to sound their fundamental notes whenever vibrated. Hugely disappointed with the quarter bells, Airy proposed several solutions, including turning the second and third bells, which were too sharp, to alter how they were struck.

To Whewell, Airy explained the cause of the defect, outlining how the theory of harmony required "a good cadence at the end of each quarthour, by a third, a fourth, or a fifth. A drop of octave upon the hour bell. In all cases, a less harmonic interruption between the first complet and the second complet of each quarthour, thus dividing each quarthour not unpleasantly." At Westminster, the "cadences of 1 and 6 are bad, and in almost all, there is that horrible discordance of 20 and 27, which would kill a nervous man."[50] Airy reviewed the discordant vibration ratios with Whewell, who later remarked that Airy's fear of "the bad effect of the ratio 20:27 in the Westminster chimes was unfounded. For, put 26 2/3 instead of 27, and the relation becomes a perfect fourth." However, there remained "the important fault that the chime-bells are harmonized to their highest bell instead of to their lowest."[51] Whewell thought that Airy's original calculation of 27 was correct, reckoning the series to be D, G, A, B, C, D, E, F, but feeling that the E was not an exact fourth. He thought the imperfection only slight but agreed that "our Cambridge melody is much better than the Westminster one."[52]

Airy's report on the Westminster bells caused immediate concern and renewed efforts to create a harmonious set of chimes for Parliament. On June 5 he wrote to Werner telling him that the government had decided to stop using Big Ben for a limited period of time and wanted a temporary arrangement for the quarter bells. He asked Werner for a composition of three quarter bells, with the D bell striking the hour, which could "produce a tolerably musical effect."[53] William Cowper (1811–88), First Commissioner of the Board of Works, offered to write to the engineer at the Palace of Westminster, Faber James, to arrange for a temporary solution while a more permanent measure was found.[54] Although he did not think James "at all musical," Airy agreed, confident that "when a musical succession of bells shall be arranged he will be able to give the proper arrangement to the machinery for correctly playing the music." Airy had his own "musical advisor" who was helping him compose a set of chimes. They produced several scores with the aid of a pianoforte, but Airy confessed it might be safer to mark each quarter with strikes of the F♯ bell and the hour with the D. Another arrangement was to have a "Ting-Tong" of B and F♯ at each quarter and a D for the hour. Eventually he advocated the "Ting-Tong," believing the alternatives "would not ... be endured by the Public."[55]

Predictably the bell's founder, Warner, denied that the notes were defective.[56] Reverend William Taylor, who had composed the chimes for the bells, believed they did strike the correct arrangement, but felt "inclined to distrust my own judgement, as I know of no higher authority in scientific matters than the Astronomer Royal."[57] He suggested the bells sounded different at varying locations and attributed any error to this. In response, Airy would "only say that the two professional musical gentlemen who accompanied me were in perfect agreement as to the notes given by the Bells, and that I myself remarked the perfect accord of the notes of the Bells with the notes on the violoncello." He asserted that the only "method of arriving at the truth" was for either Warner or Taylor to examine the bells in person, accompanied by an independent musical authority. Airy was slightly alarmed that "Taylor's experiences seem to imply that Messrs Warner rely solely on the dimensions of the bells for the notes," rather than accurate acoustic measurement.[58]

On June 27, Faber James reported that James Turle, the organist of Westminster Abbey, had proposed an alternate melody for the bells, which Airy requested be sent to Greenwich where he would have it played.[59] Turle had heard the bells from Westminster Abbey and believed them to be A, G, F, and C.[60] On receiving and playing Turle's score, Airy approved of the tune, providing the notes corresponded accurately to the bells, and sanctioned its use until the Westminster bells were corrected to play the chimes of Great St. Mary's.[61] Turle was delighted, believing this preferable to Airy's proposed "Ting-Tong," which he found "anything than pleasing to my ear."[62] Airy therefore advised the Board of Works that a "trial be made *by ear*" between Turle's tune and the "Ting-Tong," and that the chief commissioner of the Board of Works assess which was most pleasing.[63] Cowper, however, did not like the idea and believed himself unqualified to judge, while he had "no one in my office with a musical ear" who could help: civil servants were evidently not appointed for their musical abilities.[64] Airy therefore recommended William Pole be invited to decide and asked the Board of Works to give him access to the clock tower. A self-taught mathematician and University College London's professor of engineering, Pole was also an enthusiastic musician, being organist at a Wesleyan chapel in Birmingham from the age of seventeen and organist of St. Mark's Church on North Audley Street in London from 1836 to 1844 and 1850 to 1866. After reporting on musical instruments at the 1851 Great Exhibition, Pole studied music at St. John's College Oxford before completing a doctorate in music at Oxford in 1867.[65] Airy trusted Pole, explaining that "though a Professor of Engineering," he was "also a musician of the highest order," capable of solving "the doubtful questions on the tones of the Bells."[66] Pole himself boasted "much experience in both the theory and practice of the art," having "been Organist of a large London church for many years, and have taken the degree of Mus. Bac. at Oxford."[67]

Airy instructed Pole to examine the Westminster bells but requested he do so at various locations and distances, specifying that they "should sound (beginning with Big Ben) E (lower scale), B, E, F♯, G♯," but that "my professional musicians, on trying the notes by unisons with a violoncello, found, E (lower scale), D (a little sharp), E (very sharp), F♯ (a little too sharp), B." This measurement had been obtained in the "Clock Room, or in its neighbourhood," however, Airy remarked that "on going to the Belfry, the highest bell actually gave A in the next higher scale." Airy wanted Pole to verify these observations and advised him to find an assistant to strike the bells with a six-foot hammer while he listened first from the belfry, then the clock room, and on the stage above the bells. He warned that on the striking of a quarter or hour Pole might "hear the tremendous beating, which exceeds anything that I ever heard."[68]

On attempting his investigation, Pole was denied entry to the tower but, from beyond the palace, thought the quarter chimes sounded B, E, F♯, and G♯. At a certain distance these were in fact pleasing, but they were discordant at closer range, leading Pole to state that he did "not wonder at any difficulty found in determining the note when *close to the bell*, the chief note being . . . confused with a multitude of other sounds harmonic or otherwise. I fancy that distance is the only way of eliminating these disturbing elements."[69] He also forwarded some of his earlier observations made at the Institution of Civil Engineers in November 1859 concerning musical pitch, where he had raised the subject of the Westminster bells. Pole thought the B bell "decidedly imperfect" and flat in pitch, but the other three chime bells to be good. However, he found the arrangement displeasing, as the "great Bell had a more serious musical defect, as, in addition to its principal note E, and the harmonics properly accompanying it, it sounded also a false note, not contained in the harmonic, or indeed in any musical scale, namely something between C♯ and D." The "effect to these living near was excruciating," claimed Pole, "the defect in tone being universally complained of even by those who were unable to define its musical nature."[70]

Airy sent Pole Hopkins's index of St. Mary's chimes but rejected his poor opinion of the Great Bell.[71] On August 18, at the second attempt,

Pole secured access to the belfry and confidently reported on the notes struck. He found that the quarter bells sounded at 872, 758, 674, and 507 vibrations per second, with the great bell at 337.[72] Using tuning forks, Pole determined the great bell struck an E, with quarter bells of B, E, F♯, and G♯, later revising Big Ben's rate of vibration to 337.5, and 516 (B), 675 (E), 770 (F♯), and 874 (G♯) for the chimes. He attributed earlier discrepancies over the notes to changes in standard of pitch over time, as was immediately under discussion at the Society of Arts. Therefore, the great bell was both an old F and a new E. The problem was that notes were relative to C-above-middle C and this was not a fixed tone. France had only just established its *diapason normal*, giving a C of 522 vibrations per second, but most English operas used 537, while the recent Society of Arts committee determined C to be 528. Nevertheless, Pole confirmed that the bells had defects, with the G♯ sounding like a low B♭ and the B giving a D♭. In both cases it was hard "to distinguish its fundamental note at all," even at a distance. If sounded alone, the B was fine, but "when chimed with the others it gives the impression of being too flat, although this is not borne out by the number of its vibrations." Pole also reiterated that the great bell sounded more like a very flat D and was, if heard up close, "extremely harsh and disagreeable."[73]

Pole's confirmation that the bells did not strike their intended notes supported Airy's fears that the Westminster chimes were neither harmonious nor a replication of St. Mary's chimes. Little could be done about this, but the controversy had provided Airy with an irksome lesson in how difficult it could be to control sound and obtain a musical consensus. In *On Sound*, Airy recalled this problem of reconciling mathematical theory with practical musical experience when discussing how to calculate harmonic vibrations. However, he asserted that of "all parts of theoretical music, the theory of Temperament is the most troublesome. We cannot pretend here to go into it to the extent which would be useful to the professional musician." Airy here drew directly on his experiences of the Westminster and Cambridge chimes, outlining how bells could indicate joy or triumph by completing an octave in descending order, as long as each bell was well related to the others. Both the rise and the fall of an octave were acoustically "exciting," with the effect depending "on some unknown physiological cause." Airy subsequently treated the St. Mary's chimes to a mathematical analysis, noting how they had been replicated for the Palace of Westminster. He included Hopkins's calculation of their tones, showing the relations of each note in the quarter and hour chimes in terms of their vibrations. Students could see, numerically, what a successful harmony looked like. A similar mathematical investigation of

God save our gra-cious Queen Long live our no-ble Queen
2 2 2 3 1 2 2 2 2 3 1 2
C C D B C D E E F E D C
48 48 54 45 48 54 60 60 64 60 54 48

God save the Queen | | Send her vic-to-ri-ous Hap-py and
2 2 2 2 | 4 | 2 2 2 3 1 2 2 2 2
D C B C | | G G G G F E F F F
54 48 45 48 | | 72 72 72 72 64 60 64 64 64

Glo-ri-ous Long to reign o-ver us G-o-d save the Queen.
3 1 2 2 1 1 1 1 3 1 1 1 ½ ½ 2 2 2
F E D E F E D C E F G a G F E D C
64 60 54 60 64 60 54 48 60 64 72 80 72 64 60 54 48

FIGURE 4.6 Airy's figures to show the proportion of vibrations in the playing of "God Save the Queen," with notes shown. Airy, *On Sound*, 231. Public domain.

"God Save the Queen" revealed a melody that Airy claimed to be "meagre, if considered only with reference to the relations of adjoining notes" (fig. 4.6). "Adeste Fideles" offered a musically superior contrast, as the "grandeur of this piece of music," continued Airy, was "fully explained by the consideration of the numerical relations of the vibrations corresponding to the notes." These musical relations were, Airy believed, of value to both mathematics and music undergraduates. As he put it, in reference to St. Mary's chimes, "there could be no better education for a young Cambridge musician, than to learn habitually to associate the tones of the several bells with the numbers that we have attached to them, and to repeat these numbers on hearing the sounds of the bells."[74] Within earshot of Airy's own Trinity College was a practical lesson for musicians and mathematicians alike.

Nevertheless, the difficulty of establishing the precise notes of bells was something Airy never really resolved. As late as 1881 the dean of Lincoln, Joseph Blakesley (1808–85), wrote to the now retired Airy, asking for advice concerning the bells of Lincoln Cathedral. After recently installing a new clock in the cathedral, several charitable neighbors donated a set of quarter bells which sounded "what are called the Cambridge, or the Westminster Chimes." However, the cathedral's congregation could not reach a consensus over if this tune was actually produced. A graduate and tutor of Trinity College Cambridge, as well as a past member of the elite university debating society, the Apostles Club, Blakesley reported that the cathedral organist

> declares that the bells are out of tune with each other. . . . The maker of the bells, on the other hand, declares that they are absolutely perfect. Our musical members of the Cathedral fancy that one of the bells fails a little: but they are not agreed as to whether it is a little too sharp or a little too flat. I think that they are perhaps not quite all in harmony, but incline to think that what there is objectionable arises from either the strength of the blow struck by the hammer on them severely being ill adjusted; or from the interval between the successive notes being of undue length. But what makes me trouble you in the matter is, that they appear to me sometimes quite in tune and sometimes slightly out, at different hours of the night. I have not slept well for some time . . . the phenomena has puzzled me much; for I cannot think it lies namely in the condition of my own ear. Is it possible that the variation of temperature could produce a disturbance in the harmonic relations of the bells?[75]

Airy's response was uncertain, but he imagined there to be a physical flaw with the bells which Blakesley's "ears perceive when in their best state. . . . I do not suppose that any fault of hearing could cause" the variation. He doubted the error was due to temperature changes as the bells were probably made of the same metal. Airy's only hypothesis was that it was a tuning fault. He advised the dean to "severely cross-examine the maker of the bells on the grounds of his assertions of correctness, and the organist on the grounds of his opposite assertion." Furthermore, he asserted that the only way that mathematical accuracy could "be secured—and trusted" was by comparison to a violin, which was more perfect than any other instrument. He instructed Blakesley

> to look to the very simple mathematics of the case. You must purchase of Macmillan "Airy on Sound". For exhibition of the agreement of theory and fact, I have taken as the first specimen the *St Mary's Chimes* . . . and you may there see the various consonances which ought to be made *perfect*.[76]

Airy's advice demonstrates how he conceived of his treatise *On Sound* not as a purely mathematical text but as a practical manual for resolving musical problems. It was an example of how mathematical knowledge could be of national value. However, this episode also reveals that, beyond the application of mathematical theory, Airy still struggled to reconcile varying interpretations of musical pitch. Sound remained a troublingly subjective sense to measure, despite its apparently mathematical character.

Hearing and Seeing Time

Airy's guiding principle at the Royal Observatory was to improve the accuracy of all observations, mobilizing increasingly precise instruments to reduce potential human error. Nowhere was this of greater urgency than in the recording and disseminating of time. In particular, the use of electromagnetic and telegraphic apparatus offered new ways of taking accurate readings from nature, especially of astronomical transit observations.[77] While Airy introduced trials to calculate the personal error of each astronomical observer and conceived of a galvanic electric system for registering precise star transits to give Greenwich time, he also took care to ensure trustworthy means for disseminating these measurements. "The ruling feature of his character was undoubtedly Order... he had the greatest dread of disorder creeping into the routine work of the Observatory," recalled his son Wilfrid in 1896.[78] This fear of disorder was evident in Airy's attitudes toward how best to display accurate time and, in this responsibility, his earlier experiences of sound shaped his approach to timekeeping practices.

During the 1830s and 1840s, the Royal Observatory's preferred method for distributing time was by the daily dropping of a large ball, positioned within sight of ships, so that crews could check that their marine chronometers were precisely regulated to Greenwich time for calculating longitude at sea. At 12:55 p.m. the time-ball, usually on top of a building, would be raised to warn viewers to prepare, before dropping at 1 p.m. After the Royal Navy's experimental time-ball at Portsmouth in 1829, one was installed at Greenwich in 1833, followed by another at Mauritius in the same year, St. Helena in 1834, the Cape of Good Hope in 1836, and Bombay and Madras in the 1840s.[79]

From 1849, Airy was in contact with the telegraph superintendent of the South Eastern Railway Company, Charles Walker (1812–82), concerning the dissemination of time by a telegraphic line from the Royal Observatory to Lewisham station and then throughout the railway network. This scheme was in place by early 1852, with an hourly impulse sent from Greenwich along telegraph cables across the nation. This impulse could be used to drop time-balls, regulate clocks, ring bells, or fire guns, providing accurate signals of Greenwich time across Britain and the empire.[80]

Airy's early preference was for this electric connection to regulate visual signals. In July 1852, Greenwich's 1 p.m. time-ball was controlled telegraphically, and from August the apparatus regulated a system of sympathetic clocks throughout the observatory. On the observatory's exterior, the newly installed Shepherd Gate Clock offered similar accuracy, and this clock soon provided an impulse to drive a sympathetic clock at London Bridge station. Airy oversaw the extension of this time system by a network of further visual signals. The Electric Telegraph Company erected a time-ball in the Strand in 1852, with another constructed at Deal in 1855. In the same year four clocks were installed in London post offices, all controlled from Greenwich. The biggest prize of all came in 1862 in the form of the Westminster Clock at the newly built Houses of Parliament. Although this was not directly controlled from Greenwich, it was adjusted in reference to a daily signal from the Royal Observatory, ensuring Greenwich time was accurately displayed from the world's most famous timekeeper. Yet not all time signals were visual. Beginning in 1861 a time-gun was fired at 1 p.m. from Edinburgh Castle, followed by time-guns in Newcastle and North Shields in 1863. There was also a time-gun at Dover Castle from the mid-1850s, while the Westminster Clock disseminated Greenwich time by the chimes of its bells as well as the hands on its face.[81] Preceding all of these, at Cape Town a pair of Dutch naval guns had signaled noon, each day, since 1806.

In November 1859, Edinburgh Town Council informed Airy that it wanted to install a cannon, electrically controlled from Greenwich, to fire daily at 1 p.m. This preference for a gun as well as a time-ball was due to the poor weather and regular lack of visibility in the city.[82] Airy believed that distributing time in Edinburgh by telegraphic apparatus was a praiseworthy "application of science to objects of great utility." Although he recognized that the existing Calton Hill time-ball was obscured from Leith by warehouses and also supposed that "the Leith merchants have difficulty in fixing on *any* place for a visible signal . . . and prefer an audible signal," Airy warned that the scheme lacked accuracy (fig. 4.7). A sonorous time signal did not have the precision of a visual clock or time-ball. Nevertheless, though he feared the time-gun would not secure reliable readings for marine chronometers, he conceded that the "peculiar value of a cannon-fire at Leith arises from this, that the shipping is so much concentrated that the error of transmission of sound is not great. This does not apply on the Thames or the Mersey."[83]

This plan for a sonorous time signal was the brainchild of the banker John Hewat, with the support of Charles Piazzi Smyth (1819–1900), Astronomer Royal for Scotland from 1846 and Regis Professor of Astron-

FIGURE 4.7 The Calton Hill time-ball, disseminating time across Edinburgh. Author's photograph, 2018.

omy at the University of Edinburgh. An evangelical, Smyth became increasingly well known for his alternative approaches to measurement, leading a campaign throughout the 1860s to replace Britain's existing system of imperial weights and measures with a set of standards based on the dimensions of the Great Pyramid at Giza. Attracting support from radical evangelicals, Smyth's "Pyramidology" claimed theological authority, asserting that the Great Pyramid had been divinely inspired and built in reference to a sacred measure. His continuing obsession with Pyramidology left him somewhat isolated from London's science elites.[84]

Both Smyth and Hewat had witnessed the time-gun in operation at the Palais Royale in Paris, which used the sun's rays to ignite gunpowder and to set off the charge. Hewat, a member of Edinburgh's Chamber of Commerce, canvassed for an audible time signal to accompany the existing city time-ball.[85] Importantly, this was not about replacing visual signaling with audible time distribution but of supplementing it. In the same year as establishing the Edinburgh time-gun, Smyth continued to closely regulate the accuracy of the city's time-ball on Calton Hill, which had been in operation on the summit of the Nelson Monument since 1854. It was Smyth's intention for the Edinburgh Royal Observatory to provide "both a visible and an audible signal."[86]

By October 1860 Edinburgh's merchants and Town Council had agreed to install the time-gun, and Airy consented to provide a regular signal "for the daily promulgation of accurate time." He nevertheless maintained his reservations over accuracy, encouraging Hewat to consider if there were "objections to the cannon-signal or audible signal, from which a visible signal is free." As the gun provided no warning of its impending boom, Airy explained that the signal might be missed as "very few masters of ships could have their eyes upon their chronometers at the time of the gun-fire." This lack of warning "would deprive it of nine-tenths of its utility," and the only solution would be for preliminary discharges to be fired before 1 p.m. Surmising that the "signal of sound is scarcely so sudden as that of sight . . . subject to a specific error, from which a signal of sight is free," Airy recommended a reconsideration of "the advantage of a visible signal in preference to an audible signal."[87] A new visible signal, ideally a semaphore carried by a one-hundred-foot high mast, was Airy's favored solution. Despite the claims of local newspapers, Airy clarified that he was not responsible for any time-guns and deliberately avoided the practice.[88]

Airy's appeal fell on deaf ears. Hewat reported that there was a strong public desire in Edinburgh for an audible signal at the castle, with a committee of local citizens having raised money for the project. Suggesting Airy subscribe, Hewat reassured the English Astronomer Royal that the "distance being easily ascertained and the deduction made for the time the sound of the gun occupied in travelling a sufficient facility for noting the rating of Time pieces will be given."[89] The Edinburgh Castle time-gun came into operation in 1861 and, despite Airy's reservations, proved popular, securing highly positive newspaper reviews. On first hearing the boom, the *Scotsman* reported the audible time signal to be a triumph, attracting a large crowd of spectators. The journal praised Smyth and Hewat's scheme, as did the *Edinburgh Evening Courant*.[90] Much to Airy's irritation, he read in a later article in the *Scotsman* of how audible signals were accurate to a fraction of a second, and that there was "high scientific support of audible signals of true time," providing calculations were made for the velocity of sound.[91] Away from the elite scientific publications and audiences that Airy was accustomed to dealing with, the popular press, with its broad public readership, was perfectly satisfied with Edinburgh's sonorous signaling.

In an effort to appease the agitated English Astronomer Royal, Smyth wrote to Airy, attaching his lecture from the Royal Scottish Society of Arts, "On the Methods Adopted to Secure Extreme Accuracy in the Edinburgh Castle Time-gun Signal."[92] Read in July 1861 and published in 1864, Smyth here made explicit the measures he had taken to secure accuracy for the audible signal, without which he confessed the cannon would be but "an idle toy." Attention had been paid to establishing a reliable telegraphic connection between Edinburgh Observatory and the castle. Using Herschel's calculations on the velocity of sound from his 1830 *Encyclopaedia Metropolitana* article, Smyth had performed experiments on the propagation of the time-gun signal in Edinburgh to account for temperature and humidity fluctuations. With an assistant, Smyth went to various locations across the city between June 8 and 13, armed with one of Mr. Hislop's time-gun maps, produced specifically to allow observers to calculate the time based on their distance from the cannon (fig. 4.8).[93] From these locations they observed the flash of the cannon and then timed how long the sound took to reach them, before comparing these observations with the propagation time specified on Hislop's map (fig. 4.9). Smyth concluded from these "creditable observations" that, even with wind variations, the cannon provided immense accuracy. Allowing for monthly temperature changes throughout the year, he calculated that there would never be more than 1/10th of a second's variation in the reception of the time signal. In practice, he was confident that the cannon should "be regarded as sensibly exact throughout the year" and that any greater accuracy could only be "a matter of scientific curiosity."[94] He felt Airy's calls for superior accuracy to be pedantic at best.

Much to Airy's annoyance, other towns looked to Edinburgh, rather than Greenwich, as a model to follow. As early as February 1859, the *Gateshead Observer* put forward the case for a time-ball on the River Tyne. As accurate chronometers were essential to maritime navigation, it was vital to have a signal which provided "the same confidence as if it were that at Greenwich." The publication argued that since a chronometer error of

Professor C. P. Smyth *on the*

Remarks by Mr Balbirnie on the Report of the Time-Gun at Edinburgh Castle, as heard in Leith Docks.

Date.	Time from Fall of Ball to Sound of the Gun.	Winds.	General Remarks.	Report of the Gun.
1861.	SEC.			
June 8.	10	E.N.E.	Light breezes, and cloudy.	Low.
,, 9.	Sunday.	...	... ...	...
,, 10.	10	E.S.E.	Light winds; hazy weather.	Faint.
,, 11.	10	W.S.W.	Light winds; clear weather.	Good.
,, 12.	10	W. by S.	Brisk wind; clear.	Faint.
,, 13.	10	Variable.	Variable light breezes and calms; hot sultry weather.	Good.
,, 14.	11	East.	Moder. breezes, and hazy.	Good.
,, 15.	10	E. by S.	Light winds, and hazy.	Faint.
,, 16.	Sunday.	...	... ...	...
,, 17.	11	East.	Fresh wind, and clear.	Faint.
,, 18.	10	E. by N.	Light winds, and clear.	Good.
,, 19.	Could not see the Ball. Could not hear the Gun.	East.	Light winds, with showers.	...
,, 20.	11	E.S.E.	Light winds, and hazy.	Very faint.
,, 21.	10	East.	Light winds, and hazy.	Faint.
,, 22.	No observation taken.	Variable.	Light winds, hazy, and rain.	...
,, 23.	Sunday.	...	... ...	...
,, 24.	11	W.S.W.	Brisk wind, and clear.	Good.
,, 25.	11	S.S.E.	Light winds; cloudy.	Faint.
,, 26.	11	E. by S.	Light winds, and hazy.	Light.
,, 27.	10	E.S.E.	Light winds, and clear.	Faint.
,, 28.	10	W. by S.	Fresh breezes, and clear.	Faint.
,, 29.	10	W.N.W.	Light winds, and calms.	Faint.
,, 30.	Sunday.	...	... ...	...
July 1.	No observation taken.	...	... ...	...
,, 2.	10	N.W.	Fresh breezes, and clear.	Faint.
,, 3.	10	W.N.W.	Light winds, and clear.	Good.
,, 4.	10	South.	Light winds, and hazy.	Faint.

FIGURE 4.8 Observations made on the propagation of the Edinburgh time-gun's sound signal. Smyth, "On the Methods Adopted to Secure Extreme Accuracy in the Edinburgh Castle Time-Gun Signal," 150.

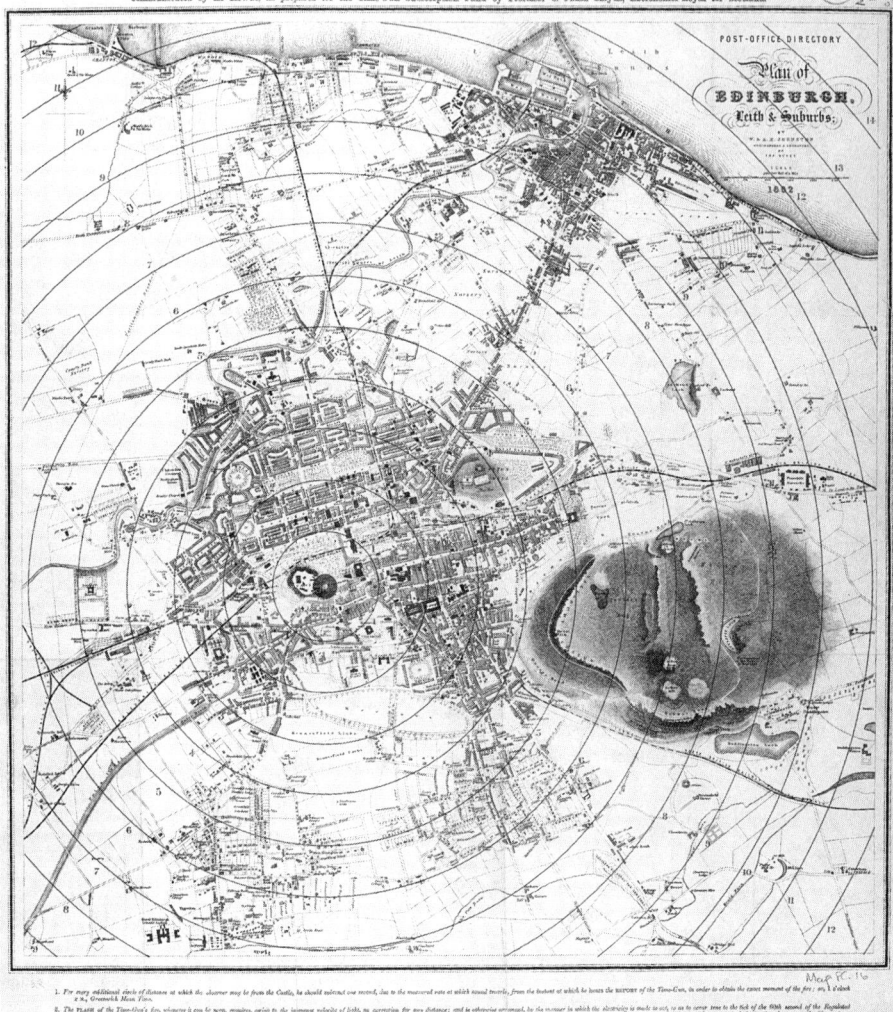

FIGURE 4.9 Hislop's time-gun map of Edinburgh and Leith allowed users to calculate how long the audible signal took to reach them in different locations around the city. For each circle from the sound's origin, a second had to be added to calculate accurate Greenwich time. Edinburgh University Library Centre for Research Collections, Map.PC.16, "Hislop's Time-Gun Map of Edinburgh and Leith," 1862.

four seconds could alter a ship's course by a mile, the Tyne's signal should not differ from Greenwich time by more than 1/30th of a second.[95] In 1863, following Edinburgh's success, the *Newcastle Daily Chronicle* raised this question again, publishing an article by the Tyne Commissioner James Mather which compared visual and audible time signaling. Mather informed readers that Greenwich "now gives forth the true time, received from astronomical data that never err, and then, demonstrating it daily to the eyes of men, flashes it on lightning wings to all parts of the kingdom." This, he continued, had to be available on the Tyne to regulate the chronometers of local vessels. Responsible for 400,000 tons of shipping a year, an eleventh of the nation's trade, the River Tyne Commissioners appointed a committee which proposed a time-ball on the High Level Bridge and a time-gun on the old Norman keep of Newcastle, providing the "diffusion of this true knowledge" of time. Mather praised Hewat as the first to propose a time-gun and Smyth for overseeing the project, and he commended the Edinburgh time-gun's accuracy:

> When its instant flash cannot be observed, its sound is a true guide to manners; for by the law laid down by Herschel, that the velocity of sound, in a temperature of 62 deg., is 1,125 feet per second, and with 1.14 feet difference for every degree of variance from 62 deg., adding for an increase, deducting for a decrease, the exact moment of discharge is fixed.[96]

Yet Airy remained unconvinced of the accuracy of acoustic signaling. After reading Mather's newspaper article, he wrote to the commissioner emphasizing the cheapness and supreme accuracy of a time-ball, but admitted that a problem with this was that "it can be seen only by persons, who have made up their minds to see it."[97] In comparison, a

> time-gun cannot be observed quite so accurately (say that there may be an error of ½ second or more) . . . But then it can be heard . . . and it will be heard with a minor degree of accuracy by persons in no way prepared for it. Its influence will be very much more extensive than that of the time ball.[98]

This was a valuable difference. While audible signaling could not secure safe navigation and chronometer regulation, Airy believed that sound's ability to permeate houses and businesses promoted Greenwich time with nonmaritime audiences. The popularity of the Edinburgh time-gun appeared to have persuaded Airy that the audible, though inaccurate, could have advantages over visual signaling. Sound was, in this way, a popularizer of Greenwich time. Airy surmised that in

> proposing an institution which must be based on popular support, it is indispensable to secure popular sympathy, provided that no permanent disablement for the higher purposes be incurred by attention in the first instance to the popular purposes. This is just what would happen in adapting for the present the time-gun. A trifling degree of accuracy is best, but it may at any time be regained by erecting a ball-drop apparatus.... If the Tyne Commission do not object to the expense of the gunpowder, I should recommend, for the present, the time-gun.[99]

By the end of 1863 the River Tyne Commission had a time-gun positioned on the banks of the river.[100] Although Smyth initially offered a time signal to Newcastle from Edinburgh, he was so enraged on learning that Mather had been in communication with Airy that he refused to provide a connection. As a result, from 1874, Airy had to provide a direct link from Greenwich to the Tyne.[101]

It must have been with frustration that Airy read a condemnation of visual signaling in the *Reader* in December 1863. This journal argued that the divide between popular audible signals and apparently elitist visual signaling was in fact a division between the North and South. It condemned southerners for ignoring the North's superior practices, lamenting that the only accurate sound in London was Big Ben. London's churches, on the other hand, disseminated inaccurate time, sounding "repeated lying tongues from a hundred steeples." Critical of those responsible for time regulation in the South, clearly directed at Airy, it praised Newcastle and Sunderland's adoption of Edinburgh's practices. While Glasgow had had a time-ball since 1855, the *Reader* claimed that "somehow there was not enough in the time-ball system itself to fully interest the practical inhabitants of the great Western city."[102] However, in the autumn of 1863 the Glasgow time-gun came into use, with maps produced for the city in a similar fashion to those Hislop published for Edinburgh.[103] Under Smyth's control from Edinburgh Royal Observatory, this time-gun confirmed the idea that audible signaling was socially popular. The United Telegraph Company's engineer, Nathaniel Holmes, informed Smyth of the successful inauguration of the Glasgow time-gun on October 1, observing that the cannon "went off famously today amidst loud cheering from a large concourse of spectators who occupied the windows of adjacent houses." He was relieved on the gun's firing "not to hear any sound of broken windows."[104] Robert Symington, of the Scottish Telegraph Construction and Maintenance Company, agreed that the Glasgow time-gun was "becoming popular" but advocated a bigger charge to project a greater sound.[105] This Glasgow time-gun meant that Edinburgh, Glasgow, Newcastle, and Sunderland were all running audible time signals. The *Reader* wondered why London was so behind the North, condemning Greenwich for failing to see the value of audible time. In contrast, it reported there was a scheme in Paris for an electrical system to regulate all public clocks in the city, under the celebrated clock maker Auguste-Lucien Vérité's (1806–87) direction. The journal preferred the Scottish audible system, but it warned that London was falling behind not only Edinburgh but also Paris.[106]

Smyth was well aware of this southern skepticism toward time-guns. Following difficulties in getting the Electric and International Telegraph Company to "spare a circuit" to connect Edinburgh Royal Observatory with the Glasgow time-gun in 1863, Nathaniel Holmes told Smyth bluntly that the problem was that "guns are unpopular in the South."[107] When Smyth subsequently asked Walter White (1811–93), assistant secretary and librarian of the Royal Society, if he thought a time-gun would enhance the timekeeping practices of London's foremost learned society, White replied curtly that they did not "in this part of London suffer any inconvenience from the absence of a time-gun and hitherto I have never had occasion to go out to look at the drop of the Strand Ball," largely because he could hear the chimes of St. James's clock every fifteen minutes.[108] Even without a time-ball, members of the society managed "to get through our duties with punctual observance," White reported.

At the same time, although the Edinburgh time-gun's signal was generally reliable, Smyth struggled to maintain standards. In 1872 he found that a clumsy artilleryman, responsible for loading the gun, had tampered with the trigger mechanism, preventing its firing and, as a result, decided to fire the cannon manually, causing a delay of almost a minute and a half.[109] Again in January 1876, the clock maker Jamie Ritchie, who had built the clock for regulating the Edinburgh time-ball, wrote to Smyth explaining that the master gunman at the castle had been absent from a firing, leaving his deputy in charge who had accidently set the cannon off by hand.[110] Ritchie recommended Smyth produce a written manual to be kept at the castle to prevent future mishaps, but the problem remained that, unlike time-balls, the use of a time-gun was contingent on the skill of artillerymen to load the cannon with charge without undermining the integrity of the apparatus. By April 1876, Smyth complained to Edinburgh Town Council of over eight misfires in under two months and advised the council to act fast to "prevent the City losing the advantage of its audible time-signal."[111] He faced similar problems with the time-gun established at Dundee, which was proving unreliable by 1872. On learning that the cannon fired twenty seconds late most days, Smyth admitted the news was "most painful to me," as he took such care to send a time signal to Dundee that was accurate to one-tenth of a second of Greenwich time. He thought it "better to have no signal at all" than an inaccurate projection that would undermine confidence in the time disseminated.[112] On investigating Dundee's inaccuracy, Smyth found the fault was a combination of a poor connection to the gun and an artilleryman who had taken it upon himself to fire the gun on his own authority.[113] Once again, the accuracy of sonorous time signaling depended on reliable gunners following precise instructions.

Conclusion

Between 1859 and 1860, within earshot of the Society of Arts' pitch committee meetings at the society's premises on John Adam Street, the bells of the Palace of Westminster provided musicians and mathematicians with a very practical lesson on how hard it could be to build a musical consensus. While the committee debated what an ideal standard pitch should be, Airy was coming to realize that getting professionals from within the music industry to agree on what a note sounded like was difficult. It was hard to get various authorities to agree to a certain frequency of pitch as a standard, but it was equally difficult to get them to concur on what that frequency actually sounded like. For all the idealism of a standard pitch and the measurability of sound, the ear was a subjective organ. Airy, accustomed to the ordered, highly disciplined practices of astronomy, was troubled that the world of music appeared so chaotic.

The problem, for Airy at least, was that acoustic time signals remained popular, with a gun installed at Liverpool in 1867 and proposals for guns at Calcutta in 1870 and at Bristol, Middlesbrough, Hartlepool, and Darlington in 1873.[114] Nevertheless, by 1898 the Admiralty had three times as many time-balls as time-guns registered across the globe, and this was largely due to the perceived uncertainties over the speed of sound through atmospheric variations.[115] Rival schemes for disseminating time engendered contrasting interpretations of accuracy, of which debates over audibility and visuality were central. At stake were alternate enterprises for how society should receive time: either with absolute precision, by eye, or with convenience, by ear. In a sense, this dichotomy between seeing and hearing involved much broader questions about society's relationship with industrial time: was it something populations should be disciplined to observe, prioritizing astronomical accuracy, or something transmitted aurally into homes and factories by guns and bells, which compromised accuracy in exchange for a more extensive audience? For Airy, the empire's discipliner-in-chief, visual signaling was clearly superior, as this alone offered the accuracy required to secure maritime navigation. And, when examining Airy's discussions over time-dissemination practices in the context of his broader work and sonorous experiences, it is clear that, by the 1860s, attaining accuracy in sound was something Airy thought hard to achieve.

Although keen to place acoustics within the study of mathematics, especially for undergraduate wranglers, as an example of how the discipline could be applied to solve physical problems, Airy and his mathematical colleagues knew that, in practice, building a musical consensus was challenging. Musical authorities, including instrument makers, bell founders, tuners, and musicians, could not always reach a consensus over pitch, regardless of mathematical theory. If Airy could not trust the ear for time-disseminating practices, then this was because he knew what a capricious organ the ear could be. Placing his work on time regulation within its musical contexts helps to explain why Airy was so loath to depend on aural signaling. However, it is equally clear that, beyond Britain's scientific communities, the Astronomer Royal struggled to exert authority in this matter. Throughout the debates regarding time-guns, it becomes apparent that Airy's calls for accuracy were aimed at scientific and maritime audiences, concerned with exact measurement and the precise timekeeping on which navigation depended. In contrast, his adversaries appealed to a more general public who were increasingly invested in the business and daily routine of cities and towns. These were readers who would find the spectacle of a time-gun a far more convincing symbol of order and civic pride than the extreme accuracy of an astronomical observatory's time-ball.

PART III
Materialism and Morality
Religious Authority and the Science of Sound

CHAPTER 5

Musical Matter

Religious Authority, John Tyndall, and the
Challenge of Materialism, 1859–1914

> *From the edge of the sea came a ripple and whisper. Beyond these lifeless sounds the world was silent. Silent? It would be hard to convey the stillness of it. All the sounds of man, the bleating of sheep, the cries of birds, the hum of insects, the stir that makes the background of our lives—all that was over.*
>
> H. G. WELLS, *The Time Machine*

While the controversy over pitch in 1859 and 1860 constituted a rebuke to those professing scientific knowledge who sought cultural authority over sonorous matters, this did not inhibit philosophical inquiries into sound. Indeed, during the 1860s and 1870s, there was a revival of interest in acoustics. When, in 1873, John William Strutt (1842–1919), 3rd Baron Rayleigh, informed the Cambridge physicist James Clerk Maxwell (1831–79) that he was writing a treatise on acoustics, Maxwell suggested the title "Theory of Sound" and reflected that "You speak modestly of a want of Sound books in English. In what language are there such, except Helmholtz, who is sound, not because he is German but because he is Helmholtz. The next book is Herschel whom you may regard as a German." He advised Rayleigh to aim the treatise at "organists desirous of scientific knowledge."[1] Rayleigh embraced Maxwell's suggestion and published his seminal *Theory of Sound* in 1877 which became the standard textbook on the subject.[2] Nevertheless, Maxwell's endorsement of Herschel's "Sound" is revealing, echoing Faraday's own promotion of this treatise in 1835. For over forty years, Herschel's text had remained Britain's premier work on sound. However, all this changed in 1862 with the publication of physicist Hermann Helmholtz's (1821–94) *Die Lehre von den Tonempfindungen als physiologische Grundlage für die Theorie der Musik*, which initiated a revival of sonorous investigation throughout Europe. Helmholtz's work provided a physical definition of musical tones and a physiological explanation for their sensations.[3] Having begun his acoustic inquiries in 1856 with the aim of reducing the principles of musical harmony to a few basic laws and connecting the physical nature of sound with the aesthetics of music, Helmholtz taught that complex musical tones resulted from the combination of simple tones and harmonic overtones.[4] This study was part of a new mid-nineteenth-century understanding that musical systems (aesthetics) varied over time and between cultures. It was a reconciliation of the universal, law-like account of sound sensation with a belief in the musical aesthetic as a cultural product.[5]

In Britain, something of an acoustic renaissance followed Helmholtz's work. At the Royal Institution, John Tyndall (1820–93) orchestrated new sonorous research, culminating in the publication of his *Sound* in 1867. Airy published his mathematical *On Sound* shortly after, with Rayleigh's *Theory of Sound* following in 1877, and then Edmund Gurney's (1847–88) *The Power of Sound* in 1880. Not since the early 1830s had British experimentalists and natural philosophers paid such attention to the science of sound. In the early 1860s, Herschel's work provided a rare account of sonorous phenomena, but by the end of the decade, readers could consult Tyndall or Airy for English accounts of sound, while from the 1880s, they might also read Gurney, Rayleigh, or Ellis's 1875 translation of Helmholtz. Traditional historical narratives of the science of sound often portray celebrated works, such as Gurney's and Helmholtz's, as representing an expansion of sonorous knowledge in which acoustics became increasingly scientific.[6] There appears an almost Whiggish expansion of philosophical understanding surrounding the sonorous. As Steven Turner argued, Helmholtz's "research of the mid-1850s created the science of physiological acoustics in its modern form."[7] If we were to reconcile this with Herschel's lamentable experiences over pitch and harmony and Airy's difficulties in working with musicians, then we might conclude that while those of mathematic training and scientific knowledge did not exert authority over music, they did define it, but even this would be a reductionist analysis. Those claiming to be scientific did not possess any inherent authority to inform audiences over what sound was or why music was aesthetically pleasing. To provide physical explanations for music risked disenchanting it by removing its spiritual, supernatural quality, contributing to what Max Weber called the "disenchantment of the world."[8] The enrapturing melodies of the supernatural were becoming the rationalized product of harmonic laws and this, understandably, brought such accounts into conflict with traditional Church authorities.[9] As historian David Pantalony put it, "it seemed improbable, even offensive to some, that musical sounds could be analysed in the same way that a chemical compound could be reduced to elements."[10]

In nineteenth-century Britain, the leading protagonist to promote an apparently materialistic account of sound and music was Tyndall. He saw it as his duty to convince the British public that science, rather than religion, should instruct society in matters of natural philosophy. During the 1860s and 1870s he campaigned for scientists, rather than churchmen, to be acknowledged as the sole authorities on nature. For Tyndall, so much of nature was a mystery that he felt theology's claims to absolute knowledge were inconsistent with objective scientific investigation. He found miracles and the idea that prayer could produce physical effects particularly abhorrent, as such beliefs encouraged society to trust the clergy, rather than scientists, in moments of crisis.[11] Tyndall's primary aim was to undermine the Church's scientific authority and, when it came to acoustic phenomena and music, he had a point. In sermons and religious tracts there was much eagerness to provide spiritual accounts of how sound operated and metaphysical explanations of why music was euphonically pleasing. What Tyndall confronted was a religious consensus that music and sound were sacred.

Musical and sonorous phenomena were common subjects for nineteenth-century religious commentators. Most obviously, the place of music in worship was a source of continual discussion. However, there were also religiously informed readings of what the nature of sound, including speech and music, was. From dissenting Unitarians to well-to-do Anglicans, theological commentators endeavored to define the physical nature of sound and explain why music was aesthetically pleasing. This was the spiritual consensus that scientific materialists like Tyndall challenged when they provided increasingly physical explanations for such phenomena. And yet these religious authorities were often just as keen to fashion themselves as "scientific" as Tyndall. Many boasted impressive philosophical or mathematical credentials. They did not reject scientific accounts of sonorous phenomena but, on the contrary, mobilized them to explain how sound worked. They were, however, selective in who they cited, avoiding natural philosophers who appeared overtly antireligious. Importantly, while scientists of Tyndall, Rayleigh, and Helmholtz's caliber wrote for a very limited, specialist readership, clergymen directed their words to much broader social audiences. If we are to move away from historical accounts of nineteenth-century sonorous knowledge centered on scientific elites, then it is to sermons that we should look. These were perhaps the most shared cultural experiences for the Victorians, and it was to the Church, rather than to Helmholtz or Tyndall, that the majority of society looked for knowledge of sound and music. It is only by placing philosophical works on sound within these religious and musical contexts, where a broad public possessing varying degrees of education and expertise constituted a very different audience from those of scientific lectures and treatises, that we can fully understand the complexities of nineteenth-century authority within the natural sciences.

The Sound of John Tyndall

Throughout late 1859, religious, scientific, and musical authorities discussed the question of musical pitch in the Society of Arts' handsome neoclassical building on John Adam Street. In the midst of these meetings, just a few streets away, the publisher John Murray released a new work which was to have profound implications for all natural philosophy. On November 24, 1859, Charles Darwin's *On the Origin of Species* went on public sale. Several months later, at the 1860 BAAS meeting in Oxford, the challenge that this work presented to traditional Church authority in the natural sciences became very apparent. Within the newly built Oxford University Museum, a bitter dispute broke out between the bishop of Oxford, Samuel Wilberforce (1805–73), and Thomas Huxley (1825–95) which was to become iconic of religion's perceived retreat from natural philosophy. Wilberforce believed Darwin to be guilty of empirical inaccuracies but now found himself on the receiving end of an unexpectedly brutal attack from Huxley. Though the precise details of the disagreement are unrecorded, they contested man's relationship to monkeys.[12] It was not just Darwin's treatise that concerned religious commentators at this time. In March 1860, seven authors, six of whom were clergymen, published a volume of Biblical criticism. This work, *Essays and Reviews*, rejected the passive acceptance of Biblical doctrines unless validated by external evidence. A product of the liberal Anglican Church and including favorable references to Darwin and calls for a critical study of the Bible, it offended many conservative churchmen.[13] These were trying times for traditional Church authority. In this context of growing unease between religious and scientific communities, knowledge of sonorous phenomena did not escape scrutiny.

New evolutionary understandings of nature had ramifications for Victorian musical culture. The philosopher and composer Joseph Goddard (1833–1911) had published an evolutionary account of music, "The Moral Theory of Music," featured in *The Musical World* in March 1857, but it was Herbert Spencer's (1820–1903) article the following October in *Fraser's Magazine*, much of which was repeated in his subsequent *Origin and Function of Music*, which was most influential.[14] Sound and music also played a central role in Spencer's "Progress: Its Law and Cause," published in the *Westminster Review* in the same year, in which Spencer advocated an evolutionary account of nature and society, with everything in the universe developing from the simple to the complex. This progression was a universal law, stretching from astronomy and organic life to the human mind and society. Language was a prime example, with the lowest form of speech being the singular exclamations of animals. Music was also a product of this progressive evolution from the homogenous to the heterogeneous, with Spencer claiming that while "barbarous tribes" performed "monotonous chant," the Israelites had sung and danced with primitive instruments under Moses's direction following the defeat of the Egyptians at the Red Sea. Since then, music had continually developed into something complex and diverse.[15] Although ancient Greek music was more heterogeneous, Spencer continued,

> there existed nothing but melody: harmony was unknown. It was not until Christian church-music had reached some development, that music in parts was evolved; and then it came into existence through a very unobtrusive differentiation. Difficult as it may be to conceive *à priori* how the advance from melody to harmony could take place without a sudden leap, it is none the less true that it did so.[16]

Spencer believed that the practice of employing two choirs singing alternately the same piece had hastened this development. He argued that since this breakthrough the increase of musical heterogeneity had been undeniable, proven by the "many forms of sacred music, from the simple hymn, the chant, the canon, motet, anthem, &., up to the oratorio; and the still more numerous forms of secular music."[17]

Like Spencer, Darwin also found music an intriguing evolutionary subject, recording his initial ideas in 1838 and developing these into a theory of music in 1844, which he eventually published in his *The Descent of Man, and Selection in Relation to Sex* in 1871. This thesis rejected Spencer's claims that music had evolved from speech, instead contending that music was a separate evolutionary function of sexual selection, itself preceding language.[18] With Darwin and Spencer's evolutionary interpretations of sound, those convinced of music's metaphysical value could be forgiven for thinking that spiritual accounts of the sonorous were under attack. Increasingly, music was placed in what many interpreted as a materialistic vision of nature. It was John Tyndall, however, who was the foremost acoustic authority to fashion such an approach: he came to epitomize a growing scientific challenge to religious teachings over music's divine character. Along with Huxley and Spencer, Tyndall was a member of the notorious X Club, a London-based philosophical society which demanded scientists have sole authority over the investigation of nature.[19] Formed in 1864, the X Club was committed to developing naturalistic conceptions of man, nature, and society, which were consistent with recent scientific findings; these "scientific naturalists" believed that only natural laws, as opposed to those supernatural, operated in the world.[20] While not as hostile toward religion as Huxley or Spencer, throughout the 1860s and 1870s Tyndall cultivated what was widely read as a materialistic interpretation of nature, in which all phenomena could be attributed to physical laws. His own earlier investigations into sound and music very much informed Tyndall's natural philosophy.

Born in County Carlow, Ireland, Tyndall studied mathematics, physics, and chemistry at the University of Marburg from 1848, before working in Gustav Magnus's laboratory in Berlin in 1851. On returning to England to pursue a career in science, at the Royal Society he met Charles Wheatstone, who became an early supporter of the German-trained experimentalist. After delivering an authoritative Friday Evening Lecture in the summer of 1853, the Royal Institution offered Tyndall its prestigious chair in natural philosophy, which he held until 1887. Michael Faraday had been influential in this appointment, being impressed with Tyndall's experimental knowledge and practices, and they collaborated at the Royal Institution until Faraday's death in 1867.[21] The topic of Tyndall's career-establishing Friday Evening Lecture had been an acoustic one, examining the sounds produced through heated metals, which Trevelyan, Faraday, and Forbes had investigated during the 1830s. Tyndall had first approached this subject under Magnus in Berlin during 1853, before developing these experiments into a Royal Institution lecture in the same year.[22] He observed that a Saxon, M. Schwartz, had first discovered this sonorous phenomenon in a smelting works, during the process of solidifying silver by placing it on an anvil to hasten its cooling. Although Trevelyan had come across this phenomenon several years later, it was he who realized its importance for the study of heat as a demonstration of the transformation of energy into sonorous vibrations.[23] Tyndall outlined Forbes's theory that the vibrations resulted from temperature differences between two surfaces in contact: there was "a repulsive action exercised in the transmission of heat from one body into another which has a less power of conducting it." Forbes's belief that this repulsion was "a new species of mechanical agency in heat" had excited Tyndall during the summer of 1853 because it appeared to offer "a chance of becoming more nearly acquainted with the intimate nature of heat itself." However, Tyndall's subsequent inquiries undermined each of Forbes's laws regarding the production of tones from the movement of heat. Tyndall's production of a musical note from a sheet of hot iron on top of cold iron proved Forbes's first assertion, that vibrations never happened between two substances of the same nature, wrong. Forbes's second law, that only metallic substances produced tones, was equally erroneous, with Tyndall achieving sounds with heated metal on cold cut quartz as well as with rock, glass, and earthenware. His results also showed that Forbes's third and final law, that vibrations took place with an intensity proportional to differences in the conducting powers of the metals, to be equally false. Tyndall asserted that to produce a sound, the fundamental requirement was that heat caused an expansion in a very small space, at the point of contact with the colder material: it was vital that the edge in contact was thin and concentrated to produce the sonorous phenomenon.[24] He felt this evident from the comparison of effects of a heated silver crown on a cold silver rest. Usually this produced no effect, but if the coin's edge was hammered thin and put in contact, then sound was made.

Tyndall enjoyed demolishing Forbes's work, privately recording how he had made "utter smash of an investigation of a very eminent man," but eventually judged the subject valuable only as a Friday Evening Discourse. Nevertheless, when Tyndall demonstrated these experiments at the Royal Institution, he won a positive response, impressing its managers, Faraday, and Charles Anderson (1790–1866), the lecture-room assistant, who concluded that Tyndall "promises well." Evidently Tyndall's performance was not just about sonorous phenomena and heat but about building his own credentials in this prestigious location. Following this success, in January 1854, Tyndall revised his lecture for the Royal Society. He noted in his journal that he followed this paper with some experiments in the society's library, with the fellows "amused at the manner in which I have 'demolished Forbes,' as they express it. It is just what [Forbes] would like to do himself."[25] Two months later, Forbes wrote to Tyndall, expressing discontent with his own paper, but the exchange contributed to a growing animosity between the two natural philosophers. Tyndall included his performance of Trevelyan's experiments in his extensive *Heat: A Mode of Motion*, published in 1863, as an illustration of the "conversion of heat into mechanical energy."[26]

Most of Tyndall's later work on sound was conducted in his position as scientific advisor to the Elder Brethren of Trinity House. This institution, founded in 1514, was responsible for maintaining Britain's maritime navigation through an extensive network of lighthouses as well as sound signals for communicating through fog. For example, Faraday, having occupied the position before Tyndall, had advised on the use of electromagnetically operated fog bells during 1857, before touring the French coast in October 1859 to report on the effectiveness of fog bells in use between Calais and Boulogne.[27] Faraday performed several experiments with the three-cwt. fog bell on the western pier of Boulogne harbor, which had "a species of reflector consisting of Iron with a regular face of cement" to aid its projection. He had the bell sounded while he was in a small boat outside the harbor and recorded its performance under varying conditions. He found that the signal was subject to great interference and could be heard with differing degrees of intensity at various locations.[28] Nevertheless, Trinity House wanted Faraday's guidance on the installation of a fog bell on Start Point in Devon, proposing a sounding board be erected over and around its landward side and intended to project sound up to three miles out to sea. Faraday agreed that such a scheme might work and recommended plate iron for the soundboards, as opposed to corrugated iron which "might interfere with the numerous vibrations."[29]

Building on Faraday's inquiries, Trinity House commissioned Tyndall to compare how sound traveled through fog, rain, and snow.[30] Just east of Dover, he conducted experiments on an array of sound signals, including guns, horns, steam-driven whistles, and an American steam siren.[31] He tested these various signals positioned at 235 and then 40 feet above sea level and took recordings off the coast aboard the paddle steamers *Irene* and *Galatea*. The results were confusing, with the gun occasionally traveling further than the siren; his observations suggested that, contrary to common thought, fog and rain conveyed sound effectively. Indeed, on July 3, 1872, Tyndall found sound signals traveled poorly over a calm sea, so the unreliability of sonorous communication was not determined by precipitation, contradicting Herschel's claim that over "smooth water . . . sound is propagated with remarkable clearness and strength."[32] Tyndall accounted for this by explaining that "Beams [of heat] of equal power were falling on the sea, and must have produced copious evaporation. That the vapour generated should so rise and mingle with the air as to form an absolutely homogeneous medium was in the highest degree improbable." This effect produced partial echoes within the atmosphere, he argued, as hot air rose "in invisible streams, breaking through the superincumbent air now at one point, now at another, thus rendering the air *Flocculent* with wreaths and striae. At the limiting surfaces of these we should have the conditions necessary to the production of partial echoes and the consequent waste of sound."[33]

Tyndall resumed these sonorous trials at Greenwich and Westminster, measuring how the sound of the Palace of Westminster's bells traveled over the Serpentine at night.[34] He then attempted to replicate this phenomenon under controlled laboratory conditions in the Royal Institution, examining how the sound of a high-pitched whistle moved through sheets of descending heavy gas and layers of ascending heated air. Tyndall placed a sensitive flame on the opposite side of the laboratory from the whistle and measured the communication of sound by its disruption of the flame, finding that the layers of air and gas prevented the flame from being affected. The conclusion from this was that maritime sound signals were contingent on the homogeneity of air. More importantly, though, Tyndall's laboratory trials confirmed that alternating sheets of denser and lighter air caused an echo, with the sound reflected between these atmospheric variations, from which he deduced that mariners were mistaken in their beliefs that clouds reflected sound. Rather, a sound's echo was contingent on variations in density, or the formation of what Tyndall referred to as "invisible acoustic clouds." He concluded that often during his experiments, "the echoes reached us, as if by magic, from the invisible acoustic clouds with which the optically transparent atmosphere was filled. The existence of such clouds in all weathers . . . is one of the most important points established by this inquiry."[35] Tyndall delivered these results to the Royal Society in 1874, emphasizing the urgency of his investigation into sound's velocity over sea: in the previous ten years at least 273 ships had been lost off the coast of Britain due to poor visibility.[36] In this context of maritime disasters, sound knowledge took on considerable urgency.

As well as sonorous investigations, Tyndall repopularized sound as a lecture subject at the Royal Institution, conceiving of new experimental displays to illustrate acoustic phenomena. Helmholtz's 1862 *Tonempfindungen* convinced Tyndall to publish these lectures in 1867 as *Sound: A Course of Eight Lectures Delivered at the Royal Institution of Great Britain*. Demonstrations were central to both Tyndall's lectures and their following publication, with the subject "treated experimentally throughout" so as to make it accessible to those "who do not possess any special scientific culture." In his opening lecture on sound's propagation, Tyndall illustrated how sound waves traveled as impulses, with each particle moving only slightly, through the game of solitaire, which consisted of a row of glass balls resting in a groove (fig. 5.1). When a ball with force was added, the impulse traveled through the row, moving only the final ball. He then scaled this up by lining five assistants in a row, touching each other so that when he pushed the first, the energy was passed through each in turn without moving them, with only the final assistant falling over (fig. 5.2). These novel arrangements helped to make sound visible in new ways. In

FIGURE 5.1 Tyndall's example of a game of solitaire showed how a sonorous impulse moved through particles of air. Tyndall, *Sound*, 3. Public domain.

FIGURE 5.2 Five assistants provided Tyndall's second demonstration of the communication of a sonorous impulse. Tyndall, *Sound*, 4. Public domain.

his second lecture, as part of his efforts to relate the physical processes of sound with the emotional effects of music, Tyndall produced experiments to show optically the vibration of musical notes. He placed a thin strip of sheet copper on the tip of a tuning fork, which he put into vibration with a bow and then used this to mark the path of the vibration on a piece of smoked glass. The effect could also be achieved with chalk on a blackboard.[37]

Along with the oscillation of a tuning fork, Tyndall delivered a thorough analysis of the vibration of a string, including by analogous demonstrations with waves in a water tank and by placing paper "riders" on a vibrating string to reveal nodal points (fig. 5.3). He repeated Wheatstone's "beautiful experiment" of conveying the music of a piano being played

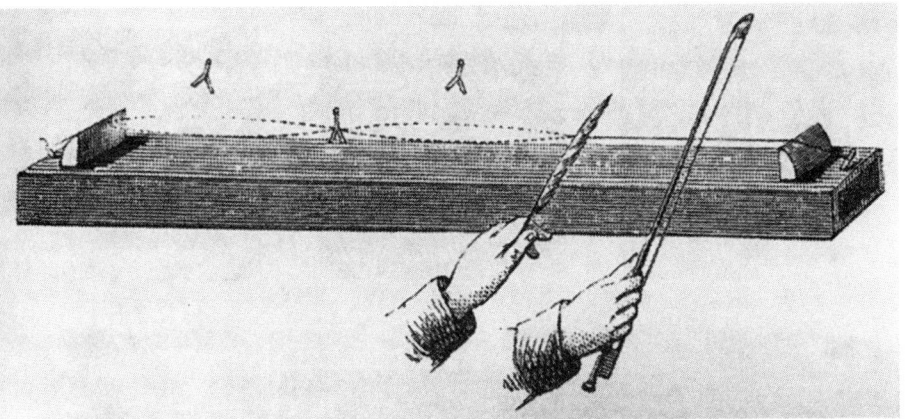

FIGURE 5.3 Tyndall's use of paper "riders" to illustrate the nodal points of a vibrating string. Tyndall, *Sound*, 103. Public domain.

two floors down into the lecture theater: using a tin tube protecting a series of deal rods in contact with a violin's soundboard, he produced the phenomenon illustrated by the Enchanted Lyre in the same venue in 1828. In his fourth lecture, Tyndall introduced Wheatstone's kaleidophone in his analysis of the transverse vibration of rods, plates, and bells (fig. 5.4). After producing several pleasing effects, he conducted trials with Chladni's acoustic plates.[38] His fifth lecture discussed resonance, with Tyndall repeating Wheatstone's experiments on the phenomenon and also presenting an account of Wolfgang von Kempelen and Robert Willis's work to replicate human speech.

In his final lecture, Tyndall examined musical intervals and the relations between different ratios of vibrations. This subject had two sides, "one physical, the other aesthetical," with the link between ratios and euphonically pleasing sounds long known; Tyndall observed that these ratios were not "determined by scientific knowledge" but "chosen empirically, and in consequence of the pleasure which they gave, long before anything was known regarding their numerical simplicity."[39] Considering why smaller ratios of vibration were aesthetically superior to those more complex, Tyndall suggested two solutions: the first "metaphysical" and the second "physical." Pythagoras provided the metaphysical; he supposed that as there were seven notes in a scale, the intervals expressed

> the distances of the planets from their central fire; hence the choral dance of the worlds, the "music of the spheres," which according to his followers, Pythagoras was the only mortal privileged to hear. And

FIGURE 5.4 The figures produced by the movement of the vibrating rods of Wheatstone's kaleidophone, exhibited during Tyndall's Royal Institution lectures on sound. Tyndall, *Sound*, 135. Public domain.

might we not in passing contrast this glorious superstition with those that have taken hold of human fantasy in our day? Were the character which superstition assumes in different ages, an indication of man's advance or retrogression, assuredly the nineteenth century would have no reason to plume itself in comparison with the sixth B.C.[40]

This relation between musical ratios and *"pleasure"* had therefore been a concern since ancient times and ascribed with celestial causation. Nevertheless, the physical explanation for dissonance and consonance was, Tyndall claimed, now at hand thanks to Helmholtz, who argued that dissonance was the result of a rapid succession of beats on the ear. Slower beats were less disagreeable, according to Helmholtz, who emphasized the importance of overtones to a vibration's pleasing sonorous qualities; a pure sound would be dull and flat, so to be euphonically satisfying required combinations of overtones and fundamental tones. This took Tyndall and his audience "to the verge of the Physical portion of the science of Acoustics."[41] The question that remained, then, was whether music possessed any characteristics beyond those purely physical. In other words, what was the metaphysical worth of the sonorous? And it was this problem that would draw Tyndall into religious controversy over the nature of sound.

Acoustics in Tyndall's Materialism

Tyndall's reference to the "superstition" of the nineteenth century was not a mere passing remark. Beyond his growing reputation as an experimentalist, Tyndall was best known to contemporary Victorian audiences for his criticism of Christianity and apparently materialistic interpretation of nature. Indeed, Tyndall's first point in his lectures on sound was that, like all sensations, it was the product of "molecular motion," with varying movements, such as those optical or sonorous, acting on different nerves.[42] Between 1860 and 1875 sound and music contributed to Tyndall's materialistic account of nature, and nowhere was this more obvious than in his claims on the role of imagination in science. This was already evident in his article "Physics and Metaphysics," featured in the *Saturday Review* in August 1860. Here he reviewed how light and heat could both be understood as mechanical phenomena in the same way as sound. While sound involved vibrating air, these other phenomena were, he contended, the vibrations of an invisible ether. As all human knowledge of the external world consisted of "an interchange of motion" detected by sensory organs, it was probable that emotion was, likewise, materially stimulated: sound provided a prime example for this. Tyndall described the journey of sonorous impulses from a vibrating body to the ear and auditory nerve. This was purely "the excitement of the nerves," yet such vibrations affected consciousness. Music provided Tyndall with an example of how this material action could be confused for spiritualism, there being occasions when human emotions might "be taken for a direct spiritual emanation entirely independent of 'brute matter.'"[43] Tyndall employed his knowledge of sound to demonstrate this. He defined how notes were produced from musical instruments; these were very material actions. Helmholtz had showed that

> certain incidental notes commingle in each case with the principal one, and produce a composite result. The "harmonics" of a string are known to be due to minor vibrations which superpose themselves upon the principal ones, as small ripples cover parasitically the surfaces of large sea-waves. The notes of the true simple wave and its parasites are heard at once, and it is the variation of the latter which produces differences in the *timbre* of a musical instrument or of the human voice.[44]

Here was a material explanation for the quality of a note, illustrating the mechanical processes of harmonic waves. Tyndall then asserted that such harmonics, in a church, could affect the emotional state of a congregation or worshipper. A woman praying might suddenly appear overcome with radiance, her "soul ... shining through her features," with her emotions miraculously altered, yet this was no miracle but had "a proximate mechanical origin." Her moment of divine feeling, Tyndall asserted, was merely a sensual response to a material process. In church, as in all places, "the expression which is to stir the soul, to kindle love or quench it, exists in space as a purely mechanical affection of matter."[45] As all thoughts were products of physical processes, Tyndall argued that there was nothing divine about music or its emotional power.

These questions of sense and thought had implications for the production of knowledge. Imagination was central to Tyndall's epistemological approach, as it was only through such creativity that a natural philosopher could envisage the original formation of life. Similarly, imagination was crucial to understand the vibrations of heat and light or the existence of unobservable atoms. Man's imaginative powers reconciled what could be known from the senses with a broader belief in the unity of nature.[46] In his address to the Mathematical and Physical Section of the BAAS at Norwich in 1868, Tyndall developed these arguments, stressing how mathematics and physics were inseparably bound because, "no matter how subtle a natural phenomenon may be, whether we observe it in the region of sense, or follow it into that of imagination, it is in the long run reducible to mechanical laws." Once again, to these material mechanical forces, Tyndall ascribed the production of consciousness. Love, for instance, was the "motion of molecules of the brain," probably in the opposite direction of molecules producing the emotion of hate. Though unobservable, both were material processes, but the question of "why" was unanswerable and, for Tyndall, marked the limits of scientific knowledge.[47] Tyndall's address sparked much criticism. In a sermon attacking the speech, the Scottish minister and theologian Joseph Taylor Goodsir (1815–93) argued that Tyndall had unwisely moved into a metaphysical discussion. While agreeing with Tyndall's "scientific statements," Goodsir denounced "the entire philosophic and religious light" of his observations. Regardless of forces and molecules, Goodsir's view of the cosmos was built on the "ancient . . . eternal basis" of the Bible, which was evidence of a Creator. He felt Tyndall's understanding of the universe was poor in comparison to Whewell's *Astronomy and General Physics* and Herschel's *Preliminary Discourse*. Based on Herschel's claim that knowledge was limited by the senses, Goodsir summarized that "the eye is for seeing; and no better nor more clearly, no otherwise than it will permit, can a man see. The ear is for hearing; and no better nor more audibly, nor otherwise than it will permit, can man hear. Thus doth God rule over the bodies of man."[48] God governed the senses and ordered human imagination: both were beyond purely material explanation.

Undeterred, Tyndall continued this discussion in his address, "Scientific Use of the Imagination," to the 1870 BAAS meeting in Liverpool, as he again employed sound to define the relationship between thoughts and material processes. To move between sensory experiences and the construction of general laws of nature required imagination, "the mightiest instrument of the physical discoverer."[49] Tyndall produced several examples to illustrate this creative power, the first and most prominent of which was of acoustic knowledge. The observation of a drop of rain falling into a pond and producing outwardly expanding ripples could help a philosopher envisage how sound traveled, with a sonorous impulse comparable to this watery motion thanks to man's intellectual faculties. From this, Tyndall continued, it was possible to understand how sound moved through air and material bodies. As he put it,

> We know the phenomena and laws of vibrating rods, of organ-pipes, strings, membranes, plates, and bells. We can abolish one sound by

another. We know the physical meaning of music and noise of harmony and discord. In short, as regards sound we have a very clear notion of the external physical processes which correspond to our sensations.[50]

In the formation of this body of knowledge, natural philosophers required a combination of sensible experience and imagination, given that the eye "cannot see the condensations and rarefactions of the waves of sound. We construct them in thought, and we believe as firmly in their existence as in that of the air itself." Furthermore, by imagination, knowledge of sound could become an understanding of other natural phenomena. "Having mastered the cause and mechanism of sound," Tyndall contended, it was possible to comprehend the mechanism of light and "extend our enquiries from the auditory nerve to the optic nerve." He then proceeded to account for the similarities and disparities between light and sound undulations, both purely mechanical forces that necessitated a "definite, tangible, vibrating body." The same seemed true of all forces. Tyndall concluded that the "scientific imagination, which is here authoritative, demands as the origin and cause of a series of ether waves a particle of vibrating matter quite as definite, though it may be excessively minute, as that which gives origin to a musical sound. Such a particle we name an atom or a molecule."[51]

Tyndall's religious views have been much debated, with reductionist efforts to label him as an atheist, agnostic, or pantheist failing to provide a sufficient explanation of his beliefs. While his parents were Irish Protestants with Quaker roots and Methodist sympathies, Tyndall sometimes lamented his own loss of faith, was sensitive to the spiritual dimension of religion, and maintained a constant dislike of Catholicism.[52] Stephen Kim argued that Tyndall was actually a man of faith but rejected the cultural authority of the Church of England, demanding it keep out of scientific questions and resenting the privileged position from which clergymen spoke on matters of nature.[53] Furthermore, Kim claimed that Tyndall cultivated a framework of "transcendental materialism," defining this as the combination of a materialistic interpretation of nature with an idealistic belief in the spiritual.[54] Unlike Huxley and Spencer, Tyndall often stressed the importance of faith and, while opposing organized religion, did not reject God. Likewise, historian Ursula DeYoung has shown how Tyndall emphasized that science could not fully explain everything in nature. Along with Faraday's influence, Thomas Carlyle (1795–1881) and Johann Wolfgang von Goethe (1749–1832) both encouraged Tyndall's wondrous interpretation of the natural world. Carlyle had called for a national rebirth in response to what he believed was a devolution of Victorian society following the loss of its traditional medieval structure, which found sympathy with the scientific naturalists of the X Club. Tyndall was especially taken with Carlyle's writings, and his early awe of the philosopher amounted to something nearing hero-worship. Following his own loss of belief in established protestant religion, Carlyle's insistence that there was an unknowable, yet undeniable, superior intelligence appealed to Tyndall. Yet Tyndall remained adamant that nature should be studied without religious interference, despite acknowledging there were questions beyond science which were mystical.[55]

While Tyndall's early BAAS discussions were controversial, it was his presidential speech to the 1874 BAAS meeting in Belfast which caused the greatest sensation. Though very similar to his earlier papers, as president, his address reached much broader audiences. Again, Tyndall claimed that nature acted through particles beyond the capabilities of human vision, echoing Maxwell's 1873 BAAS paper on the conservation of energy, but he now rejected Maxwell's assertion that the existence of apparently manufactured atoms suggested the existence of a "Maker." Tyndall instead claimed all sensation to be molecular and all beauty in nature attributable to natural selection, as Darwin had shown regarding the attraction of insects to brightly colored flowers. He emphasized the importance of the doctrine of the conservation of energy, which projected a vision of the universe as a balanced system of indestructible matter, and he extended this material account to the senses and thought. Organs had developed from vague matter into complex mechanisms for receiving sensations through experience: both the eye and the ear were the products of the slow learning of tissue. Likening this process to a musical education, Tyndall invited his audience to consider "the musician, who by practice, is enabled to fuse a multitude of arrangements, auditory, tactual, and muscular, into a process of automatic manipulation. Combining such facts with the doctrine of hereditary transmission, we reach a theory of instinct." The ear, then, had evolved in a manner comparable to the cultivation of a musician. As Spencer taught, the brain's appreciation of music was an evolutionary product, developing over time in response to experiences. This idea carried racial implications, with Tyndall alleging that through experience, "it happens that faculties, as of music, which scarcely exist in some inferior races, become congenital in superior races." In conclusion, Tyndall declared that life was "developed out of matter" and that present scientific knowledge was "in favour of 'spontaneous generation.'"[56]

There was a context to Tyndall's materialist claims. After Huxley's presidential speech in 1870, widely interpreted as inherently materialist, the following three presidential addresses were decidedly antimaterialist, with William Thomson in 1871 and the physiologist William B. Carpenter in 1872 both referring to a Creator, while the chemist Alexander Williamson criticized efforts to reduce all nature to inanimate matter when he served as president in 1873. So when Tyndall spoke in 1874, he wanted to break this succession of moderate presidential addresses which promoted nature as divinely created. As much as with Anglican churchmen, promoters of Darwinian evolution, like Tyndall, found themselves in confrontation with what Crosbie Smith has described as a "North British" group of mathematicians, physicists, and engineers, who radically reformed the physical sciences around the concept of energy. From the early 1850s William Thomson and Macquorn Rankine, joined later by Scottish natural philosophers Maxwell, Peter Guthrie Tait, and Fleeming Jenkin, developed a new "science of energy," eventually known as thermodynamics, in which energy and its conservation replaced traditional understandings of a universe governed by mechanical forces. As Smith puts it, "the universe would now be understood neither in terms of action-at-a-distance forces nor in terms of discrete particles moving through void space, but as a universe of continuous matter possessed of kinetic energy." This science of energy was very much a product of this North British network, developing within a Scottish Presbyterian framework. With the Scottish Free Kirk increasingly tending toward Biblical literalism after the death of its leader Thomas Chalmers in 1847, and the growth of evolutionary materialism and Biblical criticism following the anonymous publication of *Vestiges of the Natural History of Creation* in 1844, promoters of this science of energy represented a more "moderate" Presbyterian interpretation of nature. Rather than a god of power, such as Faraday conceived, at the center of the science of thermodynamics was a divine architect. The two guiding laws of this new science were that, primarily, all energy in the universe was constant and transformable and, secondly, that the total amount of transformable energy was finite and in decline, as the universe gradually cooled toward its inevitable death. Given the universe's eventual termination, it seemed logical that it had had a beginning and therefore a Creator who had brought molecules into existence.[57] As Maxwell surmised at the 1873 BAAS meeting in Bradford, while the properties of molecules did not change, "the exact equality of each molecule to all others of the same kind gives it . . . the essential character of a manufactured article, and precludes the idea of its being eternal and self-existent." Furthermore, Maxwell surmised that science was "incompetent to reason upon the creation of matter itself out of nothing.

We have reached the utmost limit of our thinking faculties when we have admitted that because matter cannot be eternal and self-existent it must have been created."[58]

This scheme of nature rivaled the scientific naturalism of Tyndall and the X Club. In his treatise, *Heat*, Tyndall ignored the second law of thermodynamics, that the amount of transformable energy in nature is always diminishing, because it implied the universe's certain end. Instead, Tyndall focused on the first law of thermodynamics, that of conservation, in which all transformable energy was finite. Revealing Tyndall's unease over the cosmological ramifications of the dissipation law, he preferred a cyclical account of nature. In part, this was because Tyndall feared the principle of energy entropy represented an attempt to reintroduce a Biblical creation-to-apocalypse narrative into science. In contrast to Thomson's religious commitment to an end of the world, Tyndall favored Immanuel Kant's model of an infinite universe, as presented in his 1755 *Allgemeine Naturgeschichte*, in which gravity created planetary systems and celestial collisions sustained a cyclical process of renewal. An end suggested a beginning, and therefore a Creator, which meant that materialist philosophy was dependent on a cyclical account of the cosmos. Taking the organic processes of life and death, in which energy was recycled, materialists such as Tyndall supposed that while energy might be lost in one place, it was created in another part of the universe.[59]

This challenge of reconciling the conservation of energy with an evolutionary universe was prominent throughout Tyndall's Belfast address, but the degree to which the speech was materialistic has been debated. DeYoung asserted that Tyndall was exercising science's cultural authority, as opposed to religion's, and that his address was not a materialist manifesto.[60] Similarly, Ruth Barton concludes that Tyndall's Belfast address was an attempt to contextualize materialism within his own idealistic metaphysics. While religious authorities certainly interpreted the address as an atheistic attack on Christianity—and Tyndall was advocating an account of nature in which all could be reduced to mechanical laws—Tyndall also accepted that such knowledge was limited and some things had to remain mysterious.[61] However, regardless of Tyndall's intentions, his speech provoked an aggressive response from the pulpit. William Binns, preaching a sermon at the Unitarian Church in Birkenhead, accused the BAAS president of making "the blunder of treating as rubbish things that were divine." Finding Tyndall less unpleasant than the morally repugnant Huxley, Binns could not agree with his "theory of materialism" which was but "a guess at truth."[62] Likewise John Quarry, rector of Donoughmore, rejected Tyndall's suggestion that all was lifeless matter, believing this destroyed all hope for humanity. Quarry lamented that so many were "deeply imbued, with the materialistic notions to which our modern scientific men have shewn so great a tendency." Yet he was cautious in labeling Tyndall a "materialist," speculating that the scientist did "not feel satisfied that in the atomic constitution of matter he has a sufficient account of the mystery of all mundane existence." Quarry was untroubled with much of Tyndall's interpretation of nature, because the fundamental question of Creation remained unanswered. At the same time, he thought materialism failed to truly explain self-consciousness and reason, with man's emotions beyond pure matter. Quarry maintained that "to assume that sensation and perception, thought and conscience and will are the outcome of mere material evolution is . . . an irrational inversion of the order of cause and effect." Humans experienced the world through sensations, but imagination could not be explained simply as a product of matter.[63]

Responding to these criticisms in 1875, in the *Fortnightly Review*, Tyndall directed his defense of materialism against James Martineau (1805–1900), the Unitarian principal of Manchester New College in London, who had rejected Tyndall's Belfast claims in an address of his own. Speaking to Manchester New College in 1874, Martineau had boasted sympathy with science but refuted the doctrine of the conservation of energy. Tyndall countered this by arguing that everything could be explained through molecular processes, including the brain's activities, because all force, including heat and light, consisted of matter.[64] If anyone doubted this, he suggested the example of an acorn's transformation into a tree by the action of solar light and heat. Tyndall here resorted to his considerable acoustic experiences, invoking the Enchanted Lyre and the motion of sonorous vibrations to illustrate this example, as he recalled Wheatstone's experiment,

> where the music of a piano is transferred from its sound-board, through a thin wooden rod, across several silent rooms in succession, and poured out at a distance from the instrument. The strings of the piano vibrate, not singly, but ten at a time. Every string subdivides, yielding not one note, but a dozen. All these vibrations and subdivisions are crowded together into a bit of deal not more than a quarter of a square inch in section. Yet no note is lost. Each vibration asserts its individual rights; and all are, at last, shaken forth into the air by a second sound-board, against which the distant end of the rod presses. Thought ends in amazement when it seeks to realise the motions of

that rod as the music flows through it. I turn to my tree and observe its roots, its trunk, its branches, and its leaves. As the rod conveys the music, and yields it up to the distant air, so does the trunk convey the matter and the motion—the shocks and pulses and other vital actions—which eventually emerge in the umbrageous foliage of the tree.[65]

As music had moved through Wheatstone's deal rods, so heat and light moved through growing wood in nature. For Tyndall, this was evidence of the staggering "potency of matter," equally capable of growing acorns or communicating sound. He declared the scientifically minded to be

> conscious of a music subtler than that of the piano, passing unheard through these tiny boughs, and issuing in what Mr Martineau would opulently call the "clustered magnificence" of the leaves. Does it lessen my amazement to know that every cluster, and every leaf—their form and texture—lie like the music in the rod, in the molecular structure of these apparently insignificant stems? Not so.[66]

The movement of sound was in this way directly analogous to light and heat. However, unlike Whewell, Herschel, and Somerville's religiously informed accounts of a united nature, Tyndall employed a decidedly materialistic framework.

Tyndall proceeded to unite the inorganic (the sea, sun, and rain) and organic (plants and animals) in nature within the analysis of material forces, asserting all botanic and animal growth to be the result of natural forces, or molecules, at work. While the precise movement of molecules within a growing acorn were unknowable, the movements of a vibrating rod were examinable by both ear and, through clever experiments, by eye. Like the acorn, the human egg supported Tyndall's assertions. No more than 1/120th of an inch in diameter, the egg was "matter," yet through molecular action could develop into a complex network of organs. The ear was a good example of this, consisting of a "tympanum, cochlea, and Corti's organ—an instrument of three thousand strings, built adjacent to the brain, and employed by it to sift, separate, and interpret, antecedent to all consciousness, the sonorous tremors of the external world."[67] This had all been organized, Tyndall claimed, through the "mysterious" processes of matter.

Tyndall's "Materialism and Its Opponents" aroused further hostility, with an anonymous clergyman alleging in 1879 that Tyndall wanted to sever all links between nature and scripture. It was impossible, the clergyman alleged, to attempt to separate faith in matter from faith in God: they were the same, as all natural laws were decreed by divine intelligence. Beyond heat, light, and sound, there was another "invisible immaterial energy" that Tyndall had overlooked, and this was that of the soul. In describing how an acorn became a tree due to the material action of light and heat, Tyndall ascribed too much potency to inert matter. Rejecting Tyndall's analogy between the sonorous vibrations of deal rods and the material growth of acorns, the clergyman provided the alternative comparison of building a house. This work could not be a purely material action but required both the materials of construction and intelligent design. Tyndall's vision of a world of "chaos of matter, of hopeless, hapless, unprogressive, dead materials," was ignorant of the mystery of Creation: not all was reducible to vibratory forces.[68] Away from Tyndall's scientific networks and readership, his purely physical account of sound's power encountered substantial resistance.

Spiritual Sounds and Material Melodies

While Tyndall lectured to fashionable London audiences at the Royal Institution and wrote for philosophically inclined readers, pulpits and sermons remained leading places of instruction on matters musical for churchgoing society. Religious authorities continued to deliver rival sources of sonorous knowledge throughout the 1860s, 1870s, and well into the twentieth century, speaking to more socially diverse audiences than Tyndall and other scientific authorities writing in treatises. From well-to-do gentry to illiterate laborers, the sermon remained an accessible source of knowledge of nature. As much as Tyndall might shed light on sound's material properties, ministers of varying Christian denominations continued to offer spiritual lessons on the divine nature of sound and music's metaphysical value. A robust religious consensus surrounding this idea of music's sacred character continued throughout the nineteenth century. Especially influential was Reverend Hugh Reginald Haweis's (1839–1901) popular *Music and Morals*, published in 1871 and in print through sixteen editions until the 1930s. Haweis alleged that music was fundamentally connected to the emotions, having a unique power to alter human consciousness.[69] These sentiments made a common subject for sermons. On the opening of a new organ at the Holy Trinity Church at Burton-upon-Trent in October 1872, for instance, W. Weldon Champneys, dean of Lichfield, marveled at He who had created the ear in his sermon, "The Immortality of Music." Champneys explained how sound traveled in waves through the air, how the waves were received by the ears and then conveyed to the brain. Not only was all of this divinely created, but so too was music's power to evoke sensations and pleasure. As Champneys put it,

> wonderful it is that the *air* which is our *life* should be also the instrument of our pleasure. If there were no air, there would be neither *life* nor *music*. The air, which we cannot see, is that which, put into motion by the string or pipe, makes those sounds, which when they come in a single chain of notes are *melody*, and when in notes that agree is *harmony*.[70]

Likewise, when Canon F. W. Farrar spoke on music and religion at Westminster Abbey in July 1882, he claimed that there was "little of what can accurately be called music in Nature; for music is the divine prerogative of human and angelic beings; and Nature furnishes only the rude elements of it," these being sonorous phenomena. Music was the opposite of everything that was "loose, ill-regulated, disorderly" and revealed "the beauty and order of the Universe."[71] Several years later, in 1890, the precentor John H. Mee attributed a sacred value to the musician, above that of the scientist, claiming that the "achievements of the intellect in metaphysics and in science are undoubtedly very wonderful indeed, but it is the artistic powers of the mind that have the noblest field, viz., man himself. To influence the actions and thoughts and feelings of his brother man is the task of the Poet, the Orator, the Painter and the Musician."[72] For all of Tyndall's materialism, the religious consensus that music was of metaphysical value endured, echoing the sermons of the mid-nineteenth century and the writings of Somerville and Whewell.

Yet even the persisting authority that religious commentators wielded over the spiritual character of music was under increasing pressure by the late 1870s. In 1878, Henry Charles Lunn, a member of the 1859–60 pitch committee, picked up on these growing tensions in a lecture delivered at the Royal Academy, promoting the establishment of a national school for the training of vocalists. He wanted musicians to pay greater attention to empirical practices and "Modern Science," emphasizing that science and art were mutually beneficial to each other, because both involved the study of divine work.[73] Art and science observed

> nature from different and distinct points of view, yet both may gather and display the evidences of the same truth; for surely the principle of order which the man of science searches out, are equally and alike the issue of Divine law.[74]

Art, as much as science, had to be cultivated to build knowledge of God, and music was, according to Lunn, the most valuable of all arts for revealing nature's beautiful order. Nevertheless, the art was to first be studied materially, given that

> Music *per se* is not a spiritual thing; it is not even a human thing. It is a quite material thing, independent of man in its eternity.... Invisible, yes, but sternly material; and the spiritual parts of ourselves can never, at its best, be *transfused* there through, until we know and obey the material aspect of inorganic form.[75]

It was essential, Lunn continued, to treat the art of music as something material, rather than spiritual, to ensure aesthetic quality. Nevertheless, he recognized that music was unlike the visual arts because it lacked natural standards. In painting, for example, there were always visible standards in nature as a means of comparison, separate from man's judgment. In music, however, it was unclear what the standard should be. Indeed, this had led to a division over how to pursue the art, between "Purists," who aimed for mechanical accuracy, and "Sensationalists," who aimed for feeling. Yet any science of music would, Lunn contended, have to combine feeling with mechanical accuracy, uniting the head, heart, and hand in its construction, conception, and execution. He hoped that one day an institution, comprised of "Physiologists, Acousticians, Physicists, and Musicians," would teach and develop vocal music, uniting the technical and aesthetical. Lunn invited "scientists to join with us in friendly aid; then we may unite in raising the masses from experiment to certainty, from fancy to fact, from personal taste to a knowledge of the irrefutable laws organized in Nature by the God who made it."[76] To safeguard the future of music, it was important that it be judged both in terms of the spiritual passions it evoked and its employment of material laws. In conclusion, Lunn surmised that

> We have a Carpenter, a Huxley, a Spencer, a Bain, and others, all of whom have told us somewhat of the human mind. Surely these men are worthy of attention, and their thoughts worthy of application in so subtle a region as that of invisible sound. It is to these mental philosophies we must look for supplying to us the counterpart in the *mind* of what Acousticians, such as Prof. Helmholtz, have supplied to us in atmospheric force,—this, the outside acting agent; that, the mind, the receiving faculty: the two rightly associated making the highest Art-result.[77]

In this way, Lunn promoted a material understanding of music, yet devised this through a framework in which the spiritual was still central.

Two years later, Edmund Gurney (1847–88) raised even more troubling questions over music's physical and aesthetic nature in terms of its mystery and spiritualism. Gurney had attended Trinity College Cambridge from 1866, coming fourth in the Classical Tripos Examinations, and had been appointed a college fellow in 1872. Following the deaths of his parents during his childhood and then the tragic drowning of his three sisters in a boating accident on the Nile in 1875, Gurney focused his attention on the psychology and philosophy of music. After studying medicine at University College London, he returned to Cambridge and published *The Power of Sound* in 1880, which argued that musical excellence was understood by a special faculty of musical intuition, unique from other emotions.[78] He explained that since Helmholtz's *Tonempfindungen*, the "indispensable *material* of musical phenomena" had been known and "widely popularized," but little had been done "to apply scientific treatment to the musical phenomena themselves." It was this ground between the physical and the aesthetic that interested Gurney, but he was aware that such work appeared threatening to those who regarded music as something divine. He recognized that many an "ardent believer in the spiritual character of the art" would "fear to find the domain of genius measured by mechanical rules, or the feelings whose indescribable and mysterious nature no one, I think, can have realised more deeply than myself, docketed off under cut-and-dry psychological formulae."[79] Gurney did not want to unveil all harmonic mystery. Indeed, part of his analysis was to "define the boundary of the vast-region" that was beyond scientific understanding, and his conclusion included an acknowledgment of "the hopelessness of penetrating Music in detail."[80] Yet he was keen to uncover the relationship between music and its obvious influence over human emotion.[81] As Gurney put it, "the effect of Song on the masses is like a glimpse of infinite spiritual possibilities; and owing to the fewness of the moments where even the suggestion of a universal kinship in lofty sentiment appears possible, such occasions seem to have a very singular and impressive significance in human life."[82] Gurney's work presented a direct challenge to religiously informed readings of music as a metaphysical entity.

Nor was this growing controversy over what music was, its relationship to the human emotions, and its ability to transcend the purely material confined to academic disputes and elite scientific treatises. Following the failure of the National Training School for Music, established in 1873 to provide free musical training to exceptional students, in 1882 Prince Leopold, Duke of Albany, delivered a speech at a meeting in Manchester Free Trade Hall, calling for the establishment of a new college of music with royal patronage. This new institution found support from the Duke of Edinburgh and Prince Christian of Schleswig-Holstein, with each delivering speeches following Leopold's address. Yet it was Leopold, Queen Victoria's fourth and final son, who drew on a question which plagued music throughout the 1860s and 1870s: was it a series of mathematical laws, or was it a divinely ordered spiritual phenomenon? Leopold had more than a passing interest in the art. A hemophiliac and epileptic, he had attended Christ Church College Oxford, attending lectures on history, music, and fine art as well as the sciences. After graduating in 1876, he set about promoting the arts, serving as the president of the Royal Society of Literature from 1878 and vice-president of the Society of Arts from 1879, and he began to advocate for the establishment of a national conservatory of music.[83] Not long before his early death in 1884, Leopold's proposal was realized as the Royal College of Music at South Kensington, established by Royal Charter in 1882.[84] In front of his Manchester Free Trade Hall audience, which included Charles Hallé, Leopold expressed his own conviction that music was a fundamentally Christian art. As the meeting's opening speaker he felt it his duty to explain the nature of music, but he found human thought completely unable to do so and merely observed that

> Music must be felt; its nature cannot be explained. "He who hath not music in his soul" will be none the wiser if I tell him the number of vibrations which is supposed to constitute harmony as distinguished from discord; and he who hath music in his soul will resent his ethereal art being brought down from Heaven to earth, and reduced to the rigid laws of a mathematical science.[85]

Leopold wanted music kept sacred: not just idealized and set beyond the material world, but left as something divine. In his address, mathematical science appeared, by royal decree, as something contradictory to the heavenly.

Leopold had certainly identified a very tangible confrontation between religious and materialistic interpretations of music. However, it would be simplistic to describe this as a clearly demarcated disagreement between science and faith. Often religious and scientific authorities transcended what, in the twentieth century, would become demarcated as professional boundaries between scientists and clergymen. As much as Gurney and Tyndall challenged religious accounts of music, other figures remained, until the very end of the century, as eager to combine science and theology in their understanding of sound. The mathematician and Church of Ireland clergyman, George Salmon (1819–1904), for example, addressed the nature of music in 1892, comfortably uniting the religious with the scientific. From 1858 he lectured in mathematics, becoming Trinity College Dublin's Regius Professor of Divinity in 1866 and serving as president of the Mathematics and Physics Section of the BAAS in 1878.[86] He was also a competent musician and, in 1892, united his love of music with his scientific interests in a sermon preached at the opening of Southwell Minster's new organ. For evidence of God's desire that man was made to work together in order and obedience, Salmon looked not to the day's great engineering projects but to an orchestral chorus. The performance of vocal music required timing, rhythm, and harmonic knowledge. The power of song transcended scientific explanation: there was, he declared, so much "in man that cannot be weighed or measured, and how many faculties of his nature display themselves in the course of his evolution, which the most thorough knowledge of his material structure could not have enabled the most skilful anatomist to predict." Citing William Paley's (1743–1805) classic design argument over how each organ was so well constructed to perform its required work as to suggest a Creator, Salmon claimed the ear to be an especially wondrous mechanism.[87] Clearly, man had been gifted with ears to appreciate sound and faculties to enjoy the laws of harmony because such a Creator wanted music to be a cultivated pleasure.

Conclusion

Religious-scientific tensions surrounding sound and music continued into the twentieth century, but in the nineteenth they were accompanied by genuine desires to unite philosophical understandings of the sonorous within a theological framework. In many respects, this revealed the lingering influence that religious authorities continued to exert over sound. As late as 1911, Edward Curling, vicar of St. Leonard's Church in Downham, Lancashire, had a series of sermons published exploring both Creation and music. A graduate of Brasenose College Oxford, he had studied theology and classics between 1870 and 1874 and become an associate of Edward Pusey and Henry Liddon at a moment of great confidence for Oxford's High Churchmen. Curling was, by the 1900s, aware of Anglicanism's fading philosophical authority amid growing scientific materialism. In his sermon, "Creation and Evolution," he criticized those who rejected Genesis, who believe "that Science—and especially the theory of Evolution—has made it impossible for grown-up men and women to believe in a Creator." Warning that the "view Science and Religion need not be enemies . . . is treated as a last resource of Christianity" and was therefore dangerous, Curling felt that scientific knowledge, including that of Darwin or germ theory, still failed to explain the universe's original creation. In the end, convinced that Huxley and Spencer were mistaken over creation, he reminded listeners that Darwin had had faith, but it had faded along with his physical decline.[88] In a later sermon, "On God's Gift of Music," he expanded on this reasoning, attacking scientific materialism. Curling this time asserted that there

> is a fundamental thought of creation—"It is He that hath made us, and not we ourselves." It matters not whether we hold that we must go back, for the first traces of our bodily frames, to some tiny germ far ages since. That wonderful germ must have come from somewhere. We cannot without self-contradiction believe in a universe without a cause, a world of beauty and order without an orderer.[89]

Music was, therefore, one of the greatest examples of God's Creation, given to man so that he could praise his Architect.[90] This art exerted so much influence on the human mind and such power to express feeling beyond language, that it was inseparable from the

> consciousness of incompleteness, the longing after something higher and nobler than earth can give, the hidden struggle with evil, the vague yet real sense of awe and mystery which makes the materialist's view of life well-nigh impossible to many minds—where shall we find these things set forth in any art so powerfully as in some of Beethoven's sympathies, or in his wonderful Mass in D? Or again, the beauty of perfect harmony, the subordination of each individual in the interests of the whole, yet the need and nature of each separable part, for the general effect—where can these Christian ideas be found more aptly symbolised than in Bach's Passion music, or in the better-known *Elijah* of Bach's devout-disciple, Felix Mendelssohn?[91]

Ordered and beautiful, music was, according to Curling, a divine gift offering a rare taste of heaven; it was beyond the comprehension of all materialist understandings of nature.

Throughout the nineteenth and early twentieth century, it was not just the place of music in worship that concerned religious commentators but the nature of music itself and its medium of sound. Clergymen were just as eager as scientists to provide knowledge of the sonorous and mobilize it within a vision of a divinely ordered universe. Through religiously informed accounts of sound and music, the church was the traditional source of authority in matters sonorous, and it was with such spiritual, idealistic accounts of sound that scientific authority had to compete in Victorian Britain. Sonorous phenomena, especially those musical, were part of broader debates over who should convey knowledge of nature to society. While Tyndall's writings were obviously central to these discussions, to avoid a history of the science of sound in the nineteenth century that is limited to scientific elites, printed sermons and religious tracts provide rich sources for evaluating what more diverse audiences thought about what sound was and why music was enjoyable. Ironically, many of those delivering these sermons were themselves part of Britain's sociopolitical elites, but they were speaking to congregations which included the rich and poor, Anglicans and dissenters, and generally a wide array of social classes. Along with the musicians, instrument makers, composers, and critics who vied with those of scientific reputation for authority over the sonorous, ministers and theologians provided spiritually inscribed acoustic and musical knowledge. Traditionally, the science of sound was understood within the religious framework which shaped the works of Whewell, Herschel, Faraday, and Somerville as much as the sermons of churchmen. The 1860s and 1870s undoubtedly marked the beginnings of a rupture between faith and science, but spiritual authorities continued to provide musical and sonorous knowledge to diverse social audiences well beyond the nineteenth century.

EPILOGUE

Musical Spiders and Sounds Scientific in the Modern Age

> See deep enough, and you see musically; the heart of
> Nature being everywhere music.
>
> THOMAS CARLYLE, *On Heroes, Hero Worship and the Heroic in History*

December 1880 saw a disturbing article in the science journal *Nature*. "The Influence of a Tuning-fork on the Garden Spider" revealed a grim series of experiments which directly linked a tuning fork's vibrations with a spider's ability to detect animal life. Charles Vernon Boys (1855–1944), of the Physical Laboratory at South Kensington, recalled that on watching spiders spinning webs in his garden the previous autumn, he had been intrigued at how the insects responded to the vibrations of an A tuning fork. Boys found that by touching any leaf connected to a web with his vibrating fork, the spider turned to face the direction of the disturbance. After measuring the approaching vibrations with its forefeet, the spider rapidly advanced on the fork. "If the fork is not removed when the spider has arrived it seems to have the same charm as any fly," observed Boys, "for the spider seizes it, embraces it, and runs about on the legs of the fork as often as it is made to sound." When he brought the vibrating fork directly at a spider when positioned centrally on its web, the insect instantly dropped from a thread. Boys went further and subjected the spider to more trials. "By means of a tuning-fork a spider may be made to eat what it would otherwise avoid," he boasted, "I took a fly that had been drowned in paraffin and put it into a spider's web and then attracted the spider by touching the fly with a fork." Quickly the spider realized this was not food and left it, but by continually touching the fuel-soaked insect with the fork, the experimentalist "compelled the spider to eat a large portion of the fly."[1]

At first inspection, these experiments might seem like the trivial pursuits of a warped mind. But on closer inspection they are revealing of the interconnections between sound, music, and the natural sciences in nineteenth-century Britain. Boys was not some philosophical amateur but a reputable physicist who would go on to win the Royal Society's Royal Medal in 1896 and its Rumford Medal in 1924. He worked at the South Kensington Physical Laboratory in the Science Schools, part of the scientific and cultural center of "Albertopolis" which would, in 1890, become the Royal College of Science. Boys, however, was not a fan of South Kensington's laboratory provisions, where it was difficult to perform delicate experiments due to passing urban traffic; he preferred the peaceful laboratories of Oxford.[2] The place of publication, Nature, established in 1869, was also significant. Both the South Kensington Laboratory and Nature were at the forefront of the changing relationship between science and the British state during the 1870s. Following calls at the 1868 BAAS meeting for the government to be more active in cultivating science, a committee of leading scientists lobbied Parliament for a formal inquiry into the state of British science. In 1870, the Liberal prime minister, William Gladstone, responded by appointing the Royal Commission on Scientific Instruction and the Advancement of Science, with the 7th Duke of Devonshire as its chairman and Norman Lockyer, Nature's founding editor, as secretary. This inquiry provided impetus to a series of Parliament-backed scientific endeavors, including a report on government-funded science and arts departments in 1872, university reforms in 1873, and an account of the nation's museums in 1874. For six years, the appointed commissioners held eighty-five meetings, with Lockyer's Nature providing a commentary on its proceedings: this was no mere publisher of natural curiosities but an outspoken organ in the changing relationship between society and the natural sciences, in which experimental and mathematical knowledge were taking on increased cultural authority. Given that the Royal Commission had also recommended that the Royal School of Mines and the Royal College of Chemistry unite into a single science school at South Kensington, both Boys's institutional affiliation and his choice of publication were, by 1880, highly credible venues, which had been prominent within the commission's reorganization of state-science relations.[3] This invoking of governmental authority marks an appropriate place to conclude a book that has been concerned with how scientific practitioners fashioned themselves as authoritative purveyors of acoustic knowledge.

The inclusion of sonorous experiments on spiders, from a respected scientific practitioner, within an influential science journal, emphasizes that, as late as 1880, sound was still providing important opportunities for investigating nature. This demonstration of the influence of a vibrating tuning fork over a spider would not have been out of place in Brewster's *Letters on Natural Magic*, half a century earlier, in which understandings of sonorous phenomena presented natural philosophers with powerful means for manipulating the senses and exerting control over the human mind. Boys emphasized the value of such sonorous inquiry for the natural sciences, asserting that his techniques afforded "a method which might lead a naturalist to notice habits otherwise difficult to observe." Within this experimental framework, musical apparatus like tuning forks played a central role, just as they had done in Wheatstone and Faraday's philosophical investigations during the 1820s and 1830s. Arguably more significant, however, was Boys's relating of sonorous vibrations to animal life. To the humble spider, what separated a living fly from a dead one appeared reducible to the mechanical action of a vibrating tuning fork. Viewed in the context of John Tyndall's earlier observations, made in 1875, on the comparison between musical sounds traveling through a wooden rod and an acorn's growth, in which vibration featured as but one process within a materialistic world, the spider experiments were potentially troubling. Boys's ability to trick a spider into dining on the inedible suggested a mechanical explanation for the senses and habits of the insect. Boys extended this to questions musical, supposing that his experiments had ramifications for "the supposed fondness of spiders for music."[4] During the nineteenth century there had been a continual preoccupation within the study of natural history on this relationship between animals and music, which Charles Darwin exemplified in his 1871 *The Descent of Man*. Boys's tormenting of his arachnid specimens built on an intellectual tradition which informed new understandings over race, species, and creation. Through sonorous experiments, musical instrumentation, and growing concerns over the connections between animal life and vibration, Victorian musical and scientific cultures were inseparably bound.

By the 1900s, the ways of thinking about sound and hearing were very different from what they had been at the beginning of the nineteenth century. This may seem obvious, given that Alexander Graham Bell patented the telephone in 1876, Thomas Edison secured a patent for his microphone in 1877, and the same decade saw the construction of several primitive sound-recording machines. As scholars such as Lisa Gitelman, Greg Milner, Melissa Van Drie, Tony Grajeda, John Mowitt, and Mara Mills have compellingly shown, this "Edison Era" presented Western societies with new sonorous experiences from an increasingly expanding range of devices, from gramophones and radios to phonographs and théâtrophones.[5] Together with trains, trams, automobiles, factories, and a multitude of other "noises," these contributed to a changing sonic environment in which, as Karin Bijsterveld has demonstrated, mechanical sound was ascendant.[6] However, this change went beyond technological innovations, no matter how dramatic. The science of acoustics had grown into an extensive field, while sonorous experiments were a valuable part of Britain's scientific culture, not just for investigating nature but for displaying it to diverse audiences. Equally, physiological and psychological understandings of the ear and the experience of hearing were increasingly sophisticated. In producing science and in sharing knowledge of nature, the ear had a unique epistemological value. Knowledge of sound also informed grand scientific accounts of nature, be it the divinely ordered universe of Herschel, Somerville, and Whewell or the materialistic, mechanical interpretations of Tyndall and Spencer.

Yet when it came to music, by 1900, science was just one of a variety of alternate authorities with claims to explain, define, and regulate musical phenomena. If we want to know what society thought about music, about its aesthetics, and about why it sounded good, it is not enough to confine our investigation to scientific elites working in psychology, physiology, and physics. Without situating the works of Tyndall, Helmholtz, Gurney, and Rayleigh within their broader sociocultural contexts, our understanding of the nineteenth-century relationship between science and music is severely limited. Music was, above all, a social product over which scientists, psychologists, physiologists, and physicists did not wield exclusive authority. To analyze what people thought about what they heard, we should also look to musicians and clergymen. Throughout this book, which has provided a cultural history of nineteenth-century science and music, it has been overwhelmingly clear that as soon as the science of sound is placed within Britain's religious, musical, social, and philosophical contexts, we are immediately dealing with questions of authority. During the 1820s and 1830s, Wheatstone and Faraday combined musical and scientific knowledge seamlessly in London's workshops, laboratories, lecture theaters, and showrooms. In the mid-century, Somerville, Whewell, and Herschel conscripted musical knowledge into their visions of a harmonious universe. These experimental and mathematical projects were in fact promoting an ideal in which science and music were inseparably bound; in which to be an authority on nature could also make one an authority on music and, arguably, vice versa. In practice, this vision of science proved difficult to realize as scientific authorities struggled to exert influence over music.

For the performance of music, standardization, that most essential of Victorian scientific concerns, remained firmly in the hands of musical, religious, and military communities. Musical pitch became the measure of science's limits in Victorian Britain. While controversies over standards for electricity, heat, and time could be resolved in the laboratory and observatory, a standard for music remained elusive. Despite Herschel's campaign for C512, it was Britain's musical elites, including those from within the Church and army, who determined how the nation's music would be ordered. While Herschel strove to convince audiences of the virtues of a mathematical standard, Airy labored to secure *any* kind of consensus with instrument makers and musicians over how notes were heard. It was not just that sound was an unreliable medium of communicating time, but that to the Astronomer Royal, accustomed to the ordered regime of the Royal Observatory at Greenwich, musical practice appeared a chaotic business, with the ear a variable and subjective organ. It is significant that both Airy and Herschel were involved in unresolvable musical controversies, because they were almost certainly the most eminent mathematical authorities in mid-Victorian Britain. Similar tensions were evident as Tyndall conscripted sound and music into his materialistic account of nature. Though not in dialogue, such a naturalistic understanding of the sonorous contrasted sharply with theological interpretations of music as projected from the pulpit. It was exactly this sort of religious commentary on nature that Tyndall rebuked. Discussions over the materiality of music were inseparable from the wider debates between theologians and scientists over who had authority to explain nature to society. Wheatstone, Whewell, Somerville, Airy, Herschel, and Tyndall all wrote extensively on sound and music, but all found the musical world a challenging place to influence. The utopian union between music and science, promised in the 1820s and 1830s, largely failed to materialize in Victorian Britain.

Nevertheless, as late as 1913, Alfred Parr emphasized these romanticized links between music and astronomy, observing that Herschel and Somerville were both typical of those who embraced both disciplines, as had composer Camille Saint-Saëns, an "ardent amateur astronomer." Writing in the *Westminster Review*, Parr noted that the "quaint" ancient conception of the "Music of the Spheres" afforded a romantic framework for conceiving of theories of "Universal Harmony."[7] Just a year later, however, such sentiments were rudely shaken, as the parameters between music, natural philosophy, and religion were thrown into chaos. In many respects, the romantic vision that scientific knowledge would induce a utopian future was irretrievably lost in the First World War. In 1914, the nineteenth century came to a brutalizing stop. In the trenches of Flanders and France, philosophical questions over machinery and industrialization were abandoned in the pursuit of national survival. As Peter Sloterdijk dramatically put it in his 2009 *Terror from the Air*, "The 20th century dawned in a spectacular revelation on April 22, 1915, when a specially formed German 'gas regiment' launched the first, large-scale operation against French-Canadian troops in the northern Ypres Salient using chlorine gas as their means of combat."[8] For Sloterdijk, it was this mobilization of laboratory science in the form of chemical gas warfare that marked the end of the nineteenth century. While this declaration perhaps places too much emphasis on a single weapon, it does encapsulate a broader sense that the war marked the end of an era, both technologically, socially, culturally, and scientifically.[9] At this moment, worries over what was material became starkly immaterial as sound quickly moved from a subject of philosophical discourse to a crucial weapon in the breaking of the Western Front's stalemate.

Amid the industrial slaughter, the ear became an instrument of war as never before and was subjected to unpreceded sonic experiences. As historian Yaron Jean explains, the First World War provided a "bridge between the soundscapes of the nineteenth and the soundscapes of the twentieth century."[10] This was a loud war where soldiers had to live with continuous shell fire. By 1918, over 2,800 artillery batteries were firing more than 10,000 shells a month along Europe's frontlines while, at the same time, smoke and trenches limited vision. Beneath the ground, miners depended on sound to identify enemy engineers at work in subterranean tunnels. With the emergence of large-scale aerial warfare, new methods of acoustic defense were fashioned, including geophones, double trumpet sound-locators, acoustic visors, sound mirrors, and listening wells. New apparatus accompanied refined modes of listening to identify incoming enemy aircraft with the ear, before they became visible to the eye.[11] At sea, the German U-boat threat was managed with the latest sonorous technology, securing the imports on which the Allied war effort depended. Although numbering only twenty-eight vessels in 1914, over 300 U-boats were constructed throughout the war, with the ear key to locating and tracking these as they lurked deep underwater. Between 1914 and 1916 British scientists worked to develop new sonorous equipment for this purpose; then American naval engineers joined them in this endeavor.[12] By the end of the conflict, hearing was at the very center of the Allies' final victory. During the last hundred days of the war, planes flew loudly over the frontline to conceal the assembling of rumbling tanks. Listening devices accurately located German artillery positions which could then be quickly neutralized. In 1918, the Australian General John Monash, architect of the decisive Allied offensive at Amiens, explained how this total war was in fact comparable to the careful coordination of an orchestra. As he put it, the "perfected modern battle plan is like nothing so much as a score for an orchestral composition, where the various arms and units are the instruments, and the tasks they perform are their respective musical phrases."[13]

These militaristic applications of sound and acoustic technology had philosophical ramifications. Roland Wittje has compellingly described how the science of sound underwent a radical transformation between the late nineteenth century and the 1920s. From a field closely connected to understandings and performances of classical music, acoustics was dramatically reorganized into one of electrical engineering and media technology or, rather, "electroacoustics." In his 2016 *Age of Electroacoustics: Transforming Science and Sound*, Wittje traced this change from the acoustics of the 1860s and 1870s, dominated by the writings of Helmholtz and Rayleigh and in which sound was a subdiscipline of physics with links to physiology and musicology, into the highly industrialized discipline which materialized in the 1920s. The First World War played a crucial, if complicated, role within this transformation. Although the emergence of electroacoustics predated 1914, with the development of psychophysics and experimental psychology in the late nineteenth century, the war stimulated an unprecedented frenzy of techno-industrial research into sound, which meant that the soundscape of the 1920s was far removed from that which had preceded the war. Wittje concluded that while acoustics had been an evolving field before 1914, the industrial-military mobilization of sound for wireless communication, aircraft locating, artillery sound ranging, and submarine warfare accelerated the cultivation of electroacoustics. By the 1920s, technologies for wireless radio broadcasting and for the recording, propagating, and amplifying of sound contributed to a redefinition of the science of acoustics away from the bourgeois ties to classical music as high culture which had traditionally characterized the discipline.[14]

Between 1914 and 1918 it mattered less if music was spiritual or mechanical: what was important was how sound could be mobilized as a weapon. In the fields of Ypres and the Somme, it became painfully apparent that science and industry had not ushered in a utopian future but unprecedented horror. Just consider the declining interest in John Ruskin's (1819–1900) writings, for example. In late Victorian Britain few philosophical positions held so much influence as Ruskin's attack on industrial capitalism, in which machine labor was rejected in favor of a return to an idealized medieval way of life. In 1906, for instance, more Labour MPs named Ruskin as a guiding political figure than any other single author; of forty-five MPs, seventeen acknowledged Ruskin's significance, compared with sixteen who cited the Bible.[15] The inspiration for the Arts and Crafts Movement of William Morris, Ruskin's teachings held that true value was to be found in artisanal craft rather than precise industrial production. Identifying a conflict between man and machine, in his celebrated *Stones of Venice* (1851–53) Ruskin alleged that,

> You must either make a tool of the creature, or a man of him. You cannot make both. Men were not intended to work with the accuracy of tools, to be precise and perfect in all their actions. If you will have that precision out of them, & make their fingers measure degrees like cogwheels, and their arms strike curves like compasses, you must unhumanize them.[16]

Although Ruskin died in 1900, his ideas resonated throughout Edwardian Britain. By 1918, however, they had lost much of their popularity.[17] In a war in which guns, ships, chemicals, and tanks were decisive, the impossibility of rejecting machinery, precision, and factory production became staggeringly clear. Men and women had to become tools capable of precision and measurement. The demise of idealism was not to be found solely on the pages of Darwin or Tyndall but on the dehumanizing battlefields of the Great War. Arguably, this was the ultimate realization of the fears over industry and mechanism that informed nineteenth-century philosophical discourse, and music was very much a part of this. New scientific understandings of sonorous phenomena, mechanical productions of music, attempts to subject musical practice to industrial standards, all contributed to a growing sense that music was not spiritual or metaphysical but coldly material. As scientists and natural philosophers increasingly conscripted the ear into their investigations of nature, so they threatened to rationalize and disenchant the highest of all art forms, and one that was deemed inseparable from human existence.

This book began by claiming that the history of sound and music is important for understanding British scientific culture between 1815 and 1914. However, it concludes by going even further and asserting that the history of sonorous knowledge offers rich insights on British society and its conflicted relationship with the natural sciences throughout the nineteenth and early twentieth centuries. Often musicology has been seen as the poor relation of historical study. Yet the history of music provides a unique perspective on the development of nineteenth-century science and the existential challenges that it produced. While sonorous knowledge has been historically important in its own right, when placed within its musical contexts, it reveals how complicated the relationship between society and science was throughout this period. Indeed, the connections between the musical and the scientific carried greater philosophical urgency than any other art or discipline because of music's socially diverse character. The publication of Darwin's *On the Origin of Species* in 1859 is frequently invoked as initiating a new era in which authority moved irrevocably toward those of scientific credentials, as spiritual commentators gradually retreated from questions of natural philosophy. This is, of course, a simplistic interpretation, but historians of science and religion have nevertheless provided a wealth of well-merited studies on the theological and social implications of the physical and biological sciences. But in the same year that Darwin published his controversial treatise, mathematicians wrangled with musicians, clergymen, soldiers, and composers over what Britain's national musical pitch should be.

If it took time for the tensions between science and religion to become realized in relation to questions of evolution and astronomy, then these were already well established in matters of music. When a natural philosopher produced new acoustic experiments or sonorous explanations and related these to musical phenomena, he or she was immediately dealing with a subject in which almost everyone in society had a stake. Music transcended social hierarchies, from the elite repertoires of the opera and concert to the popular songs of music halls, workplaces, and pubs, and to the religious chants and hymns of chapels and cathedrals. Scientific knowledge of sound was, therefore, immediately relevant in a wide range of cultural contexts, and was the object of audiences far beyond the laboratory and lecture theater. Any comment on acoustics, especially if extending to musical matters, at once engendered philosophical and social concerns. More than any other subject, music permeated the boundaries between art and science and, as a result, marked the limit of pre-1914 scientific authority. This relationship between music and science was never far from making the immaterial material. It raised profound anxieties over the human condition, the existence of a soul, and the place of God in the universe. At stake were not just questions of aesthetics but of whether all was physical matter or if there were something beyond the material world of sensory experience. Music was of unique importance in these discussions because no other natural phenomenon had the same power to affect the senses and arouse emotional responses. Who had authority over the musical was a crucial social problem, because to demystify music, the highest of all arts, was to demystify nature itself.

Acknowledgments

First and foremost, I would like to thank David Trippett for making this book possible. Not only did he take the bold gamble of appointing me to his project, "Sound and Materialism in the Nineteenth Century," but he has provided constant guidance throughout. My fellow project members Melissa Van Drie, Melle Kromhout, and Stephanie Probst have been amazing, teaching me an awful lot and offering phenomenal support. Together my colleagues have offered a great deal of friendship and fun over the past three years; without them, I simply could not have written this. I also am grateful to the European Research Council for funding our project so generously. Cambridge University's Faculty of Music has been a lovely place to work and I have appreciated all the kind help from its academic and support staff, especially our project coordinators, Elaine Hendrie and Veronika Lorenser. I conducted a great deal of the work for this project at 32 Sherlock Close, where I was lucky enough to have the encouragement and entertainment of Amie Varney, Chlöe Gamlin, Tilda, and Bertram. I also owe a great deal to Silke Muylaert, who will always have my enduring appreciation.

As ever, I owe immense debts to Christina de Bellaigue, Philip Boobbyer, Grayson Ditchfield, Katherine Fry, Jane Garnett, Ben Griffin, Graeme Gooday, Alexandra Hui, Peter Mandler, Simon Schaffer, Jim Secord, Crosbie Smith, Geoffrey Tyack, and William Whyte, as well as Robert Hall, Ben King, Rachel North, and Keith Shepherd. This work has been considerably enhanced through conversations with Daniel Belteki, Geoffrey Cantor, Graham Dolan, Giorgio Farabegoli, Alexandra Hui, Julia Kursell, Peter McMurray, Roger Parker, and Peter Pesic. Thanks to Viktoria Tkaczyk and Birgitta Mallinckrodt at the Max Planck Institute, where I have enjoyed collaborating with the "Epistemes of Modern Acoustics" Research Group, and especially for all the immensely productive discussions with Fanny Gribenski; the time I spent in Berlin with these scholars has had much influence on the shape of this book. I must also thank Jenny Bulstrode, Oliver Carpenter, Sheila Cavanagh, Stephen Courtney, Perry Gauci, Michael Hall, Horatio Joyce, Ayla Lepine, David Lewis, Harry Mace, Tim Marshall, Christopher McKenna, Lucy Rhymer, Peter Roberts, Otto Saumarez-Smith, and the now legendary "Little Christopher Yabsley" for their constant friendship and support. Without all the enthusiasm they provide, the process of research and writing would be highly laborious. I apologize for anyone I have accidently forgotten to mention. My work has been enriched by discussions with students at Oxford, Cambridge, and Emory, both undergraduates and postgraduates, and I thank them all.

Cambridge is a great place for conducting research and I appreciate all the help staff have provided at the University Library, especially in Rare Books and Manuscripts, as well as at the Institute of Astronomy Library, the Wren Library, the English Faculty Library, the Seeley Historical Library, Newnham College Library, Girton College Library, the Casimir Lewy Library, St. John's College Library, the Whipple Library, and the Pendlebury Library. I have also enjoyed help from Oxford University's Bate Collection, Claire Horscroft at Hove Public Library, the British Library, King's College London Library, the Royal Institution, the Royal Society Library, the Royal Society of Arts Library, Edinburgh University's Centre for Research Collections, Alison Metcalfe at the National Library of Scotland, Karen Moran at the Edinburgh Royal Observatory Library, Liverpool University's Sydney Jones Library, the Institute of Civil Engineers Library, the Science Museum, and at Oxford's Bodleian Libraries, in particular at the Weston Library. Additional thanks go to everyone at Trinity and St. Edmund's Colleges in Cambridge, the History Faculty, and the university's History and Philosophy of Science Department.

This book owes much to the excellent feedback at a number of conferences and seminars. In particular, thanks go to all those who provided constructive input at Cambridge's Faculty of Music Colloquia, the History and Philosophy of Science departmental seminar in Cambridge, Cambridge Modern British History Seminar, Oxford Modern British History Seminar, the Victorian Society, the Science Museum Research Seminar, the Royal Historical Society Oxford's Long Nineteenth Century Seminar, the IHR's Parliaments, Politics, and People Seminar, the 2018 History of Science Society conference in Seattle, the "Life and Work of Sir George Biddell Airy" workshop at Cambridge University Library, the 2018 "Measurement at the Crossroads" conference in Paris, the "Spectres de L'Audible" and "Sensing the Sonic" conferences at Paris and Cambridge during 2018, the "Acoustics of Empire" conference, the 53rd Annual Meeting of the Royal Musical Association, and "After Idealism: Sound as Matter and Medium in the Nineteenth Century," which was also held in Cambridge in 2017.

Finally, I am grateful to my family and friends in Devon for everything they have done for me and all the patience they have shown over the past few years with my work and eccentric habits. Thanks go to my parents, Steve and Louise Spencer, my brother Alexander Teague, Montgomery Spencer ("the best of all of us"), Alfred Spencer, Robert White, Laura Treloar, Peter Exell, everyone at Sharpham Vineyard, the Fagg family, especially Lizzie and Steve, Charlotte, Ellie, and Jamie, and all of the Teague family, especially Steve and Pauline. This book is dedicated to my dear old Nan, the late Estelle White, who was always terrific fun.

Notes

Introduction

1. As argued in Parker and Rutherford, "Introduction. London Voices, 1820–1840: A 'Luminous Guide,'" 1–14, 25.

2. Pesic, *Music and the Making of Modern Science*, 9–20, 121–31, and 151–60; on ancient harmony, see Barker, *Science of Harmonics in Classical Greece*; Creese, *Monochord in Ancient Greece Harmonic Science*; on early modern acoustics, see Gouk, *Music, Science and Natural Magic in Seventeenth-Century England*, 224–57.

3. Media theorist Friedrich Kittler argued that during the nineteenth century music changed from a metaphysical concept to a discipline firmly grounded in mathematical knowledge, as explored in Kromhout, "'Antennas Have Long Since Invaded Our Brains,'" 89–104, 92–94.

4. Grattan-Guinness, "Mathematics and Mathematical Physics from Cambridge, 1815–1840," 84–111.

5. On Chladni, see Jackson, *Harmonious Triads*, 13–44.

6. Picker, *Victorian Soundscapes*, 4–6.

7. Russell, *The Wave of Translation*, 97–98.

8. Tyndall, *Sound*, 50–51.

9. Hui, *Psychophysical Ear*, xiii–xvii.

10. On the dissemination of scientific knowledge to musical audiences, see Petersen, "Craftsmen-turned-scientists?," 212–31.

11. Scientific credibility is not purely dependent on a body of knowledge's validity but is built through persuasion and rhetoric, as shown in Shapin, *Never Pure*, 18; Marsden and Smith, *Engineering Empires*, 7.

12. Gibson, "The British Sermon, 1689–1901," 3–30, 9, 25; Knight, "Parish Preaching in the Victorian Era," 63–78, 72.

13. Cantor, *Religion and the Great Exhibition of 1851*, 14.

14. Thormählen, "From Dissent to Community," 159–78.

15. Grande, "On Tongues and Ears," 137–58.

16. Pink, "Order and Uniformity, Decorum, and Taste," 215–26, 222–24.

17. Gatens, *Victorian Cathedral Music*, 1–6, 20–23.

18. Often shaping these sermons were evangelical values, including a call to conversion, activism, a focus on the Gospels, and a commitment to the Bible as the absolute truth and word of God. See Bebbington, *Evangelicalism in Modern Britain*, 2–3.

19. Gillin, "Prophets of Progress," 928–56.
20. Aveling, *Recreations, Physical and Mental*, 24, 25–26, 49–50.
21. Discussed in Morrell and Thackray, *Gentlemen of Science*, 30–33; on the economy and natural order, see Daunton, *Progress and Poverty*, 495.
22. Hahn, *Pierre Simon Laplace*, 172.
23. Morus, *When Physics Became King*, 192–93.
24. Desmond, *Politics of Evolution*, 1–12; Secord, *Visions of Science*, 7–8.
25. Quoted in Niven, ed., *Scientific Papers of James Clerk Maxwell*, 2:376.
26. Jamie James, *Music of the Spheres*, 3–4.
27. Jackson, *Harmonious Triads*, 75–78.
28. Hegel, *Hegel's Aesthetics*, 891. Thanks to David Trippett for discussions on German Romantic philosophy and materialism.
29. E. T. A. Hoffmann quoted in Charlton, ed., *E. T. A. Hoffmann's Musical Writings*, 237, 239.
30. Giacomo Leopardi quoted in Caesar and D'Intino, eds., *Zibaldone*, 123.
31. Hanslick, *On the Musically Beautiful*, 1–2.
32. Trippett, *Wagner's Melodies*, 5–6, also see 330–92; Lange, *History of Materialism*.
33. Richard Wagner quoted in Trippett, *Wagner's Melodies*, 356.
34. Chua, "Vincenzo Galileli," 17–29, 20.
35. Jackson, *Harmonious Triads*, 79–82.
36. E. T. A. Hoffman quoted in Jackson, *Harmonious Triads*, 79; also see Pinch and Bijsterveld, "'Should One Applaud?,'" 536–59.
37. If Romanticism was chiefly characterized as emotional, spiritual, organic, and liberal, then it is equally evident that such values were influential over the mechanistic, which emphasized reason, mathematics, and determinism. See Tresch, *Romantic Machine*, 3, 5; also see 223–51.
38. See Yeo, *Defining Science*; on laboratories, see Schaffer, "Physics Laboratories and the Victorian Country House," 149–80.
39. Morus, *When Physics Became King*, 1.
40. Latour, "Visualization and Cognition," 1–40, 4, 6–7; on inscription, see Latour and Woolgar, *Laboratory Life*, 45–53.
41. See Foucault, *Discipline and Punish*, 200–206; Hirst, "Foucault and Architecture," 52–60, 56.
42. Otter, *Victorian Eye*, 1–4.
43. Musselman, *Nervous Conditions*, 17–19.
44. Schaffer, "Transport Phenomena," 71–91, 71, 74.
45. Tyndall quoted in Schaffer, "Transport Phenomena," 76.
46. Lightman, *Victorian Popularizers of Science*, 37, 167–218.
47. Morus, "Seeing and Believing Science," 101–10; Morus, "Worlds of Wonder," 806–16; Morus, Schaffer, and Secord, "Scientific London," 129–42.
48. Morus, *Frankenstein's Children*, 70–98; Lightman, *Victorian Popularizers of Science*, 8–9, 18–19; Secord, *Victorian Sensation*; Secord, *Visions of Science*, 3–4.
49. Morus, "Seeing and Believing Science," 101–10; on eye-witness accounts of spectacles, see Schaffer, "Natural Philosophy and Public Spectacle," 1–43; Shapin, "The House of Experiment," 373–404.
50. Morus, "Illuminating Illusions," 37–50, esp. 40.

51. Smith, *Sensing the Past*, esp. 41–58; for example, see Schickore, *The Microscope and the Eye*.

52. For examples, see Cantor, *Optics after Newton*, 114–46; Wittje, "Electrical Imagination" 40–63.

53. Meulders, *Helmholtz: from Enlightenment to Neuroscience*, 153–78, 179–99; Steege, *Helmholtz and the Modern Listener*; Lenoir, "Helmholtz and the Materialities of Communication," 184–207; Beyer, *Sounds of Our Times*, 55–82.

54. Sterne, *Audible Past*, 3; on society, noise, and technology, see Bijsterveld, *Mechanical Sound*.

55. Sterne, *Audible Past*, 3; on aural culture and modernity, see Emily Thompson, *Soundscape of Modernity*.

56. Sterne, *Audible Past*, 2.

57. Hui, Kursell, and Jackson, "Music, Sound, and the Laboratory," 1–11, 5; on sound and laboratories, see Schmidgen, "Silence in the Laboratory," 47–61; Kursell, "'A Gray Box,'" 176–97.

58. Gooday, "Architectural Acoustics," 101–14; on Sabine's equation for reverberation, see Emily Thompson, "Listening to/for Modernity," 253–80.

59. Hankins and Silverman, *Instruments and the Imagination*, 86–108; Tresch and Dolan, "Towards a New Organology," 278–98; on musical and scientific instruments, see Dolan, *Orchestral Revolution*, 11–13.

60. Jackson, *Harmonious Triads*; Pesic, *Music and the Making of Modern Science*; Erlman argues that the concept of musical resonance has been valuable to modern philosophy. As he put it, "scientific thinking and musical sensibilities merged on a higher plane, mutually shaping each other and advancing sophisticated arguments about the foundations of knowledge and personhood"; in Erlman, *Reason and Resonance*, 26.

61. Davies and Lockhart, "Fantasies of Total Description," 1–26, 3.

62. Tkaczyk, "The Making of Acoustics around 1800," 27–55.

63. Gieryn, "Boundary-work and the Demarcation of Science," 781–95, 781, 784–85.

Chapter 1

1. See Lightman, "Refashioning the Spaces of London Science," 25–50.

2. Dickson, "Charles Wheatstone's Enchanted Lyre," 125–44, 131–33.

3. Compare with theatrical acoustic demonstrations in mid-nineteenth-century Paris: Pantalony, *Altered Sensations*, 37–60, 145–48.

4. Brewster, *Letters on Natural Magic*, 14–15.

5. Morrison-Low, "Brewster, Sir David (1781–1868)."

6. Professor John Millington gave a description during his lectures in Chancery-Lane; see "X," "The Invisible Lady," 69–71.

7. Brewster, *Letters on Natural Magic*, 152–53, 157; Professor John Millington exhibited this experiment in 1807 in Chancery-Lane.

8. Brewster, *Letters on Natural Magic*, 170–78, 167.

9. See Wright, *The Jews-Harp in Britain and Ireland*, 16–17; Fox, "Introduction," 15–39, 15–21.

10. Anon., "Eulenstein"; Anon., *Sketch of the Life of C. Eulenstein*, 12–14.

11. Anon., "Eulenstein"; Anon., *Sketch of the Life of C. Eulenstein*, 17–18.
12. Anon., *Sketch of the Life of C. Eulenstein*, 20.
13. Anon., "The Fine Arts."
14. Anon., *Sketch of the Life of C. Eulenstein*, 52, 58.
15. Anon., *Sketch of the Life of C. Eulenstein*, 1–3.
16. Neuffer, "To Eulenstein, Musician on the Jew's Harp," 162–63.
17. Anon., *Sketch of the Life of C. Eulenstein*, 47–48.
18. Anon., "Music—the Jew's Harp," *Hull Packet and Humber Mercury* (Hull, England), issue 2262, March 25, 1828.
19. Anon., "The Jew's Harp," *Caledonian Mercury* (Edinburgh, Scotland), August 16, 1830.
20. Anon., "Mr Eulenstein's Jew's Harp," *Liverpool Mercury* (Liverpool, England), issue 1121, October 26, 1832.
21. Anon., "Public Amusements," *Bristol Mercury* (Bristol, England), issue 2244, April 20, 1833.
22. Anon., "Eulenstein," *Bristol Mercury* (Bristol, England), issue 2242, April 6, 1833.
23. Quoted in Stratford, *Gloucestershire Biographical Notes*, 190.
24. Metz, "The Scientific Instruments of Charles Wheatstone," 19–29, 20; Bowers, *Sir Charles Wheatstone FRS*, 5–7, 9–10.
25. Charles Wheatstone, "New Experiments on Sound," in *Scientific Papers of Sir Charles Wheatstone*, 1–13, 1–3, 4–6, 7.
26. Bowers, *Sir Charles Wheatstone*, 8–9.
27. Anon., "The Enchanted Lyre and the Diaphonicon," *New Monthly Magazine and Literary Journal* 6 (London, 1822): 200–201.
28. Wheatstone, "New Experiments on Sound," 8–9.
29. Olleson, "Busby, Thomas (1754–1838)"; Busby, *A Grammar of Music*, v, vii.
30. Busby, "The Acoucryptophone," 9–10.
31. Anon., "The Enchanted Lyre and the Diaphonicon," 200–201.
32. Anon., "Musical Intelligence: The Enchanted Lyre," *Repository of Arts, Literature, Fashions, Manufacturers &c.* 12 (London, 1821), 173–75, 174.
33. Anon., "The Enchanted Lyre," *Literary Gazette* 5 (London, 1821), 586.
34. Anon., "Enchanted Lyre," *The Circulator of Useful Knowledge, Amusement, Literature, Science, and General Information* (London, 1825), 284.
35. Anon., "Musical Intelligence: The Enchanted Lyre," 173–75, 175.
36. Charles Wheatstone, "Description of the Kaleidophone, or Phonic Kaleidoscope; a New Philosophical Toy, for the Illustration of Several Interesting and Amusing Acoustical and Optical Phenomena," in *Scientific Papers of Sir Charles Wheatstone*, 21–29, 23–24, 27, 21–22.
37. Anon., "The Kaleidophone, or Phonic Kaleidoscope," *Mechanics' Magazine* 8 (London, England), August 11, 1927, 49–52.
38. Dickson, "Charles Wheatstone's Enchanted Lyre," 125–44, 129, 131.
39. Bowers, "Faraday, Wheatstone and Electrical Engineering," 163–73, 165; Bowers, *Sir Charles Wheatstone*, 20.
40. Cantor, Gooding, and James, *Michael Faraday*, 28, 10, 47, 52, 11–12.
41. Gladstone, *Michael Faraday*, 33, 21.
42. "Letter 21: Faraday to Benjamin Abbott, 12 and 14 May, 1813," in James, ed., *Correspondence of Michael Faraday*, 1:52–54, 53.

43. Bowers and Symons, *Curiosity Perfectly Satisfyed*, 25.

44. Bowers and Symons, *Curiosity Perfectly Satisfyed*, 65–66.

45. Bowers and Symons, *Curiosity Perfectly Satisfyed*, 118; "Letter 2835: Faraday to Frederick Gye, 22 May, 1854," in James, ed., *Correspondence of Michael Faraday*, 4:684; "Letter 2991: Faraday to Frederick Gye, 2 Jun., 1855," in James, ed., *Correspondence of Michael Faraday*, 4:871; "Letter 3009: Faraday to Frederick Gye, 20 Jul., 1855," in James, ed., *Correspondence of Michael Faraday*, 4:888; "Letter 3448: Faraday to Angela Georgina Burdett Coutts, 2 Jun., 1858," in James, ed., *Correspondence of Michael Faraday*, 5:388; "Letter 3455: Faraday to Frederick Gye, 19 Jun., 1858," in James, ed., *Correspondence of Michael Faraday*, 5:391.

46. "Letter 3020: John Tyndall to Faraday, 27 Aug., 1855," in James, ed., *Correspondence of Michael Faraday*, 4:896–99, 896.

47. Royal Observatory Greenwich Papers, Cambridge University Library (RGO) 6/703, Meteorology, correspondence, instruments and miscellaneous, June 1861–May 1865, "George Biddell Airy to Michael Faraday," (July 7, 1861), 45.

48. Faraday, "On the Sounds Produced by Flame in Tubes, &c.," 274–80, 275; in 1819, the Swiss physician Charles-Gaspard de la Rive wrote to Faraday, expressing interest in the potential of these experiments for enhancing acoustic knowledge, but Faraday did not want to pursue the trials further.

49. Royal Institution (hereafter RI) MS F/4/H, "Miscellaneous Notebook, 1825–1850," 7, 14, 19, 21.

50. Quoted in DeYoung, *A Vision of Modern Science*, 15.

51. Anon., "Royal Institution," *The Times* (London), issue 10871, March 4, 1820, 2.

52. Bowers, "Faraday, Wheatstone and Electrical Engineering," 163–73, 165.

53. Forgan, "Faraday—from Servant to Savant," 51–67.

54. Gooding, "'In Nature's Schools,'" 105–35, 107–8, 111; Metz, "The Scientific Instruments of Charles Wheatstone," 20.

55. Charles Wheatstone, "On the Resonances, or Reciprocated Vibrations of Columns of Air," in *Scientific Papers of Sir Charles Wheatstone*, 36–46, 37, 39–40.

56. Wheatstone, "On the Resonances," 42; Anon., "Proceedings of the Royal Institution: February 15th," *Quarterly Journal of Science, Literature and Art* 25 (London, 1828): 173–74.

57. Wheatstone, "On the Resonances," 43–44; on resonance, see Jackson, "Charles Wheatstone: Musical Instrument Making," 101–24, 113–15.

58. Wheatstone, "On the Resonances," 44–45.

59. Wheatstone, "On the Resonances," 45.

60. Wheatstone, "On the Resonances," 42.

61. Anon., "Proceedings at the Friday Evening of the Royal Institution of Great Britain," *Philosophical Magazine* 3 (London, January–June 1828): 304–5.

62. Anon., "Musical Sound," *Arcana of Science and Art* (London, 1830): 65–67, 66.

63. Anon., "Royal Institution," *Dublin Literary Gazette, or Weekly Chronicle of Criticism, Belles Lettres, and Fine Arts* 1 (Dublin, 1830): 395–96, 396.

64. Anon., "Royal Institution," *Dublin Literary Gazette, or Weekly Chronicle of Criticism, Belles Lettres, and Fine Arts* 1 (Dublin, 1830): 395–96, 396.

65. Charles Wheatstone, "On the Transmission of Musical Sounds through Solid Linear Conductors, and on Their Subsequent Reciprocation," in *Scientific Papers of Sir Charles Wheatstone*, 47.

66. Francis Bacon's (1561–1626) rejection of hypothesis and Stuart patronage

made him a controversial figure for nineteenth-century natural philosophers. See Yeo, "An Idol of the Market-place," 251–98.

67. Anon., "Royal Institution," *Dublin Literary Gazette*, 396.

68. Anon., "The Aeolina," *The Times* (London), issue 13802, January 14, 1829, 2.

69. Anon., "Friday-evening Proceedings at the Royal Institution of Great Britain," *Philosophical Magazine* 7 (London, 1830): 69.

70. Anon., "Royal Institution," *Literary Gazette* (London, 1830): 369–70, 369.

71. Bowers, *Sir Charles Wheatstone*, 35–37, 40; Atlas, *Wheatstone English Concertina*, 25–30.

72. Atlas, "Ladies in the Wheatstone Ledgers," 9–10.

73. Metz, "The Scientific Instruments of Charles Wheatstone," 25.

74. "Letter 346: Charles Wilkins to Faraday, 6 Feb., 1828," in James, ed., *Correspondence of Michael Faraday*, 1:448; "Letter 347: Eleventh Duke of Somerset to Faraday, 7 Feb., 1828," in James, ed., *Correspondence of Michael Faraday*, 1:448.

75. Anon., "Royal Institution," *Dublin Literary Gazette*, 395.

76. RI MS F/4/C, "Friday Evenings, 1825–1829," 240–41, 242–43.

77. RI MS F/4/K, "Friday Evenings, 1829–1831," 119–21, 123.

78. Wheatstone, "On the Transmission of Musical Sounds," 47–63, 49–51, 53–55.

79. Wheatstone, "On the Transmission of Musical Sounds," 57–60, 62.

80. Anon., "Royal Institution," *Literary Gazette* (London, 1830): 223–24; Anon., "Friday-evening Proceedings at the Royal Institution of Great Britain," 315–17, 316.

81. In 1850 Faraday would second Smart for election to the Athenaeum. Adolph, "Smart, Benjamin Humphrey (*bap.* 1787, *d.* 1872)"; Smart, *Theory and Practice of Elocution*.

82. Morus, *Frankenstein's Children*, 20–21.

83. Aldrich, "Hickson, William Edward (1803–1870)."

84. Royal Institution (RI) Ms F/5/D, "Administrative Notebooks: Duties of Servants," 39–40.

85. Maconie, "Willis, Speech, Sound, and Music," 75–84, 79.

86. RI Ms F/5/H, "Subscribers to Separate Courses of Lectures, 1835–1841"; Anon., "Royal Institution," *Morning Post* (London), issue 18827, April 14, 1831, [1].

87. Bowers, *Sir Charles Wheatstone*, 25; Charles Wheatstone, "On the Vibrations of Air in Cylindrical and Conical Tubes," in *Scientific Papers of Sir Charles Wheatstone*, 368–71.

88. Tkaczyk, "The Making of Acoustics around 1800," 27–55, 37–44.

89. On the replication of experiments, see Shapin and Schaffer, *Leviathan and the Air-Pump*, 225–82.

90. Murphy, "White, Joseph Blanco (1775–1841)"; Tuckwell, *Pre-Tractarian Oxford*, 226–57.

91. Murphy, *Blanco White*, 151.

92. "Letter 418: Blanco White to Faraday, 18 Dec., 1829," in James, ed., *Correspondence of Michael Faraday*, 1:496.

93. Sydney Jones Library Archives, University of Liverpool (hereafter SJL), Blanco White, Mss. III/22, "Music" (ca. 1830), pages unnumbered.

94. SJL, Blanco White, Mss. III/21, "On Musical Sounds" (ca. 1830), 1–2, 1.

95. SJL, Blanco White, Mss. III/21, "On Musical Sounds" (ca. 1830), 3, 5–7.

96. SJL, Blanco White, Mss. III/21, "On Musical Sounds" (ca. 1830), 8–9, 16–17, 34–38, 41.

97. Francis Newman quoted in Daniel, ed., *Our Memories*, 24.
98. Daniel, ed., *Our Memories*, 24.
99. Anon., "Obituary," *The Times* (London), issue 34759, December 13, 1895, 6.
100. Robert Brown quoted in Daniel, ed., *Our Memories*, 81–82.
101. On Chladni, see Jackson, *Harmonious Triads*, 13–44.
102. Savart, "Researches on the Elasticity of Regularly Crystallized Bodies," 141–46, 141, 143, 145–46; Anon., "On the Structure of Metals, by M. Savart," *Quarterly Journal of Science, Literature, and Art* 28 (London, 1830): 163–67.
103. King's College London Archives (henceforth KCL), Wheatstone 5, "Rock-Crystal," in "Various: 8 Lectures on Sound," (ca. 1835).
104. Anon., "Royal Institution," *Morning Post* (London), issue 20371, March 22, 1836, 5; KCL, Wheatstone 5, "Lecture IV" in "Various: 8 Lectures on Sound" (ca. 1835); acoustic plates could also be used to examine earthquake phenomena, as argued in Gillin, "Seismology's Acoustic Debt," 65–82.
105. Myles Jackson has observed that the location of Wheatstone's experiments differed significantly from Chladni's earlier demonstrations of the same principles. While London provided Wheatstone with immense opportunities to popularize acoustic plate experiments, there was nowhere on the Continent of similar scale for Chladni to perform; he instead traveled the German lands, Holland, Brussels, and Paris to promote the phenomenon. See Jackson, "Charles Wheatstone: Musical Instrument Making," 106.
106. Charles Wheatstone, "On the Figures Obtained by Strewing Sand on Vibrating Surfaces, Commonly Called Acoustic Figures," in *Scientific Papers of Sir Charles Wheatstone*, 64–83, 67.
107. KCL, Wheatstone 5, "Lecture IV" in "Various: 8 Lectures on Sound" (ca. 1835).
108. L. Pearce Williams, *Michael Faraday*, 178.
109. Cantor, "Reading the Book of Nature," 69–81, 69.
110. Cantor, *Michael Faraday: Sandemanian and Scientist*, 170–72, 194–95, 226.
111. Martin, ed., *Faraday's Diary*, 1:329–30, 332–34.
112. Michael Faraday, "On a Peculiar Class of Acoustical Figures; and on Certain Forms Assumed by Groups of Particles upon Vibrating Elastic Surfaces," in Faraday, *Experimental Researches in Chemistry and Physics*, 314–58, 315, 317–18, 319–21, 327, 331.
113. "Letter 489: Charles Wheatstone to Faraday, 23 Mar., 1831," in James, ed., *Correspondence of Michael Faraday*, 1:556–57.
114. Faraday, "On a Peculiar Class of Acoustical Figures," 335, 338–39.
115. Martin, ed., *Faraday's Diary*, 1:336–37, 340–44, 355, 357.
116. Faraday, "On a Peculiar Class of Acoustical Figures," 341, 344–45.
117. Faraday, "On a Peculiar Class of Acoustical Figures," 349, 356, 357.
118. Faraday, "On a Peculiar Class of Acoustical Figures," 358; on Fresnel, see Cantor, *Optics after Newton*, 159–72.
119. Faraday quoted in Tyndall, *Faraday as a Discoverer*, 22.
120. Anon., "Royal Institution," *New Monthly Magazine and Literary Journal*, pt. 3 (London, 1831): 319; Anon., "June 10th.—Mr Faraday on the Arrangements Assumed by Particles on the Surfaces of Vibrating Elastic Bodies," *Journal of the Royal Institution of Great Britain* (London, 1831): 2:130–31.
121. Anon., "June 10th.—Mr Faraday," 2:130.

122. "Letter 489: Charles Wheatstone to Faraday, 23 Mar., 1831," in James, ed., *Correspondence of Michael Faraday*, 1:556–57.
123. Martin, ed., *Faraday's Diary*, 1:367.
124. Faraday, *Experimental Researches in Electricity*, 1–41, 42–75.
125. Martin, ed., *Faraday's Diary*, 2:424.
126. Cantor, Gooding, and James, *Michael Faraday*, 69.
127. Hamilton, *Faraday: The Life*, 235–36 and 243–44.
128. Tweney, "Inventing the Field," 31–47, 33–36; on field theory, see Tweney, "Representing the Electromagnetic Field," 687–700.
129. Tweney, "Faraday's Discovery of Induction," 189–209, 205; Tweney, "Stopping Time," 149–64, 151–53.
130. "Letter 563: Faraday to André-Marie Ampére, 5 Apr., 1832," in James, ed., *Correspondence of Michael Faraday*, 2:32–33.
131. "Letter 589: Faraday to Eilhard Mitscherlich, 15 Jun., 1832," in James, ed., *Correspondence of Michael Faraday*, 2:57–58; "Letter 610: Hans Christian Oersted to Faraday, 23 Aug., 1832," James, ed., *Correspondence of Michael Faraday*, 2:78; Williams, *Michael Faraday*, 137–38, 179.
132. "Letter 557: Faraday to John George Children and Sealed Note, 12 Mar., 1832," in James, ed., *Correspondence of Michael Faraday*, 2:26–27.
133. Pesic, *Music and the Making of Modern Science*, 198, 213; on Faraday's visual reasoning and electromagnetism, see Gooding, "From Phenomenology to Field Theory," 40–65, 51–52.
134. This continued until at least 1846 when, a year after finding that passing a ray of polarized light through a powerful electromagnetic field caused the plane of light to rotate, Faraday challenged the common view that electricity was propagated like a wave, or fluid, traveling along a wire in opposite directions. Instead, he proposed that magnetism and electricity behaved like stretched strings, transmitting vibrations across space. While specifying that such vibrations were different from "that which occurs on the surface of disturbed water, or the waves of sound in gases or liquids," Faraday contended that heat, light, magnetism, and electricity could all be understood as vibratory motion. See Michael Faraday, "Thoughts on Ray-Vibrations. To Richard Phillips, Esq.," in Faraday, *Experimental Researches*, 366–72, 367, 370; Cantor, Gooding, and James, *Michael Faraday*, 15, 76–77, 82.
135. Marrian, "On Sonorous Phaenomena," 382–84, 383, 384.
136. "Letter 1629: John Percy to Faraday, 23 Oct., 1844," in James, ed., *Correspondence of Michael Faraday*, 3:262–64, 263–64.
137. "Letter 1639: Faraday to William Robert Grove, 31 Oct., 1844," in James, ed., *Correspondence of Michael Faraday*, 3:273–74, 273.
138. Trevelyan, "On the Vibration of Heated Metals," 321–32, 321–22, 324–25, 327.
139. RS: HS 14.383, "John Robison to John Herschel" (17 Jan., 1831).
140. Anon., "Friday-Evening Proceedings at the Royal Institution of Great Britain," *Philosophical Magazine* 9 (London, January–June 1831): 461; Michael Faraday, "Trevelyan's Experiments on the Production of Sound during the Conduction of Heat," in Faraday, *Experimental Researches in Chemistry and Physics* 311–14, 311.
141. Faraday, "Trevelyan's Experiments on the Production of Sound," 312–13.
142. RI MS F/4/K, "Mr Trevelyan's Experiments," in "Friday Evenings, 1829–31" (April 29, 1831): 130.
143. Trevelyan, "Further Notice of the Vibration of Heated Metals," 85–86.

144. Arthur Trevelyan, "Theory of Heat" (ca. 1843), Wren Library, Trinity College Cambridge (henceforth WRN) Ref.15.b.80.3, 2.

145. Arthur Trevelyan, "Theory of the Vital Principle of Animal and Vegetable Life" (ca. 1843), WRN Ref.15.b.80.3, 4.

146. Shairp, Tait, and Adams-Reilly, *Life and Letters of James David Forbes*, 66; Smart, "Forbes, James David (1809–1868)."

147. Forbes, "Experimental Researches," 429–61, 429–30, 431, 433, 435–39, 446, 443–44.

148. Morus, *When Physics Became King*, 129–30; Crosbie Smith, *Science of Energy*, 32–39.

149. Forbes, "Experimental Researches," 454–55, 459–61.

150. "Letter 2704: John Tyndall to Faraday, 24 Jul., 1853," in James, ed., *Correspondence of Michael Faraday*, 4:540–41, 541; "Letter 2707: John Tyndall to Faraday, 28 Jul., 1853," in James, ed., *Correspondence of Michael Faraday*, 4:544–45, 545.

151. Gladstone, *Michael Faraday*, 116.

152. Gladstone, *Michael Faraday*, 212–13.

153. Charles J. B. Williams, "Observations on the Production and Propagation of Sound," 25–34, 25.

154. Charles J. B. Williams, "Observations on the Production and Propagation of Sound," 26, 29.

155. Royal Society Archives (hereafter RS): HS 18.146, "Wheatstone to John Herschel" (May 13, 1835).

156. KCL, Wheatstone 5, "Lecture VIII" in "Various: 8 Lectures on Sound" (ca. 1835).

157. KCL, Wheatstone 5, untitled note in "Various: 8 Lectures on Sound" (ca. 1835).

158. KCL, Wheatstone 5, "Lecture II" in "Various: 8 Lectures on Sound" (ca. 1835). These lectures were largely based on readings of Herschel's 1830 "Sound."

159. KCL, Wheatstone 5, "Lecture II" in "Various: 8 Lectures on Sound" (ca. 1835); on the development of early nineteenth-century architectural acoustics, see Tkaczyk, "Listening in Circles," 299–334.

160. KCL, Wheatstone 5, "Lecture IV" in "Various: 8 Lectures on Sound" (ca. 1835).

161. KCL, Wheatstone 5, "Lecture IV" in "Various: 8 Lectures on Sound" (ca. 1835).

162. KCL, Wheatstone 5, "Lecture VI" in "Various: 8 Lectures on Sound," (ca. 1835).

163. Kemp, "Wheatstone's Waves," 327.

164. KCL, Wheatstone 5, "King's College London. Experimental Philosophy Professor Wheatstone," 1–4.

165. Charles Wheatstone, "An Account of Some Experiments to Measure the Velocity of Electricity and the Duration of Electric Light," in *Scientific Papers of Sir Charles Wheatstone*, 84–96, 84, 85–88, 97–98.

166. Bowers, *Sir Charles Wheatstone*, 57.

167. Otis, *Networking*, 125.

168. Marsden and Smith, *Engineering Empires*, 191, 193–97.

169. Wheatstone, "On the Transmission of Musical Sounds," 47–63, 63.

170. Buchanan, *Robert Willis*, 46, 48, 50.

171. Wheatstone, "On the Vowel Sounds, and a Reed Organ-pipe," in *Scientific Papers of Sir Charles Wheatstone*, 348–67, 363, 367; Davies, "Instruments of Empire," 145–74, 156–59.

172. Bowers, *Sir Charles Wheatstone*, 34–35.

173. Anon., "Telegraphy," *The Times* (London), issue 29160, January 24, 1878, 11.

174. Anon., "Royal Institution Lectures: Wheatstone's Acoustic Discoveries," *Illustrated London News*, issue 1923, June 3, 1876, 546.

Chapter 2

1. Thanks to Fanny Gribenski for giving me this reference to William Herschel in George Eliot's *Middlemarch*.

2. Herschel, "Sound," 747–824, 755.

3. Secord, *Visions of Science*, 3, 7–8.

4. Jamie James, *Music of the Spheres*, 3–5; on supernatural accounts of music, see Chua, "Vincenzo Galileli," 17–29.

5. Nathan, *An Essay on the History and Theory of Music*, 7, 9.

6. Anon., "An Essay on the History and Theory of Music, and on the qualities, capabilities, and management of the human voice, second edition . . . by I. Nathan," *Westminster Review* 19 (London, July 1833): 242–47, 244.

7. Anderson, "On the Motion of Solids on Surface," 315–82, 315–16.

8. Morus, *When Physics Became King*, 192–93, 202; Robert W. Smith, "Remaking Astronomy," 154–73, 160.

9. Chapman, "Astronomical Revolution," 35–77, 42.

10. Crowe, "Herschel, Sir John Frederick William, First Baronet (1792–1871)"; Grattan-Guinness, "Mathematics and Mathematical Physics," 84–111; on Cambridge mathematics, see Warwick, *Masters of Theory*, 49–113.

11. William Herschel quoted in Lubbock, *Herschel Chronicle*, 31.

12. Lubbock, *Herschel Chronicle*, 32, 39–40; Hoskin, *Discoverers of the Universe*, 12.

13. Lubbock, *Herschel Chronicle*, 59–62; Hoskin, *Discoverers of the Universe*, 13.

14. Lubbock, *Herschel Chronicle*, 35–36; Buttmann, *Shadow of the Telescope*, 4–6.

15. Dolan, "Music as an Object of Natural History," 27–46, 41–42; Loughridge, "Celestial Mechanisms," 47–76, 48, 53, 60.

16. Werrett, "Disciplinary Culture," 87–98, 88, 91, 95.

17. Case, "'Land-marks of the Universe,'" 417–38; also see Cobb, "Inductivism in Practice," 21–54; Case, *Making Stars Physical*.

18. Royal Society Archives (hereafter RS) HS 18.331, Thomas Young to John Herschel (November 21, 1824); RS HS 23.249, John Herschel to Robert Brown (December 15, 1858); RS HS 18.417, Francis Wilson to John Herschel (June 17, 1831); RS HS 18.142, Charles Wesley to John Herschel (September 1828).

19. Herschel, "Sound," 747–824, 752–53.

20. Herschel, "Sound," 755.

21. Herschel, "Sound," 777–803, 804–6, 809.

22. Herschel, "Sound," 766, 772, 773.

23. Herschel, "Sound," 789.

24. RS HS 18.144, Charles Wheatstone to John Herschel (August 23, 1825).

25. RS HS 21.146, John Herschel to Charles Wheatstone (August 21, 1833); RS HS

18.145, Charles Wheatstone to John Herschel (August 23, 1833); RS HS 18.146, Charles Wheatstone to John Herschel (May 13, 1835).

26. Swade, "Babbage, Charles (1791–1871)."

27. Secord, *Visions of Science*, 62.

28. RS HS 2.243, John Herschel to Charles Babbage (February 28, 1830); RS HS 2.244, Charles Babbage to John Herschel (March 4, 1830).

29. Patterson, *Mary Somerville*, 71.

30. Herschel, "Sound," 810.

31. Herschel, "Sound," 810.

32. RS HS 2.249, John Herschel to Charles Babbage (March 24, 1830).

33. Herschel, "Physical Astronomy," 647–734, 647, 650–52.

34. Buttmann, *Shadow of the Telescope*, 60; also see Herschel, "Light," 341–586.

35. Secord, *Visions of Science*, 81, 87, 89.

36. Herschel, *A Preliminary Discourse on the Study of Natural Philosophy*, 83–84, 248–49.

37. Herschel, *A Preliminary Discourse on the Study of Natural Philosophy*, 260, 249.

38. Higgins, *The Philosophy of Sound, and History of Music*, 17, 43–44, 1–3, 6.

39. Martha Somerville, *Personal Recollections*, 16, 32, 132, 36, and 50.

40. Martha Somerville, *Personal Recollections*, 105.

41. Patterson, *Mary Somerville*, 38; Martha Somerville, *Personal Recollections*, 128–33, 142.

42. Secord, *Visions of Science*, 114–16.

43. Mary Somerville Papers (hereafter SOM), Dep. C. 352—MSSW—5, 8; Patterson, *Mary Somerville*, 71.

44. Mary Somerville, *Mechanism of the Heavens*, vi, xii, xv, lvii.

45. Mary Somerville, *Mechanism of the Heavens*, lvii–lviii, lxix; on the mathematical overlap between different natural phenomena, see Garber, "Reading mathematics," 31–54, 40.

46. SOM, Dep. C. 300—MSH—1, John Herschel to Mary Somerville (October 28, 1831), 281.

47. Herschel, "Review of Mechanism of the Heavens," 537–59; Galloway, "Review of Mechanism of the Heavens," 1–25; Anon., "Review: Mechanism of the Heavens," *Monthly Review*, n.s., 1 (January 1832): 133–41, 137.

48. Neeley, *Mary Somerville*, 101–2, 106–8; as Walter Cannon put it, "'Analogy,' not 'unity,' was the word commonly used to describe the connection of the sciences, for unity of method was a matter of aspiration than of accomplishment," in his "John Herschel and the Idea of Science," 215–39, 238.

49. Mary Somerville, *On the Connexion of the Physical Sciences*, vii.

50. Patterson, *Mary Somerville*, 125–28.

51. Mary Somerville's annotated copy of Herschel's "Sound" is at Girton College Library, Cambridge.

52. Mary Somerville, *On the Connexion of the Physical Sciences*, 138, 147–48, 155–59.

53. Mary Somerville, *On the Connexion of the Physical Sciences*, 163, 185–87.

54. Mary Somerville, *On the Connexion of the Physical Sciences*, 250–51, 250–52.

55. Neeley, *Mary Somerville*, 9–10, 38–40.

56. Mary Somerville, *On the Connexion of the Physical Sciences*, 408.

57. Mary Somerville, *On the Connexion of the Physical Sciences*, 412.

58. Secord, *Visions of Science*, 129, 133; Martha Somerville, *Personal Recollections*, 140.

59. Secord, *Visions of Science*, 119.
60. Taylor, "Review of *On the Connexion of the Physical Sciences*," 202–3, 202.
61. SOM, Dep. C. 369—MSB—11, David Brewster to Doctor Somerville (February 29, 1834), 357.
62. Brewster, "Review of *On the Connexion of the Physical Sciences*," 154–71, 159–60.
63. Mary Somerville, *On the Connexion of the Physical Sciences*, 2nd ed., 140, 144, 168–70.
64. Patterson, *Mary Somerville*, 143–45; Mary Somerville, *On the Connexion of the Physical Sciences*, 2nd ed., 173–74.
65. Mary Somerville, *On the Connexion of the Physical Sciences*, 2nd ed., 174.
66. Mary Somerville, *On the Connexion of the Physical Sciences*, 2nd ed., 174, 176–77.
67. Mary Somerville, *On the Connexion of the Physical Sciences*, 137–62; Mary Somerville, *On the Connexion of the Physical Sciences*, 2nd ed., 148–79.
68. Mary Somerville, *On the Connexion of the Physical Sciences*, 2nd ed., 1–2.
69. Letter 4869 (2:802a): Charles Wheatstone to Faraday (June 2, 1835), in James, ed., *Correspondence of Michael Faraday*, 6:637.
70. "Letter 4877 (2:917a): Faraday to Benjamin Abbott, 16 May, 1836," in James, ed., *Correspondence of Michael Faraday*. Vol. 6, 6:641.
71. Whewell, "Review of *On the Connexion of the Physical Sciences*," 54–68, 59–61.
72. Todhunter, *William Whewell*, 59.
73. RS HS 18.173, William Whewell to John Herschel (April 3, 1828).
74. Yeo, "Whewell, William (1794–1866)."
75. Whewell, *Astronomy and General Physics*, 1–2, 5, 118, 122–23, 123–24, 125.
76. Whewell, *History of the Inductive Sciences, vols. 1 and 3*.
77. Whewell, *History of the Inductive Sciences*, 2:233–34, 237.
78. Yeo, *Defining Science*, 12–15, 184; Snyder, *Reforming Philosophy*, 23, 206.
79. Mill, *System of Logic Ratiocinative and Inductive*, 147, 223–24; Snyder, *Reforming Philosophy*, 96.
80. Like Immanuel Kant (1724–1804), Whewell struggled with the problem of accounting for knowledge which experience alone could not provide, and this is why imagination was so important for his philosophy. In 1840, Whewell admitted that he had taken Kant's understanding of the universal character of space and time as a starting point but, while Kant regarded science as an objective truth above politics and religion, Whewell believed science should embrace morality. See Yeo, *Defining Science*, 9–10, 13–14; Snyder, *Reforming Philosophy*, 43–44, 46; Schaffer, "History and Geography of the Intellectual World," 201–31, 214–15.
81. Fisch, "A Philosopher's Coming of Age," 31–66, 32, 35, 64–65.
82. Whewell, *History of the Inductive Sciences*, 2:241, 247, 258, 264–65.
83. Whewell, *Philosophy of the Inductive Sciences*, 1:310.
84. Whewell, *Philosophy of the Inductive Sciences*, 25–27, 80–82, 164–65, 171, 274, 277, 283.
85. Whewell, *Philosophy of the Inductive Sciences*, 14, 17–8, 25, 287–91.
86. Whewell, *Philosophy of the Inductive Sciences*, 314–17, 319–21, 323; on the ear's ability to analyze sound waves, see Turner, "The OHM-Seebeck Dispute," 1–24, 3.
87. Whewell, *Philosophy of the Inductive Sciences*, 323, 325, 327.
88. Gibson, "The British Sermon, 1689–1901," 3–30, 9, 25; Knight, "Parish Preaching in the Victorian Era," 63–78, 72.

89. Rubinstein, "The Social Origins and Career Patterns," 715–30, 723–24.
90. Neeley, *Mary Somerville*, 78–79.
91. Rainbow, *Choral Revival in the Anglican Church*, 202–7.
92. Gatens, *Victorian Cathedral Music*, 9.
93. Gatens, *Victorian Cathedral Music*, 1–6.
94. Yates, *Oxford Movement and Anglican Ritualism*, 9; on High Churchmanship and politics, see Nockles, *Oxford Movement in Context*, 44–103.
95. Yates, *Oxford Movement and Anglican Ritualism*, 11–12, 13–16, 26; on Oxford Movement sermons, see Morris, "Preaching the Oxford Movement," 406–27.
96. Chadwick, *Spirit of the Oxford Movement*, 91–93.
97. Adelmann, *Contribution of Cambridge Ecclesiologists*, 1–2, 3, 4–7, 12–20, 49–52.
98. Pawley, "Wordsworth, Christopher (1807–1885)."
99. Wordsworth quoted in Strudwick, *Christopher Wordsworth*, 8.
100. Wordsworth, *Sacred Music*, 5.
101. Wordsworth, *Sacred Music*, 6, 9, 10, 12, 18.
102. Miller, *Church Music*, 7–8.
103. Miller, *Church Music*, 8, 13–14, 15–16.
104. Curthoys, *The Cardinal's College*, 201–37, esp. 202–9.
105. Walter Kerr Hamilton, *Church Music*, 3, 5, 9.
106. Chandler, "Liddon, Henry Parry (1829–1890)."
107. Chandler, *Life and Work of Henry Parry Liddon*, 64, 69–71.
108. Gresley, *A Sermon on Church*, 7, 13–14, 17.
109. Austen, *Scepticism of the Nineteenth Century*, 4–5, 28, 71, 73; also see Helmstadter and Lightman, eds., *Victorian Faith in Crisis*; Morris, "Preaching the Oxford Movement," 406–27, 416–17.
110. Nockles, *Oxford Movement in Context*, 105–7; Seccombe, "Jebb, John (1805–1886)."
111. Rainbow, *Choral Revival in the Anglican Church*, 26–33, 34; on Wesley, see Gatens, *Victorian Cathedral Music*, 128–46.
112. Gatens, *Victorian Cathedral Music*, 39–43.
113. Rainbow, *Choral Revival in the Anglican Church*, 4, 6.
114. Gatens, *Victorian Cathedral Music*, 9.
115. Jebb, *Divine Origin of Music*, 5–6.
116. Jebb, *Divine Origin of Music*, 9.
117. Jebb, *Divine Origin of Music*, 10, 14–15, 26, 29.
118. Hewett, *Choral Worship*, 2.
119. Hewett, *Choral Worship*, 2, 10, 13–14.
120. Brent, "Copleston, Edward (1776–1849)"; Corsi, *Science and Religion*, 21–22, 106–23.
121. William James Copleston, *Memoir of Edward Copleston*, 169.
122. Edward Copleston, *Church Music*, 14.

Chapter 3

1. Anon., "Parliamentary Acoustics," *Punch* (London), March 13, 1847, 112.
2. On sound at Parliament, see Gillin, *Victorian Palace of Science*, 30, 56–59, 140–43.
3. Haynes, *History of Performing Pitch*, 346–47.

4. "Regulator," "Concert Pitch," 85–86.

5. Gribenski, "Negotiating the Pitch," 173–92, 175–80; for a history of pitch standardization, see the forthcoming Gribenski, *Tuning the World*; on Anglo-French pitch, see Gribenski and Gillin, "Politics of Musical Standardization."

6. Jackson, *Harmonious Triads*, 183–230; Haynes, *History of Performing Pitch*, xxxiii–xxxiv.

7. On the demarcation of science, see Gieryn, "Boundary-work and the Demarcation of Science," 781–95.

8. See Sibum, "Reworking the Mechanical Value of Heat, 73–106; Schaffer, "Metrology, Metrication, and Victorian Values," 438–74, 443–44.

9. Schaffer, "Late Victorian Metrology," 457–78; Hunt, "The Ohm Is Where the Art Is," 48–63.

10. On the measurement of heat and electricity, see Wise and Smith, "Measurement, Work and Industry in Lord Kelvin's Britain," 147–73; also see Wise, ed., *Values of Precision*, 3–13; on measurement and revolutionary politics, see Alder, "A Revolution to Measure," 39–71.

11. Gooday, *Morals of Measurement*, 1–3, 23–30; Porter, *Trust in Numbers*.

12. Gooday, *Morals of Measurement*, 9–10.

13. Henry C. Lunn, "Musical Pitch," 663–65, 663.

14. Haynes, *History of Performing Pitch*, 348.

15. Anon., "An Imperial Pitchfork," *Spectator* (London), August 28, 1858, 910–11.

16. Anon., "An Imperial Pitchfork," 910–11.

17. Anon., "An Imperial Pitchfork," 910–11.

18. Anon., "An Imperial Pitchfork," 910–11.

19. On the limited extent of *laissez-faire* government, see Mandler, "Introduction: State and Society in Victorian Britain," 1–21, 1–2; Daunton, *State and Market in Victorian Britain*, 4–5; Hilton, "Moral Disciplines," 224–46, 225; Parry, "Liberalism and Liberty," 71–100, 71.

20. Anon., "Tuning-forks and Musical Pitch," *Chamber's Journal of Popular Literature Science and Arts* 34, no. 346 (August 18, 1860): 98–101, 100–101.

21. Davis, "Dilke, Sir (Charles) Wentworth."

22. British Library (henceforth BL), Add. Ms. 41771, Papers of Sir George Smart, vol. 1, Sydney Smith to William Hawes (August 21, 1844), 121.

23. Sharp, "Horsley, Charles Edward (1821–1876)."

24. Warrack, "Smart, Sir George Thomas (1776–1867)."

25. BL, Add. Ms. 63589, Papers of Sir George Smart, George Smart to William Hawes (March 26, 1856), 11; BL, Add. Ms. 63589, Papers of Sir George Smart, George Smart to William Hawes (March 19, 1856), 10; BL, Add. Ms. 63589, Papers of Sir George Smart, George Smart to William Hawes (December 1, 1834), 5.

26. BL, Add. Ms. 41771, Papers of Sir George Smart, vol. 1, George Smart to Dr Elvey (February 27, 1841), 120.

27. Mary Elvey, *Life and Reminiscences of George J. Elvey*, 70, 73, 144, 132; Newmarch, "Elvey, Sir George Job (1816–1893)."

28. Bennett, *Life of William Sterndale Bennett*, 249–54; O'Leary, "Sir William Sterndale Bennett," 123–45, 123–32.

29. Squire, "Donaldson, John (1789/90–1865)."

30. Olleson, "Potter, (Philip) Cipriani Hambley (1792–1871)."

31. Hallé quoted in Kennedy, *Autobiography of Charles Hallé*, 191; Kennedy, *Hallé Tradition*, 6–94; Kennedy, *Autobiography of Charles Hallé*, 115–64.
32. Stainer, "Character and Influence of the Late Sir Frederick Ouseley," 25–39, 26.
33. Shaw, "Ouseley, Sir Frederick Arthur Gore, Second Baronet (1825–1889)."
34. Ouseley and Dykes, *Choral Worship of the Church*, 5, 8.
35. Maitland, "Hullah, John Pyke (1812–1884)."
36. Adelmann, *Contribution of Cambridge Ecclesiologists*, 15, 50–51.
37. Haynes, *History of Performing Pitch*, 337; on tuning forks, see Jackson, "From Scientific Instruments to Musical Instruments," 201–23.
38. Boase, "Collard, Frederick William (bap. 1772, d. 1860)."
39. With the introduction of "pistons" between the key-slips to replace the organ's traditional composition pedals, Willis boasted that this organ "was the great pioneer of the improved Pneumatic movement. A child could play the keys with all the stops drawn," quoted in Sumner, *Father Henry Willis*, 18. On the organ industry, see Thistlewaite, *Making of the Victorian Organ*, 256–91.
40. Kent, "Willis, Henry (1821–1901)."
41. Thistlewaite, *Making of the Victorian Organ*, 296.
42. E. I. Carlyle, "Cowie, Benjamin Morgan (1816–1900)."
43. Hove Public Library (henceforth HPL), Testimonial and Other Letters to and concerning W. M. Pole, 1814–1900. Engineer and Musician, J. Jebb to Prof. Christie (December 4, 1843), 142; Various Testimonials (December 3, 1836), 124.
44. HPL, Testimonial and Other Letters, F. A. G. Ouseley to W. Pole (July 1, 1859), 167; HPL, Testimonial and Other Letters, F. A. G. Ouseley to W. Pole (January 25, 1860), 168.
45. HPL, Testimonial and Other Letters, F. A. G. Ouseley to the President of St John's College, Oxford (June 5, 1860), 169; Pole, *Some Short Reminiscences*, 1–13; Anon., "Mr William Pole," *The Times* (London), issue 36339, December 31, 1900, 8.
46. Wylde, *Music and Its Art-Mysteries*, 107.
47. Wylde, *Harmony and the Science of Music*; the copy of this held in St. John's College, Cambridge, was original owned by astronomer John Couch Adams.
48. Ringer, "On the Alteration of the Pitch of Sound," 275–81, 276.
49. Morgan, *Four Biographical Sketches*, 161–79.
50. Anon., "Uniform Musical Pitch," *The Times* (London) issue 23324, June 4, 1859, 12.
51. Royal Society of Arts Archives, London (hereafter RSA) PR/GE/112/12/96, "Minutes of Various Committees," 100, 116, 124, 136, 138; Richardson, *Thomas Sopwith*, 272.
52. Anon., "English Committee on Musical Pitch," *Spectator* (London), June 18, 1859, 639–40, 639.
53. In favor of considerably reducing the pitch were J. H. Grenfell from Penzance, William Lockyer, John Howard of Trinity College Dublin's University Choral Society, H. Perkes of the Worcester Harmonium Society, Graeff Nicholes, S. Reeves, J. H. Frobisker of the Halifax Piano Forte, Harmonium and Music Repository, and John Ireland. RSA PR/GE/121/10/5, Musical Pitch, J. H. Grenfell to Foster (September 8, 1859); RSA PR/GE/121/10/5, Musical Pitch, William Lockyery (September 10, 1859); RSA PR/GE/121/10/5, Musical Pitch, J. H. Grenfell to Foster (September 8, 1859); RSA PR/GE/121/10/5, Musical Pitch, John Howard to Foster (October 15, 1859);

RSA PR/GE/121/10/5, Musical Pitch, H. Perkes to Foster (September 6, 1859); RSA PR/GE/121/10/5, Musical Pitch, Graeff Nicholes to Foster (September 13, 1859); RSA PR/GE/121/10/5, Musical Pitch, S. Reeves to Foster (January 7, 1860); RSA PR/GE/121/10/5, Musical Pitch, J. H. Frobisher to Foster (September 12, 1859); RSA PR/GE/121/10/5, Musical Pitch, John A. Ireland to Foster (September 13, 1859).

54. RSA PR/GE/121/10/5, Musical Pitch, W. Mason to Foster (September 6, 1859); RSA PR/GE/121/10/5, Musical Pitch, W. Mason to Foster (September 5, 1859).

55. RSA PR/GE/121/10/5, Musical Pitch, Charles Saldman to Foster.

56. RSA PR/GE/121/10/5, Musical Pitch, E. Sheperd to Foster (September 8, 1859).

57. RSA PR/GE/121/10/5, Musical Pitch, J. Dyson to Foster (September 6, 1859).

58. RSA PR/GE/121/10/5, Musical Pitch, Belsic Palmer to Foster (January 10, 1860); RSA PR/GE/121/10/5, Musical Pitch, Matheres Burges to Foster (September 12, 1859).

59. *Uniform Musical Pitch. Minutes of a Meeting of Musicians*, 1–8, 5.

60. RSA PR/GE/121/10/5, Musical Pitch, George Hogarth to Foster (December 2, 1859).

61. RSA PR/GE/121/10/5, Musical Pitch, W. W. Price to Foster (September 18, 1859); RSA PR/GE/121/10/5, Musical Pitch, W. W. Price to Foster (December 1, 1859).

62. RSA PR/GE/121/10/5, Musical Pitch, J. Surman to the Committee Secretary (September 8, 1859).

63. RSA PR/GE/121/10/5, Musical Pitch, W. Holland to Foster (September 12, 1859).

64. RSA PR/GE/121/10/5, Musical Pitch, Alfred Nicholson to Foster (January 4, 1860).

65. RSA PR/GE/121/10/5, Musical Pitch, H. Distin to Foster (January 5, 1860).

66. RSA PR/GE/121/10/5, Musical Pitch, Rendall Rose to Foster (January 11, 1860).

67. RSA PR/GE/121/10/5, Musical Pitch, Henry Willis to Foster (December 28, 1859).

68. "Uniform Musical Pitch," *Journal of the Society of Arts*, 2, 3, 4.

69. "Uniform Musical Pitch," *Journal of the Society of Arts*, 3.

70. Pantalony, *Altered Sensations*, 10–12, 23; Jackson, *Harmonious Triads*, 209–10.

71. "Uniform Musical Pitch," *Journal of the Society of Arts*, 4, 5; the British committee avoided questions of temperament and focused exclusively on mathematical theory. For historical discussions over the increased adoption of equal temperament to replace just intonation, see Jackson, *Harmonious Triads*, 153–57.

72. "Uniform Musical Pitch," *Journal of the Society of Arts*, 5.

73. There were different ways of counting vibrations. While the British referred to complete oscillations, the French counted each vibration as two movements. Here Donkin used the French method, referring to an A_{850} which would be A_{425} if measured in completed vibrations. In this fashion C_{512} would be C_{1024} in French vibrations.

74. RSA PR/GE/121/10/5, Musical Pitch, W. F. Donkin to Rowden (June 20, 1860).

75. "Uniform Musical Pitch," *Journal of the Society of Arts*, 5, 6.

76. "Uniform Musical Pitch," *Journal of the Society of Arts*, 2, 6.
77. Schaffer, "Metrology, Metrication, and Victorian Values," 443–49.
78. Herschel, *Familiar Lectures on Scientific Subjects*, 421–22, 426, 181, 427, 428–29, 430, 430–31, 433.
79. Royal Society Archives (hereafter RS) HS 13.203, F. A. Gore Ouseley to John Herschel (September 16, 1839); RS HS 13.204, F. A. Gore Ouseley to John Herschel (October 8, 1839); RS HS 13.205, F. A. Gore Ouseley to John Herschel (November 14, 1839); RS HS 13.206, Harriot Ouseley to John Herschel (1839).
80. RS HS 27.82, John Herschel to the Chairman of the Committee (June 14, 1859).
81. RS HS 27.82, John Herschel to the Chairman of the Committee (June 14, 1859).
82. RS HS 27.82, John Herschel to the Chairman of the Committee (June 14, 1859).
83. RS HS 27.82, John Herschel to the Chairman of the Committee (June 14, 1859).
84. John Herschel, "Uniform Musical Pitch," *Leeds Mercury*, issue 6985, August 2, 1859; Mary Somerville Papers, Dep.C.376—MSDIP—18, cutting entitled "Uniform Musical Pitch," June 14, 1859; RSA PR/GE/112/12/96, "Minutes of Various Committees," 100. I am grateful to Fanny Gribenski for making me aware of this volume.
85. Werrett, "Disciplinary Culture," 87–98, 88–89, 94–95.
86. RS HS 17.347, Thomas Thompson to John Herschel (February 10, 1860).
87. Hullah, *Grammar of Vocal Music*, viii.
88. RS HS 17.348, Thomas Thompson to John Herschel (February 13, 1860).
89. Pantalony, *Altered Sensations*, 83–96.
90. "Uniform Musical Pitch," *Journal of the Society of Arts*, 6.
91. "Uniform Musical Pitch," *Journal of the Society of Arts*, 7.
92. "Uniform Musical Pitch," *Journal of the Society of Arts*, 7, 8.
93. Anon., "Musical Pitch," *Liverpool Mercury*, issue 3988, November 23, 1860.
94. Anon., "The Society of Arts and the Exhibition of 1862," *The Times* (London), issue 23466, November 17, 1859, 7; Anon., "Uniform Musical Pitch," *Journal of the Society of Arts*, 15–6.
95. Anon., "Tuning-forks and Musical Pitch," 98–101, 98, 99–100, 99, 101, 100–101.
96. Lunn, "Musical Pitch," 664.
97. RS HS 23.249, John Herschel to Robert Brown (December 15, 1858).
98. Robert Brown, *Elements of Musical Science* 184–85, viii.
99. Herschel, *On Musical Scales*, 3, 3–4, 4.
100. Herschel, *On Musical Scales*, 5, 7–8, 17.
101. Colin Brown, "Reviews: *A Demonstration of the Musical Scale*," 436.
102. RS HS 5.143, William H. Callcott to John Herschel (February 25, 1867); RS HS 5.144, William H. Callcott to John Herschel (October 21, 1868).
103. RS HS 13.203, Francis J. Hughes to John Herschel (undated, ca. 1867).
104. RS HS 13.203, Francis J. Hughes to John Herschel (undated, ca. 1867).
105. Pole, "Explanation of the Musical Scale," 259–63, 259; also see Hullah, *Chromatic Scale*.
106. Pole, "Explanation of the Musical Scale," 260–61.
107. Airy, *On Sound and Atmospheric Vibrations*, 217–19, 222–24.
108. RS HS 1.308, George Biddell Airy to John Herschel (March 27, 1868).
109. RS HS 1.310, George Biddell Airy to John Herschel (May 26, 1868).

110. Anon., "Musical Pitch," *Journal of the Society of Arts* 18, no. 935 (London), October 21, 1870, 906–8, 906.

111. Haynes, *History of Performing Pitch*, 355–56.

112. Lunn, "Musical Pitch," 663.

113. Lunn, "Musical Pitch," 664.

114. Lunn, "Musical Pitch," 665.

115. Haynes, *History of Performing Pitch*, 356–7.

116. Sumner, *Father Henry Willis*, 33.

117. Anon., "Orchestral Tuning," *Musical Standard* (London), June 26, 1875, 409–10, 410; Schaffer, "Astronomers Mark Time," 115–45.

118. Anon., "Orchestral Tuning," 409–10, 410.

119. MacMahon, "Ellis [formerly Sharpe], Alexander John (1814–1890)."

120. Herbert and Barlow, *Music and the British Military*, 211.

121. Anon., "The Question of a Standard Musical Pitch," *Moonshine* (London), June 27, 1885, 302.

122. RSA PR/GE/121/10/6, Miscellaneous Letters Relating to Pitch, "Report by A. J. Ellis to Committee on Pitch, 17 May, 1886," 1–2, 2, 4–5.

123. RSA PR/GE/121/10/6, Miscellaneous Letters Relating to Pitch, C. Mahillon to the Secretary of the Society of Arts (April 27, 1886).

124. RSA PR/GE/121/10/6, Miscellaneous Letters Relating to Pitch, Fernand Mahillon to Alex J. Ellis (May 15, 1886).

125. RSA PR/GE/121/10/6, Miscellaneous Letters Relating to Pitch, "Report by A. J. Ellis to Committee on Pitch, 17 May, 1886," 1–9.

126. RSA PR/GE/119/39/94, Letters of Stainer on Pitch, John Stainer to Mr Wood (January 29, 1886).

127. RSA PR/GE/119/39/94, Letters of Stainer on Pitch, John Stainer to Mr Wood (February 16, 1886).

128. RSA PR/GE/121/10/6, Miscellaneous Letters Relating to Pitch, "A National Musical Pitch. Reasons for Adopting the 'Regulation' Military Band Pitch, Already Established at the School of Music, Kneller Hall, viz., C_{542}, or A_{455} Vibrations, 10 May, 1886."

129. Anon., "Royal Academy of Music," *Musical Times, and Singing Class Circular* 37, no. 636 (February 1, 1896): 97–98, 97.

130. Weinstein, "Musical Pitch and International Agreement," 341–43, 342–43; on twentieth-century international pitch negotiations, see Gribenski, "Negotiating the Pitch," 180–85.

131. Anon., "International Standard Musical Pitch," *Journal of the Royal Society of Arts* 98, no. 4810 (December 16, 1949): 74–89, 74, 75.

Chapter 4

1. Howse, *Greenwich Time and the Longitude*, 92–95; Marsden and Smith, *Engineering Empires*, 20; also see E. P. Thompson, "Time, Work-Discipline, and Industrial Capitalism," 56–97.

2. Sterne, *Audible Past*, 3.

3. Wilfrid Airy, ed., *Autobiography of Sir George Biddell Airy*, 37; Chapman, "George Biddell Airy," 103–10.

4. Wilfrid Airy, ed., *Autobiography of Sir George Biddell Airy*, 57, 11–12.

5. George Biddell Airy, "Astronomer Royal's Remarks," 339–43, 339.

6. Royal Observatory Greenwich Papers, Cambridge University Library (hereafter RGO) 6/427, Miscellaneous Letters, Airy to the Editors of the Philosophical Magazine and Journal (April 10, 1848), 168–74, 168–70.

7. RGO/6/427, Airy to James Challis (April 13, 1848), 175; RGO/6/427, James Challis to Airy (April 14, 1848), 176.

8. RGO/6/428, Airy to the Editors of the Philosophical Magazine and Journal (May 17, 1849), 161–70.

9. Stephen, "Morgan, Augustus De (1806–1871)."

10. RGO/6/433, Pure Mathematics and Miscellaneous Science, 1857–1858, Augustus De Morgan to Airy (December 31, 1857), 310–11, 310.

11. RGO/6/433, Pure Mathematics and Miscellaneous Science, 1857–1858, Augustus De Morgan to Airy (December 31, 1857), 311.

12. RGO/6/433, Pure Mathematics and Miscellaneous Science, 1857–1858, Augustus De Morgan to Airy (December 31, 1857), 311.

13. RGO/6/433, Airy to Augustus De Morgan (January 4, 1858), 312; RGO/6/433, Augustus De Morgan to Airy (January 21, 1858), 313.

14. Royal Society Archives (hereafter RS) HS 6.307, De Morgan to John Herschel (January 1, 1858).

15. RS HS 6.283, De Morgan to John Herschel (June 4, 1856).

16. De Morgan, "On the Beats of Imperfect Consonances," 129–45, 17; explored in Jackson, *Harmonious Triads*, 57.

17. De Morgan, "On the Beats of Imperfect Consonances," 141.

18. De Morgan, "On the Beats of Imperfect Consonances," 141–42.

19. De Morgan, "On the Beats of Imperfect Consonances," 142.

20. De Morgan, "On the Beats of Imperfect Consonances," 142.

21. De Morgan, "On the Beats of Imperfect Consonances," 142.

22. On later sonic discussions, see RGO/6/433, Augustus De Morgan to Airy (October 12, 1858), 247; RGO/6/433, Airy to Augustus De Morgan (October 15, 1858), 249; RGO/6/433, Augustus De Morgan to Airy (October 16, 1858), 252; RGO/6/433, Airy to Augustus De Morgan (October 19, 1858), 254–55.

23. George Biddell Airy, *On Sound and Atmospheric Vibrations*, v, 3, 23, 24–25.

24. George Biddell Airy, *On Sound and Atmospheric Vibrations*, 137–38.

25. Airy's "0^s.2" denotes two-tenths of a second.

26. George Biddell Airy, *On Sound and Atmospheric Vibrations*, 138, 211, 214.

27. Wilfrid Airy, ed., *Autobiography of Sir George Biddell Airy*, 273.

28. Airy quoted in Wilfrid Airy, ed., *Autobiography of Sir George Biddell Airy*, 274.

29. Wilfrid Airy, ed., *Autobiography of Sir George Biddell Airy*, 275, 277, 279; on British mathematics, see Flood, Rice, and Wilson, eds., *Mathematics in Victorian Britain*; Richards, "*Mathematics in Victorian Britain* by Raymond Flood; Adrian Rice; Robin Wilson. Review," 853–55.

30. Gillin, *Victorian Palace of Science*, 214–64; McKay, *Big Ben*, 89–120, 163–71; on the cultural centrality of bells, see Corbin, *Village Bells*.

31. RGO/6/599, Charles May to Airy (December 26, 1859), 435.

32. Gillin, *Victorian Palace of Science*, 256; McKay, *Big Ben*, 89–120, 163–72.

33. RGO/6/599, Airy to Charles May (December 29, 1859), 437–40.

34. RGO/6/609, Clock for New Palace Westminster, 1857 to 1861, Airy to William Cowper (May 30, 1860), 77–78.

35. McKay, *Big Ben*, 170–71.
36. RGO/6/609, "Astronomer Royal Reports on the Clock and Bells of the New Westminster Palace," (April 21, 1860), 30–48, 39–41, 43, 43–44.
37. RGO/6/609, Airy to Augustus De Morgan (March 26, 1860), 121.
38. RGO/6/609, Augustus De Morgan to Airy (March 26, 1860), 122–23.
39. RGO/6/609, Airy to F. Dent (March 23, 1860), 135; RGO/6/609, F. Dent to Airy (March 26, 1860), 136.
40. RGO/6/609, Airy to Robert Werner (March 31, 1860), 234; RGO/6/609, Airy to Robert Werner (April 12, 1860), 235.
41. RGO/6/609, "First Report by Robert Werner" (April 1860), 237.
42. RGO/6/609, "Second Report by Robert Werner" (April 1860), 238–42.
43. RGO/6/609, Airy to Robert Werner (April 27, 1860), 244.
44. RGO/6/609, Airy to William Whewell (March 24, 1860), 250; RGO/6/609, Airy to William Whewell (April 14, 1860), 254–55.
45. RGO/6/609, William Whewell to Airy (April 3, 1860), 251–53.
46. RGO/6/609, William Whewell to Airy (April 17, 1860), 256.
47. RGO/6/609, Airy to William Whewell (April 14, 1860), 254–55.
48. RGO/6/609, "The Astronomer Royal Reports on the Clock and Bells of the New Westminster Palace" (April 21, 1860), 30–48, 35–36, 36.
49. RGO/6/609, "The Astronomer Royal Reports on the Clock and Bells of the New Westminster Palace," (April 21, 1860), 36–37, 38.
50. RGO/6/609, Airy to William Whewell (April 21, 1860), 259–60.
51. RGO/6/609, Airy to William Whewell (April 23, 1860), 261.
52. RGO/6/609, William Whewell to Airy (April 23, 1860), 262.
53. RGO/6/609, Airy to Robert Werner (June 5, 1860), 246.
54. RGO/6/609, William Cowper to Airy (June 4, 1860), 81.
55. RGO/6/609, Airy to William Cowper (June 18, 1860), 83–84, 91.
56. RGO/6/609, Alfred Bonham-Carter to Airy (June 16, 1860), 89, 90.
57. RGO/6/609, Rev William Taylor to William Cowper (June 12, 1860), 86–88, 87.
58. RGO/6/609, Airy to Alfred Bonham-Carter (June 18, 1860), 93.
59. RGO/6/609, Faber James to Airy (June 27, 1860), 171; RGO/6/609, Airy to Faber James (June 28, 1860), 172; RGO/6/609, "Turle's Score-Chimes" (June 1860), 176; RGO/6/609, Faber James to Airy (July 2, 1860), 177; McKay, *Big Ben*, 169.
60. RGO/6/609, James Turle to Faber James (June 21, 1860), 97.
61. RGO/6/609, Airy to Faber James (July 3, 1860), 178.
62. RGO/6/609, James Turle to Faber James (July 5, 1860), 180.
63. RGO/6/609, Airy to Alfred Austin (June 23, 1860), 98.
64. RGO/6/609, William Cowper to Airy (June 26, 1860), 99.
65. Stanley Smith, "Pole, William (1814–1900)."
66. RGO/6/609, Airy to the Clerk of Works (August 6, 1860), 113; RGO/6/609, Airy to Alfred Bonham-Carter (August 7, 1860), 116.
67. RGO/6/609, William Pole to Airy (August 4, 1860), 211.
68. RGO/6/609, Airy to William Pole (August 6, 1860), 212–13, 212, 213.
69. RGO/6/609, William Pole to Airy (August 6, 1860), 214–15, 214.
70. RGO/6/609, attached note from William Pole entitled "Discussion on Westminster Bells, 8 Nov., 1859," (August 6, 1860), 216.
71. RGO/6/609, Airy to William Pole (August 7, 1860), 217–18.

72. RGO/6/609, William Pole to Airy (August 18, 1860), 224.

73. RGO/6/609, "Abstract of Professor Pole's Report on the Westminster Bells" (August 29, 1860), 225–26.

74. Airy, *On Sound*, 247, 227–28, 228–29, 231–35, 230.

75. RGO/7/238, J. W. Blakesley to Airy (January 29, 1881), 14.

76. RGO/7/238, Airy to J. W. Blakesley (February 4, 1881), 14.

77. Schaffer, "Astronomers Mark Time," 115–45.

78. Wilfrid Airy, ed., *Autobiography of Sir George Biddell Airy*, 2.

79. Howse, *Greenwich Time and the Discovery of the Longitude*, 83; on Airy and the time-ball network, see Holmes, "The Astronomer Royal, the Hydrographer and the Time Ball," 381–406, 385–92.

80. Howse, *Greenwich Time and the Longitude*, 95–97; also see Morus, "'The nervous system of Britain,'" 455–75.

81. Howse, *Greenwich Time and the Longitude*, 97–103, 103–4; on time-guns versus clocks, see Gillin, *Victorian Palace of Science*, 262, 267–68.

82. RGO 6/614, Galvanic Time Balls, Time Signals, and External Clocks, regulated by the Royal Observatory, 1858 June to 1860 December, Letter to Airy (November 25, 1859), 100.

83. RGO/6/614, Airy to George Hamilton (November 26, 1859), 102–3, 102, 103.

84. Brück, "Smyth, Charles Piazzi (1819–1900)"; Schaffer, "Metrology, Metrication, and Victorian Values," 438–74, 449–59.

85. Kinns, "Early History of the Edinburgh Time Ball and Time Gun," 264–90, 275–76.

86. Royal Observatory Library, Edinburgh (hereafter ROE), A.18.101, Smyth, "Time Ball and Mean Time Clock"; ROE A.20.156, Smyth, "Edinburgh, Scotland, Royal Observatory" (1882).

87. RGO/6/614, Airy to John Hewat (October 1, 1860), 104; on Airy's commitment to time balls, see Holmes, "The Astronomer Royal, the Hydrographer and the Time Ball," 388.

88. RGO/6/614, Airy to John Hewat (October 1, 1860), 104.

89. RGO/6/614, John Hewat to Airy (October 6, 1859), 106.

90. RGO/6/615, Galvanic Time Balls, Time Signals and Sympathetic Clocks Regulated by the Royal Observatory, 1861 January to 1866 February, cutting from the *Scotsman*, January 28, 1861, "The Time Gun," 47; RGO/6/615, cutting from the *Edinburgh Evening Courant*, "The Audible Time Signal," 48.

91. RGO/6/615, cutting from the *Scotsman*, "The Time-Gun Signal," 54.

92. RGO/6/615, Charles Smyth to Airy (1861), 55.

93. Hislop produced two versions of this map, the first in 1862 and the second in 1863. Edinburgh University Library Centre for Research Collections (CRC), Map. PC.16, "Hislop's Time-Gun Map of Edinburgh and Leith" (1862); CRC, Map.PC.60, "Hislop's Time-Gun Map of Edinburgh and Leith" (1863).

94. Smyth, "On the Methods Adopted to Secure Extreme Accuracy," 136–52, 137, 146–49.

95. RGO/6/614, cutting from the *Gateshead Observer*, February 19, 1859, "The 'Time Ball,'" 42.

96. RGO/6/615, cutting from the *Newcastle Daily Chronicle*, September 1863, "Time Signals and Chronometers—by Mr James Mather, of South Shields," 112.

97. RGO/6/615, Airy to James Mather (September 7, 1863), 115–16, 115.

98. RGO/6/615, Airy to James Mather (September 7, 1863), 116.

99. RGO/6/615, Airy to James Mather (September 7, 1863), 116.

100. RGO/6/615, cutting, September 11, 1863, "River Tyne Commission," 117; ROE A.10.45, Smyth, "Letter Book, 1863," Edward Barnard to Charles Piazzi Smyth (September 27, 1863), letter 19.

101. RGO/6/615, C. F. Varley to Airy (June 10, 1864), 137–38.

102. RGO/6/615, cutting from the *Reader*, December 12, 1863, "Science: On Time and Time-guns," 239.

103. ROE A.10.45, Smyth, "Letter Book, 1863," Nathaniel Holmes to Charles Piazzi Smyth (September 25, 1863), letter 15; on Glasgow's time-guns, see Black and Clarke, "Glasgow Time Signals," 256–93, 275–82.

104. ROE A.10.45, Smyth, "Letter Book, 1863," Nathaniel Holmes to Charles Piazzi Smyth (October 1, 1863), letter 27.

105. ROE A.10.45, Smyth, "Letter Book, 1863," R. S. Symington to Charles Piazzi Smyth, (October 15, 1863), letter 36.

106. RGO/6/615, cutting from the *Reader*, December 12, 1863, "Science: On Time and Time-guns," 239.

107. ROE A.10.45, Smyth, "Letter Book, 1863," Nathaniel Holmes to Charles Piazzi Smyth (September 24, 1863), letter 10.

108. ROE A.10.45, Smyth, "Letter Book, 1863," Walter White to Charles Piazzi Smyth (October 28, 1863), letter 57.

109. ROE A.12.54, Smyth, "Correspondence re Time Gun, Letters in and out, 1870, 1872, 1873, and 1876," Charles Piazzi Smyth to J. D. Maewick, City Clerk (November 5, 1872).

110. ROE A.12.54, Smyth, "Correspondence re Time Gun, Letters in and out, 1870, 1872, 1873, and 1876," James Ritchie to Charles Piazzi Smyth (January 25, 1876).

111. ROE A.12.54, Smyth, "Correspondence re Time Gun, Letters in and out, 1870, 1872, 1873, and 1876," Charles Piazzi Smyth to William Skinner, Town Clerk (April 19, 1876).

112. ROE A.12.54, Smyth, "Correspondence re Time Gun, Letters in and out, 1870, 1872, 1873, and 1876," Charles Piazzi Smyth to A. M. Cunynghame (November 14, 1872).

113. ROE A.12.54, Smyth, "Correspondence re Time Gun, Letters in and out, 1870, 1872, 1873, and 1876," A. M. Cunynghame to Charles Piazzi Smyth (November 29, 1872).

114. ROE A.12.54, Smyth, "Correspondence re Time Gun, Letters in and out, 1870, 1872, 1873, and 1876," Office of the Director General of Telegraphs in India to Charles Piazzi Smyth (April 11, 1870); RGO/6/617, Galvanic Time Balls, Time Signals and Sympathetic Clocks Regulated by the Royal Observatory, 1870 Feb. to 1875 Oct., E. D. Latham to Airy (February 3, 1873), 110; RGO/6/617, R. S. Culley to Airy (July 21, 1873), 143; RGO/6/617, Airy to R. S. Culley (July 29, 1873), 145.

115. Kinns, "Early History of the Edinburgh Time Ball and Time Gun," 277.

Chapter 5

1. Letter from James Clerk Maxwell to John William Strutt, May 26, 1873, in Harman, ed., *Scientific Letters and Papers of James Clerk Maxwell*, 2:856–57.

2. Gavroglu, "Strutt, John William, Third Baron Rayleigh (1842–1919)."

3. Hui, *Psychophysical Ear*, 55.
4. Meulders, *Helmholtz: From Enlightenment to Neuroscience*, 153–54 and 180–81; on Helmholtz's career, see Cahan, ed., *Hermann Von Helmholtz*.
5. Hui, *Psychophysical Ear*, 56.
6. Helmholtz, *On the Sensations of Tone*; Tyndall, *Sound*; Strutt, *Theory of Sound*; Beyer, *Sounds of Our Times*, 55.
7. Turner, "The Ohm-Seebeck Dispute," 1–24, 1.
8. Weber borrowed the expression from Friedrich Schiller; see Jenkins, "Disenchantment, Enchantment and Re-enchantment," 11–32, 11.
9. Chua, "Vincenzo Galileli," 17–29, 20–21.
10. Pantalony, *Altered Sensations*, 19.
11. DeYoung, *A Vision of Modern Science*, 90–94, 97; on Tyndall and prayer, see M'Cann, *The Inter-Relations of Prayer, Providence, and Science*; Groser, *On Prayer*; on science and theology, see Stanley, *Huxley's Church and Maxwell's Demon*.
12. For a revision, see Jenson, "Return to the Wilberforce-Huxley Debate," 161–79; also see Brooke, *Science and Religion*.
13. Altholz, "A Tale of Two Controversies," 50–59, 50–51, 55–56.
14. Zon, "The 'Non-Darwinian' Revolution," 196–226, 196; on Goddard, see Zon, "'Spiritual' Selection" 215–35.
15. Spencer, "Progress: Its Law and Cause," 445–85, 456, 461–62.
16. Spencer, "Progress: Its Law and Cause," 463.
17. Spencer, "Progress: Its Law and Cause," 464.
18. Zon, "The 'Non-Darwinian' Revolution," 205; on evolution and Victorian musical composition, see Allis, *Parry's Creative Process*.
19. DeYoung, *A Vision of Modern Science*, 16.
20. Barton, "John Tyndall, Pantheist," 111–34, 114.
21. DeYoung, *A Vision of Modern Science*, 14–17; Reidy, "Introduction: John Tyndall," 1–13; Eve and Creasey, *Life and Work of John Tyndall*, 38, 56–58.
22. Eve and Creasey, *Life and Work of John Tyndall*, 50, 324.
23. Tyndall, *Heat*, 96–97; Tyndall, "On the Vibrations and Tones," 1–10, 1.
24. Tyndall, "On the Vibrations and Tones," 2–3, 3–5, 7–8, 8, 10.
25. Eve and Creasey, *Life and Work of John Tyndall*, 325, 326, 325.
26. Tyndall, *Heat*, 96–100, 297.
27. "Letter 3339: Peter Henry Berthon to Michael Faraday, 3 Oct., 1857," in James, ed., *Correspondence of Michael Faraday*, 5:288–89; "Letter 3340: Michael Faraday to Peter Henry Berthon, 6 Oct., 1857," in James, ed., *Correspondence of Michael Faraday*, 5:289.
28. "Letter 3661: Michael Faraday to Robert Gordon, 24 Oct., 1859," in James, ed., *Correspondence of Michael Faraday*, 5:570–74.
29. "Letter 3676: Peter Henry Berthon to Michael Faraday, 17 Nov., 1859," in James, ed., *Correspondence of Michael Faraday*, 5:592–93; "Letter 3685: Michael Faraday to Peter Henry Berthon, 21 Nov., 1859," in James, ed., *Correspondence of Michael Faraday*, 5:600.
30. Eve and Creasey, *Life and Work of John Tyndall*, 326.
31. Tyndall, "On the Atmosphere," 183–244, 187.
32. Herschel quoted in Tyndall, "On the Atmosphere," 195; Eve and Creasey, *Life and Work of John Tyndall*, 326.
33. Eve and Creasey, *Life and Work of John Tyndall*, 327.

34. Tyndall, "On the Atmosphere," 209–13.
35. Eve and Creasey, *Life and Work of John Tyndall*, 328–29.
36. Tyndall, "On the Atmosphere," 183.
37. Tyndall, *Sound*, viii, vii, 3–4, 49, 58–59.
38. Tyndall, *Sound*, 100–104, 80–82, 132–50.
39. Tyndall, *Sound*, 286, 288.
40. Tyndall, *Sound*, 296.
41. Tyndall, *Sound*, 297–306.
42. Tyndall, *Sound*, 1–2.
43. Tyndall, *Essays on the Use and Limit of the Imagination*, 66–68.
44. Tyndall, *Essays on the Use and Limit of the Imagination*, 69.
45. Tyndall, *Essays on the Use and Limit of the Imagination*, 69–70, 70.
46. Kim, *John Tyndall's Transcendental Materialism*, 45–48, 60–64, 64.
47. Tyndall, *Essays on the Use and Limit of the Imagination*, 53, 63.
48. Goodsir, *Divine Rule Proceeds by Law*, 6, 7, 13, 22, 25, 48.
49. Tyndall, *Essays on the Use and Limit of the Imagination*, 16–17.
50. Tyndall, *Essays on the Use and Limit of the Imagination*, 18.
51. Tyndall, *Essays on the Use and Limit of the Imagination*, 18, 19–22.
52. Cantor, "John Tyndall's Religion: A Fragment," 419–36, 422–26.
53. Kim, *John Tyndall's Transcendental Materialism*, xi–xii, 8.
54. Kim, *John Tyndall's Transcendental Materialism*, 12–13. Tyndall himself referred to this mix of idealism and materialism as "scientific materialism," but fellow X Club member Thomas Hirst suggested the term "transcendental materialism" as more appropriate; see Kim, *John Tyndall's Transcendental Materialism*, 43.
55. DeYoung, *A Vision of Modern Science*, 54, 59, 60–61, 62, 64–66.
56. Tyndall, *Address Delivered before the British Association*, 26, 32, 42, 45–50, 52, 55–56.
57. Crosbie Smith, *Science of Energy*, 7, 2, 3, 5–7.
58. Quoted in Niven, ed., *Scientific Papers of James Clerk Maxwell*, 2:376.
59. Neswald, "Saving the World," 15–31, 15, 16, 27–30.
60. DeYoung, *A Vision of Modern Science*, 111–15.
61. Barton, "John Tyndall, Pantheist," 111–34, 112, 116–19, 132–34.
62. Binns, *Science, Religion, and the Bible*, 2.
63. Quarry, *Mind, Not Matter*, 3, 4, 14, 22, 20–22.
64. Tyndall, "'Materialism' and Its Opponents," 579–99, 584, 588–89.
65. Tyndall, "'Materialism' and Its Opponents," 593–94.
66. Tyndall, "'Materialism' and Its Opponents," 594.
67. Tyndall, "'Materialism' and Its Opponents," 598.
68. Anon., *Materialistic Views of Professor Tyndall and Miss Harriet Martineau Criticized*, 7–8, 13, 15–16, 17–18, 20, 33.
69. Gatens, *Victorian Cathedral Music*, 34–35, 38; on Anglican music and morality, see Martin, "Music and the Aesthetic in Worship," 647–55, 652.
70. Champneys, *The Immortality of Music!*, 4–5.
71. Farrar, *Music: In Religion and in Life*, 6, 7–8, 10.
72. Mee, *Principle of Choral Worship*, 5.
73. Lunn, *Vocal Expression*, 3.
74. Lunn, *Vocal Expression*, 5–6.
75. Lunn, *Vocal Expression*, 10.

76. Lunn, *Vocal Expression*, 12–13, 15–16, 29.
77. Lunn, *Vocal Expression*, 28.
78. Gauld, "Gurney, Edmund (1847–1888)."
79. Gurney, *Power of Sound*, vi, viii; on materialism and aesthetics, see Benjamin Morgan, *Outward Mind*, 5–6; on materialism and the senses, see Lange, *History of Materialism*, 202–30.
80. Gurney, *Power of Sound*, viii.
81. On music and morality see, Gurney, *Power of Sound*, 360–75.
82. Gurney, *Power of Sound*, 475.
83. Rigg, "Leopold, Prince, First Duke of Albany (1853–1884)."
84. See Warrack, *Royal College of Music*.
85. *Music in England. The Proposed Royal College of Music*, 4–5.
86. Tompson, "Salmon, George (1819–1904)."
87. Salmon, *Music and Religion*, 6, 9.
88. Curling, *Transfiguration*, vii–viii, 162, 162–66.
89. Curling, *Transfiguration*, 77–78.
90. Curling, *Transfiguration*, 81–82.
91. Curling, *Transfiguration*, 84.

Epilogue

1. Boys, "Influence of a Tuning-fork on the Garden Spider," 149–50.
2. Forgan and Gooday, "'A Fungoid Assemblage of Buildings,'" 153–92, 175–76.
3. Macleod, "Support of Victorian Science," 197–230, 202–5, 213–14.
4. Boys, "Influence of a Tuning-fork on the Garden Spider," 150.
5. On sound technologies, see Mowitt, "The Sound of Music," 213–24; Milner, *Perfecting Sound Forever*; Gitelman, *Scripts, Grooves, and Writing Machines*, 97–147; Van Drie, "Hearing through the Théâtrophone," 74–90; Mills, "Deafening," 118–43.
6. Bijsterveld, *Mechanical Sound*, 91–136.
7. Parr, "Bond between Music and Astronomy," 411–22, 416, 419–20, 411.
8. Sloterdijk, *Terror from the Air*, 9–10.
9. For histories of the war as a sociocultural turning point, see Fussell, *The Great War and the Modern Memory*; Hynes, *A War Imagined*.
10. Jean, "Sonic Mindedness of the Great War," 51–62, 53, 60; also see Enke, "War Noises on the Battlefield," 7–21.
11. Jean, "Sonic Mindedness of the Great War," 51–53, 57–60.
12. Manstan, *Listeners: U-boat Hunters*, 101–35; Jean, "Sonic Mindedness of the Great War," 55–57.
13. Quoted in Terraine, *White Heat*, 314.
14. Wittje, *The Age of Electroacoustics*, 6–9, 67–114.
15. Eagles, "Political Legacies," 249–62, 256, 261.
16. Ruskin, *The Nature of Gothic*, 17.
17. Eagles, "Political Legacies," 249–62, 249, 260.

Select Bibliography

Publication is in London unless otherwise stated.

Primary Material

1. MANUSCRIPTS

Cambridge, Cambridgeshire

Royal Greenwich Observatory Archives, Cambridge University Library
RGO/6/427, Miscellaneous Letters.
RGO/6/609, Clock for New Palace Westminster, 1857 to 1861.
RGO6/614, Galvanic Time Balls, Time Signals, and External Clocks, Regulated by the Royal Observatory, 1858 June to 1860 December.
RGO/6/615, Galvanic Time Balls, Time Signals, and Sympathetic Clocks, Regulated by the Royal Observatory, 1861 January to 1866 February.
RGO/6/617, Galvanic Time Balls, Time Signals, and Sympathetic Clocks, Regulated by the Royal Observatory, 1870 February to 1875 October.
RGO/6/618, Galvanic Time Balls, Time Signals, and Sympathetic Clocks, Regulated by the Royal Observatory 1875 November to 1878 December.
RGO/6/703, Meteorology, Correspondence, Instruments and Miscellaneous, June 1861–May 1865.
RGO/6/428, Mathematical and Other Papers, 1849–1850.
RGO/6/433, Pure Mathematics and Miscellaneous Science, 1857–1858.
RGO/7/238, Miscellaneous Scientific Papers of William Christie, 1878–1909.
Wren Library, Trinity College
Ref.15.b.80.3, Arthur Trevelyan, "Theory of Heat" (ca. 1843).
Ref.15.b.80.3, Arthur Trevelyan, "Theory of the Vital Principle of Animal and Vegetable Life," (ca. 1843).

Edinburgh

Centre for Research Collections, Edinburgh University
Map.PC.16, "Hislop's Time-Gun Map of Edinburgh and Leith" (1862).
Map.PC.60, "Hislop's Time-Gun Map of Edinburgh and Leith" (1863).
Royal Observatory Library
A.10.45, Charles Piazzi Smyth, "Letter Book, 1863."
A.12.54, Charles Piazzi Smyth, "Correspondence re Time Gun, Letters in and out, 1870, 1872, 1873, and 1876."

A18.101, Charles Piazzi Smyth, "Time Ball and Mean Time Clock, from 12 June 1861 to 23 January 1863."

A.20.156, Charles Piazzi Smyth, "'Edinburgh, Scotland, Royal Observatory.' History, Buildings, Instruments, and Activities" (1882).

Hove, Sussex

Hove Public Library
Testimonial and Other Letters to and concerning W. M. Pole, 1814–1900.

Liverpool

Sydney Jones Library Archives, University of Liverpool
White, Blanco, Mss. III/21, "On Musical Sounds" (ca. 1830).
White, Blanco, Mss. III/22, "Music" (ca. 1830).

London

British Library
Add. Ms. 41771, Papers of Sir George Smart, Vol. 1.
Add. Ms. 63589, Papers of Sir George Smart.
King's College London Archives
Wheatstone 5, "Various: 8 Lectures on Sound" (ca. 1835).
Wheatstone 5, "King's College London. Experimental Philosophy Professor Wheatstone, F.R.S. Will Deliver a Course of Eight Lectures on the Measure of Sound, Light, Heat, Magnetism, and Electricity" (1837), pp. 1–4.
Royal Institution Archives
MS F/4/C, "Friday Evenings, 1825–1829."
MS F/4/H, "Miscellaneous Notebook, 1825–1850."
MS F/4/K, "Friday Evenings, 1829–1831."
MS F/5/D, "Administrative Notebooks: Duties of Servants."
MS F/5/H, "Subscribers to Separate Courses of Lectures, 1835–1841."
Royal Society Archives
HS 1.308, George Biddell Airy to John Herschel, March 27, 1868.
HS 1.310, George Biddell Airy to John Herschel, May 26, 1868.
HS 2.243, John Herschel to Charles Babbage, February 28, 1830.
HS 2.244, Charles Babbage to John Herschel, March 4, 1830.
HS 2.249, John Herschel to Charles Babbage, March 24, 1830.
HS 5.143, William H. Callcott to John Herschel, February 25, 1867.
HS 5.144, William H. Callcott to John Herschel, October 21, 1868.
HS 6.283, Augustus De Morgan to John Herschel, June 4, 1856.
HS 6.284, Augustus De Morgan to John Herschel, June 8, 1865.
HS 6.307, Augustus De Morgan to John Herschel, January 1, 1858.
HS 13.203, Francis J. Hughes to John Herschel, undated, ca. 1867.
HS 13.203, F. A. Gore Ouseley to John Herschel, September 16, 1839.
HS 13.204, F. A. Gore Ouseley to John Herschel, October 8, 1839.
HS 13.205, F. A. Gore Ouseley to John Herschel, November 14, 1839.
HS 13.206, Harriot Ouseley to John Herschel, 1839.
HS 14.383, John Robison to John Herschel," January 17, 1831.
HS 17.347, Thomas Thompson to John Herschel, February 10, 1860.
HS 17.348, Thomas Thompson to John Herschel, February 13, 1860.

HS 18.142, Charles Wesley to John Herschel, September 1828.
HS 18.144, Charles Wheatstone to John Herschel, August 23, 1825.
HS 18.145, Charles Wheatstone to John Herschel, August 23, 1833.
HS 18.146, Charles Wheatstone to John Herschel, May 13, 1835.
HS 18.173, William Whewell to John Herschel, April 3, 1828.
HS 18.331, Thomas Young to John Herschel, November 21, 1824.
HS 18.417, Francis Wilson to John Herschel, June 17, 1831.
HS 21.146, John Herschel to Charles Wheatstone, August 21, 1833.
HS 23.249, John Herschel to Robert Brown, December 15, 1858.
HS 27.82, John Herschel to the chairman of the committee, June 14, 1859.
Royal Society of Arts Archives
PR/GE/112/12/96, Minutes of Various Committees.
PR/GE/119/39/94, Letters of Stainer on Pitch.
PR/GE/121/10/5, Musical Pitch, 1859.
PR/GE/121/10/6, Miscellaneous Letters Relating to Pitch.

Oxford, Oxfordshire

Mary Somerville Papers, Somerville College; held at Special Collections, Bodleian Library
Dep.C.300—MSH—1.
Dep.C.352—MSSW—5.
Dep.C.369—MSB—11.
Dep.C.376—MSDIP—18.

2. PERIODICALS

A Journal of Natural Philosophy, Chemistry and the Arts
Arcana of Science and Art
Athenaeum
Bristol Mercury
Caledonian Mercury
Chamber's Journal of Popular Literature Science and Arts
Edinburgh Journal of Science
Edinburgh Review
Fortnightly Review
Illustrated London News
Journal of the Royal Institution of Great Britain
Journal of the Society of Arts
Leeds Mercury
Liverpool Mercury
London Medical Gazette
Mechanics' Magazine
Monthly Review
Moonshine
Philosophical Magazine
Proceedings of the Musical Association
Proceedings of the Royal Society of London
Punch
Quarterly Journal of Science, Literature, and Art

Quarterly Review
The Circulator of Useful Knowledge, Amusement, Literature, Science, and General Information
The Dublin Literary Gazette, or Weekly Chronicle of Criticism, Belles Lettres, and Fine Arts
The Examiner
The Hull Packet and Humber Mercury
The Literary Gazette
The London and Edinburgh Philosophical Magazine and Journal of Science
The Morning Post
The Musical Review, Record of Musical Science, Literature and Intelligence
The Musical Standard
The Musical Times, and Singing Class Circular
The Musical Times
The New Monthly Magazine and Literary Journal
The Penny Pulpit
The Reader
The Repository of Arts, Literature, Fashions, Manufacturers &c.
The Spectator
The Times
Transactions of the American Philosophical Society
Transactions of the Cambridge Philosophical Society
Transactions of the Royal Irish Academy
Transactions of the Royal Scottish Society of Arts
Transactions of the Royal Society of Edinburgh
Westminster Review

3. PRINTED

Aglionby, Francis Keyes. *The Life of Edward Henry Bickersteth D.D. Bishop and Poet.* London: Longmans, Green, 1907.

Airy, George Biddell. "The Astronomer Royal's Remarks on Professor Challis's Theoretical Determination of the Velocity of Sound." *Philosophical Magazine*, series 3, 32, no. 216 (1848): 339–43.

———. *On Sound and Atmospheric Vibrations, with the Mathematical Elements of Music. Designed for the Use of Students of the University.* London: Macmillan, 1868.

Airy, Wilfrid, ed. *Autobiography of Sir George Biddell Airy.* Cambridge: Cambridge University Press, 1896.

Anderson, James. "On the Motion of Solids on Surface, in the Two Hypotheses of Perfect Sliding and Perfect Rolling, with a Particular Examination of Their Small Oscillatory Motions." *Transactions of the American Philosophical Society* 3 (1830): 315–82.

Anon. *A Sketch of the Life of C. Eulenstein, The Celebrated Performer on the Jews' Harps.* London: James J. Welsh, 1833.

Anon. *Church Music: A Sermon.* London: Whittaker, 1843.

Anon. *Materialistic Views of Professor Tyndall and Miss Harriet Martineau Criticized.* London: Bickers and Son, 1879.

Austen, S. C. *The Scepticism of the Nineteenth Century. Selections from the Latest Works of the Rev. William Gresley.* London: J. Masters, 1879).

Aveling, Thomas. *Recreations, Physical and Mental, Lawful and Unlawful: A Lecture.* London, 1849.

Bennett, J. R. Sterndale. *The Life of William Sterndale Bennett.* Cambridge: Cambridge University Press, 1907.

Bickersteth, Edward. *Church Music: A Sermon, Preached in the Parish Church of St Mary, Aylesbury, on Thursday, Oct. 29, 1857. On the Occasion of the First Meeting of the Vale of Aylesbury Church Choral Association.* Oxford: John Henry, 1857.

Binns, William. *Science, Religion, and the Bible: A Sermon Preached Aug. 30, 1874, on Professor Tyndall's Address at Belfast.* London, 1874.

Boys, C. V. "The Influence of a Tuning-fork on the Garden Spider." *Nature (London)* (December 16, 1880): 149–50.

Brewster, David. "Review of *On the Connexion of the Physical Sciences.*" *Edinburgh Review* 59 (April 1834): 154–71.

———. *Letters on Natural Magic Addressed to Sir Walter Scott, Bart.* 4th ed. London: John Murray, 1838.

———. *Letters on Natural Magic Addressed to Sir Walter Scott, Bart.* New York: Harper and Brothers, 1842.

Brown, Colin. "Reviews: *A Demonstration of the Musical Scale, Founded upon the Law of Vibrations of Sound.*" *The Musical Times* 14 (April 1, 1870): 436.

Brown, Robert. *Elements of Musical Science.* London: Hamilton, Adams, 1860.

———. *An Introduction to Musical Arithmetic; With Its Application to Temperament.* London: E. Lonsdale, 1865.

Busby, Thomas. *A Grammar of Music: To Which Are Prefixed Observations Explanatory of the Properties and Powers of Music as a Science. And of the General Scope and Object of the Work.* London: John Walker, 1818.

———. "The Acoucryptophone, or Enchanted Lyre," *Concert Room and Orchestra Anecdotes of Music and Musicians, Ancient and Modern.* Vol. 1. London, 1825.

Carlyle, Thomas. *On Heroes, Hero Worship and the Heroic in History.* London: Chapman and Hall, 1897.

Champneys, W. Weldon. *The Immortality of Music! A Sermon Preached on the Occasion of the Opening of the New Organ, Holy Trinity, Burton-upon-Trent, October 9th, 1872.* Burton-upon-Trent: R. R. Bellamy, 1872.

Clayton, Richard. *A Sermon by the Rev. Richard Clayton, M.A. Preached on Occasion of the Musical Festival Held in Newcastle-upon-Tyne, at St Thomas's Chapel, on Sunday, September 18, 1842.* Newcastle-upon-Tyne: E. and T. Bruce, 1842.

Copleston, Edward. *Church Music. A Sermon Preached at Usk, July 20th, 1848. At the First Annual Meeting of the Society for the Improvement of Church Music, in the Archdeaconry of Monmouth.* London: Rivingtons, 1848.

Copleston, William James. *Memoir of Edward Copleston, D.D. Bishop of Llandaff.* London: John W. Parker, 1851.

Curling, Edward. *The Transfiguration: With Other Sermons and Short Addresses for Fast and Festival (Together with Some Thoughts on Music; on Paradise; on Science and the Faith; on Tennyson; on the Day of Intercession; on the Holy Communion; Our Strength against Temptation, and in Prayer; and on the Public Use of the Athanasian Creed, etc).* London: John Ouseley Limited, 1911.

Damer, Lionel Dawson. *Church Music: A Sermon, Preached at St Mary's, Aylesbury, at the Annual Festival of the Vale of Aylesbury Church Choral Association, June 11, 1868.* London: Rivingtons, 1868.

Daniel, H., ed. *Our Memories: Shadows of Old Oxford*. Oxford: H. Daniel, 1893.
De Morgan, Augustus. "On the Beats of Imperfect Consonances." *Transactions of the Cambridge Philosophical Society* 10 (1864): 129–45.
Eden, John. *Church Music. A Sermon, Preached at the Opening of the New Organ in the Parish Church of St Nicholas, in the City of Bristol*. Bristol: J. M. Gutch, 1822.
Eliot, George. *Middlemarch*. London: Macmillan, 2018.
Elvey, Mary. *Life and Reminiscences of George J. Elvey*. London: Sampson Low, Marston, 1894.
Enke, Julia. "War Noises on the Battlefield." *German Historical Institute London Bulletin* 37, no. 1 (2015): 7–21.
Faraday, Michael. *Experimental Researches in Chemistry and Physics*. London: Richard Taylor and William Francis, 1859.
———. *Experimental Researches in Electricity*. London: Richard and John Edward Taylor, 1839.
———. "On the Sounds Produced by Flame in Tubes, &c." *Quarterly Journal of Science* 5 (1818): 274–80.
Farrar, F. W. *Music: In Religion and in Life. A Sermon Preached by the Rev. Canon Farrar, D.D., in Westminster Abbey, July 30th, 1882*. Walsall: W. Henry Robinson, 1882.
Forbes, James D. "Experimental Researches Regarding Certain Vibrations Which Take Place between Metallic Masses Having Different Temperatures." *Transactions of the Royal Society of Edinburgh* 12, pt. 2 (1833): 429–61.
Galloway, Thomas. "Review of Mechanism of the Heavens." *Edinburgh Review* 55 (April 1832): 1–25.
Gitelman, Lisa. *Scripts, Grooves, and Writing Machines: Representing Technology in the Edison Era*. Stanford, CA: Stanford University Press, 1999.
Gladstone, John Hall. *Michael Faraday*. New York: Harper and Brothers, 1872.
Goodsir, Joseph Taylor. *The Divine Rule Proceeds by Law: An Old Sermon Now First Published, with a New Preface, Bearing on the Philosophic and Theological Tendencies of Professor Tyndall's Introductory Address Lately Delivered at Norwich*. London: Williams and Norgate, 1868.
Gorham, George Cornelius, James J. Reynolds, and Harvey Marriott. *The Character of Acceptable Prayer: The Joyful Sound: And Attendance in the House of God. Three Sermons, Preached at Brampford-Speke, Devonshire, March 13, 1853, Being the First Sunday of the Celebration of Divine Service in the New Parish Church*. London: Hatchard, 1853.
Gresley, William. *A Sermon on Church Music, Preached in St Paul's Church, Brighton. September 12, MDCCCLII*. London: J. Masters, 1852.
Groser, Thomas. *On Prayer: With Special Reference to Professor Tyndall's Objections. A Sermon Preached in the Catholic Apostolic Church, Gordon Square*. London: Thomas Bosworth, 1866.
Gurney, Edmund. *The Power of Sound*. London: Smith, Elder, 1880.
Hamilton, Walter Kerr. *Church Music. A Sermon, Preached at the Church of St Mary, Sturminster Newton, on the Occasion of the First Festival of Choirs in Union with the Salisbury Diocesan Choral Association, on Thursday, April 12, 1860*. Salisbury: Brown, 1860.
Hanslick, Eduard. *On the Musically Beautiful: A Contribution towards the Revision of the Aesthetics of Music*. Trans. Geoffrey Payzant. Indianapolis: Hackett, 1986.

Hegel, G. W. F. *Hegel's Aesthetics: Lectures on Fine Art*. Trans. T. N. Knox. 2 vols. Oxford: Clarendon Press, 1988.

Helmholtz, Hermann. *On the Sensations of Tone as a Physiological Basis for the Theory of Music*. Trans. Alexander J. Ellis. London: Longmans, Green, 1875.

Herschel, John. *Familiar Lectures on Scientific Subjects*. New York: George Routledge and Sons, 1872.

———. "Light." *Encyclopaedia Metropolitana* 4 (1830): 341–586.

———. *On Musical Scales*. London: W. Clowes and Sons, 1868.

———. "Physical Astronomy." *Encyclopaedia Metropolitana* 4 (1830): 647–734.

———. *A Preliminary Discourse on the Study of Natural Philosophy*. London: Longman, Rees, Orme, Brown, and Green, 1831.

———. "Review of Mechanism of the Heavens." *Quarterly Review* 47 (July 1832): 537–59.

———. "Sound." *Encyclopaedia Metropolitana* 4 (1830): 747–824.

———. "Uniform Musical Pitch." *Leeds Mercury* 6985, August 2, 1859.

Hewett, John William. *Choral Worship; Its design and Scriptural Authority, with an Exhortion to Choristers to Live as They Sing: A Sermon Preached in the Church of the Resurrection, Dresden, Staffordshire, on the Sunday Evening before the Festival in Lichfield Cathedral, October 8th, 1862*. Newcastle: F. Crewe, 1862.

Higgins, W. Mullinger. *The Philosophy of Sound, and History of Music*. London: Wm S. Orr, 1838.

Hopkins, W. B. *Church Music. A Sermon Preached in the Cathedral Church of Norwich, at the First Annual Service of Church Choirs, on Tuesday, 28th August, 1860*. London: Bell and Daldy, 1860.

How, F. D. *Six Great Schoolmasters*. London: Methuen, 1904.

Hullah, John. *Chromatic Scale; with the Inflected Syllables Used in "Time and Tune in the Elementary School."* London: Longmans, Green, ca. 1875.

———. *Grammar of Vocal Music, for the Use of Public Schools and Classes of Adults*. London: John W. Parker, 1843.

James, Thomas. *The Prophetic Office Connected with Poetry and Music. A Sermon Preached in the Cathedral Church of Worcester, September 30, 1800, at the Annual Meeting of the Three Choirs of Worcester, Hereford, and Glocester*. Worcester: J. Tymbs, 1800.

Jebb, John. *The Divine Origin of Music: A Sermon Preached at the Third Annual Festival of Church Choirs Held in the Cathedral Church of Norwich, on Thursday, 30th September, 1862*. London: Rivingtons, 1862.

Lange, Frederick Albert. *History of Materialism and Criticism of Its Present Importance*. 3 vols. London: Trübner, 1877.

Liddell, Robert. *Plain Song, Commonly Called Gregorian Music: A Short Sermon on PSALM CL. 6*. London: J. T. Hayes, 1868.

Liddon, H. P. "The Sights and Sounds of Christendom. A Sermon Preached in St Paul's Cathedral, on Sunday Afternoon, Sept. 11th, 1870." *Penny Pulpit*, n.s., no. 452, October 27, 1870 (London: J. Paul, 1870): 194–200.

Lunn, Charles. *Vocal Expression: Empirical or Scientific? A Lecture, Delivered at the Royal Academy, Tenterden Street, on Friday, May 17, 1878*. London: Stanley Lucas, Weber, 1878.

Lunn, Henry C. "Musical Pitch." *The Musical Times, and Singing Class Circular* 13, no. 312 (February 1, 1869): 663–65.

Lyon, Ralph. *Two Sermons, Preached on the Occasion of Opening an Organ in a Parish Church*. London: J. Hatchard and Son, 1840.

Marrian, J. P. "On Sonorous Phaenomena in Electro-Magnets." *Philosophical Magazine*, series 3, 25, no. 167 (1844): 382–84.

Martin, Thomas, ed. *Faraday's Diary: Being Various Philosophical Notes of Experimental Investigation Made by Michael Faraday*. Vol. 1: Sept. 1820—June 11, 1832. London: G. Bell and Sons, 1932.

———, ed. *Faraday's Diary: Being Various Philosophical Notes of Experimental Investigation Made by Michael Faraday*. Vol. 2: August 25, 1832–February 29, 1836. London: G. Bell and Sons, 1932.

M'Cann, James. *The Inter-Relations of Prayer, Providence, and Science*. London: Simpkin, Marshall, 1866.

Mee, John H. *The Principle of Choral Worship. A Sermon Preached in Chichester Cathedral, on Behalf of the "Ouseley Memorial Fund," on Sunday, August 10, 1890*. Chichester: Wilmshurst, 1891.

Mill, John Stuart. *A System of Logic Ratiocinative and Inductive: Being a Connected View of the Principles of Evidence and the Methods of Scientific Investigation*. London, 1886.

Miller, Edward. *Church Music. A Sermon Preached in Smethwick Old Chapel, in the County of Stafford, on the Opening of a New Organ, on Sunday Afternoon, December 10th, 1848*. Chichester: William Hayley Mason, 1849.

Morgan, John. *Four Biographical Sketches: Bishop Ollivant, Bishop Thirlwall, Rev. Griffith Jones, Vicar of Llanddowror, and Sir Thomas Phillips*. London: Elliot Stock, 1892.

Morgan, J. Vyrnwy, ed. *Welsh Political and Educational Leaders in the Victorian Era*. London: James Nisbet, 1908.

Music in England. The Proposed Royal College of Music. Three Addresses Delivered by HRH the Duke of Edinburgh, HRH the Duke of Albany, and HRH Prince Christian, at the Free Trade Hall, Manchester, Dec. 12, 1881. London: John Murray, 1882.

Nathan, I. *An Essay on the History and Theory of Music; And on the Qualities, Capabilities, and Management of the Human Voice*. London: G. and W. B. Whittaker, 1823.

Niven, W. D., ed. *The Scientific Papers of James Clerk Maxwell*. 2 vols. Cambridge: Cambridge University Press, 1890.

O'Leary, Arthur. "Sir William Sterndale Bennett: A Brief Review of His Life and Works." *Proceedings of the Musical Association*, 8th session (1881–82) (April 3, 1882): 123–45.

Ouseley, F. A. Gore. *A Treatise on Harmony*. Oxford: Clarendon Press, 1868.

Ouseley, F. A. G., and John B. Dykes. *The Choral Worship of the Church. Two Sermons, Preached in St Peter's Church, Derby, on the Second Sunday in Advent, (Dec 9) 1860, (On Occasion of the Annual Collections in Aid of the Choir)*. Derby: W. Bemrose and Sons, 1861.

Parr, W. Alfred. "The Bond between Music and Astronomy." *Westminster Review* 180, no. 4 (October 1913): 411–22.

Pinnock, W. H. *Church Choirs, and Church Music, Their Origin, and Usefulness: A Sermon Preached at the Re-opening of the Organ, after being Considerably Enlarged; At the Parish Church of St Ives, Huntingdonshire: on Friday, November 17th, 1865*. Cambridge: J. Hall and Son, 1866.

Pole, William. "Explanation of the Musical Scale and Its Component Intervals." In F. A. Gore Ouseley, *A Treatise on Harmony*. Oxford: Clarendon Press, 1868, pp. 259–63.

———. *Some Short Reminiscences of Events in My Life and Works. Abbreviated from Manuscript Notes*. Privately printed, 1898.

Quarry, John. *Mind, Not Matter, the Cause of All Things. An Argument Suggested by the Address of Professor Tyndall to the British Association at Belfast*. Dublin: Hodges, Foster, 1874.

Ramsay, Edward B. *Reminiscences of Scottish Life & Character*. Edinburgh: Edmonston and Douglas, 1874.

———. *The Use of Organs in Christian Worship: A Sermon Preached in Trinity Episcopal Church, Edinburgh, October 22, 1865*. Edinburgh: R. Grant and Son, 1865.

"Regulator," "Concert Pitch." *The Musical Review, Record of Musical Science, Literature and Intelligence* 1, no. 8 (New York, 1839): 85–86.

Richardson, Benjamin Ward. *Thomas Sopwith, M.A., C.E., F.R.S. with Excerpts from His Diary of Fifty-seven Years*. London: Longmans, Green, 1891.

Ridgway, James. *A Plea for Church Music. A Sermon by the Rev. James Ridgway, B.D., F.S.A., Principal of the Oxford Diocesan Training College, Culham; Preached in the Church of St Nicholas, Abingdon, November 8, 1868, on the Occasion of the Opening of a New Organ*. Oxford: James Parker, 1868.

Ringer, Sydney. "On the Alteration of the Pitch of Sound by Conduction through Different Media." *Proceedings of the Royal Society of London* 10 (1859–60): 275–81.

Ruskin, John. *The Nature of Gothic: A Chapter of The Stones of Venice*. London: George Allen, 1892.

Russell, John Scott. *The Wave of Translation in the Oceans of Water, Air, and Ether*. London: Trübner, 1885.

Salmon, George. *Music and Religion: A Sermon Preached at the Dedication of the Cathedral Organ of St Mary the Virgin, Southwell, October 13th, 1892*. Dublin: Dublin University Press, 1892.

Savart, Felix. "Researches on the Elasticity of Regularly Crystallized Bodies." *Edinburgh Journal of Science* 1, n.s. (April–October 1829): 141–46.

Shairp, John Campbell, Peter Guthrie Tait, and A. Adams-Reilly. *Life and Letters of James David Forbes, F.R.S.* London: Macmillan, 1873.

Smart, Benjamin Humphrey. *The Theory and Practice of Elocution*. London: John Richardson, 1820.

Smyth, Piazzi. "On the Methods Adopted to Secure Extreme Accuracy in the Edinburgh Castle Time-Gun Signal." *Transactions of the Royal Scottish Society of Arts* 6 (1864): 136–52.

Somerville, Martha. *Personal Recollections from Early Life to Old Age, of Mary Somerville. With Selections from Her Correspondence*. London: John Murray, 1873.

Somerville, Mary. *Mechanism of the Heavens*. London: John Murray, 1831.

———. *On the Connexion of the Physical Sciences*. London: John Murray, 1834.

———. *On the Connexion of the Physical Sciences*. 2nd ed. London: John Murray, 1835.

Spencer, Herbert. "Progress: Its Law and Cause." *Westminster Review* 67 (April 1857): 445–85.

Stainer, John. "The Character and Influence of the Late Sir Frederick Ouseley." *Proceedings of the Musical Association*, 16th session (1889–90) (December 2, 1889): 25–39.

Stewart, Henry William. *Church Music: A Sermon, by the Rev. Henry William Stewart, A. M., Vicar of Russagh.* London: William MacIntosh, 1869.
Stratford, Joseph. *Gloucestershire Biographical Notes.* Gloucester, 1887.
Strutt, J. W. *The Theory of Sound.* London: Macmillan, 1877.
Taylor, William Cooke. "Review of *On the Connexion of the Physical Sciences.*" *Athenaeum* (March 15, 1834): 202–3.
Todhunter, I. *William Whewell, D.D. Master of Trinity College, Cambridge. An Account of His Writings with Selections from His Literary and Scientific Correspondence.* London: Macmillan, 1876.
Trevelyan, Arthur. "Further Notice of the Vibration of Heated Metals; With the Description of a New and Convenient Apparatus for Experimenting with." *Philosophical Magazine,* series 3, 6, no. 32 (1835): 85–86.
———. "On the Vibration of Heated Metals: By Arthur Trevelyan, Esq.; Including a Letter on the Same Subject by Dr W. Knight." *London and Edinburgh Philosophical Magazine and Journal of Science,* 3rd series (November 1833): 321–32.
Tuckwell, W. *Pre-Tractarian Oxford: A Reminiscence of the Oriel "Noetics."* London: Smith, Elder, 1909.
Tyndall, John. *Address Delivered before the British Association Assembled at Belfast, with Additions.* London: Longmans, Green, 1874.
———. *Essays on the Use and Limit of the Imagination in Science.* London: Longmans, Green, 1870.
———. *Faraday as a Discoverer.* London: Longman's, Green, 1870.
———. *Heat: A Mode of Motion.* 7th ed. London: Longmans, Green, 1887.
———. "'Materialism' and Its Opponents." *Fortnightly Review,* n.s., 18, no. 107 (November 1, 1875): 579–99.
———. "On the Atmosphere as a Vehicle of Sound." *Philosophical Transactions* 164 (1875): 183–244.
———. "On the Vibrations and Tones Produced by the Contact of Bodies Having Different Temperatures." *Philosophical Transactions of the Royal Society of London* 144 (1854): 1–10.
———. *Sound: A Course of Eight Lectures Delivered at the Royal Institution of Great Britain.* London: Longmans, Green, 1867.
Uniform Musical Pitch. Minutes of a Meeting of Musicians, Amateurs, and Others Interested in Music, Held at the House of the Society of Arts, when the Report of the Committee Appointed by the Council of the Society Was Received and Adopted. Royal Society for the Encouragement of Arts, Manufactures, and Commerce, published in "Uniform Musical Pitch," *Journal of the Society of Arts* 8, no. 417 (November 16, 1860): 1–8.
Vaughan, C. "Dean Vaughan." In J. Vyrnwy Morgan, ed., *Welsh Political and Educational Leaders in the Victorian Era,* pp. 407–28. London: James Nisbet, 1908.
Vaughan, C. J. *Music in Churches: A Sermon Preached in St John's Church, Leicester, at a Festival of the Church Choral Association, on Thursday morning, September 28, 1865.* Cambridge: Macmillan, 1865.
Walker, S. A. *Church Music: A Sermon Preached in St Mary-le-Port Church, Bristol, on Sunday Morning, September 24, 1865.* London: Seeley, Jackson and Halliday, 1865.
Wheatstone, Charles. *The Scientific Papers of Sir Charles Wheatstone, D.C.L., F.R.S.* London: Taylor and Francis, 1879.

Whewell, William. *Astronomy and General Physics; Considered with Reference to Natural Theology*. London: William Pickering, 1833.
———. *History of the Inductive Sciences, from the Earliest to the Present Time*. 3rd ed., 3 vols. London: John W. Parker, 1857.
———. *The Philosophy of the Inductive Sciences, Founded upon Their History*. 2nd ed., 2 vols. London: John W. Parker, 1847.
———. "Review of *On the Connexion of the Physical Sciences*." *Quarterly Review* 51 (March 1834): 54–68.
Williams, Charles J. B. "Observations on the Production and Propagation of Sound." *London and Edinburgh Philosophical Magazine and Journal of Science* 6 (January–June 1835): 25–34.
Wordsworth, Christopher. *Sacred Music: A Sermon, Preached at the Anniversary of the Choral Association of the Diocese of Llandaff, in the Cathedral Church of Llandaff, Sept., 2, 1868*. London: Rivingtons, 1868.
Wroth, Warwick R. *"The Choral Service." A Sermon Preached in the Church of St Peter, Sudbury, on the Evening of August 3, 1858, Being the Commemoration of Its Reopening and Restoration*. Oxford: John Henry and James Parker, 1858.
Wylde, Henry. *Harmony and the Science of Music; Complete in One Volume*. London: Hutchings and Romer, 1872.
———. *Music and Its Art-Mysteries*. London: L. Booth, 1867.
"X." "The Invisible Lady; Being an Explanation of the Manner in What the Experiment Which Was Exhibited in London by M. Charles and Others, Is Performed. In a Letter from a Correspondent." *Journal of Natural Philosophy, Chemistry, and the Arts* 16 (January 1807): 69–71.

Secondary Material

Adelmann, Dale. *The Contribution of Cambridge Ecclesiologists to the Revival of Anglican Church Worship, 1839–62*. Aldershot: Ashgate, 1997.
Adolph, Anthony R. J. S. "Smart, Benjamin Humphrey (bap. 1787, d. 1872)." *Oxford Dictionary of National Biography*, Oxford University Press, 2004; http://www.oxforddnb.com/view/article/25738.
Alder, Ken. "A Revolution to Measure: The Political Economy of the Metric System in France." In M. Norton Wise, ed., *The Values of Precision*, 39–71. Princeton, NJ: Princeton University Press, 1995.
Aldrich, Richard. "Hickson, William Edward (1803–1870)." *Oxford Dictionary of National Biography*. Oxford: Oxford University Press, 2004; http://www.oxforddnb.com/view/article/13220.
Allis, Michael. *Parry's Creative Process*. Aldershot: Ashgate, 2003.
Altholz, Joseph L, "A Tale of Two Controversies: Darwinism in the Debate over 'Essays and Reviews.'" *Church History* 63, no. 1 (March 1994): 50–59.
Atlas, Allan W. "Ladies in the Wheatstone Ledgers: The Gendered Concertina in Victorian England, 1835–1870." *Royal Musical Association Research Chronicle* 39 (2006): 9–10.
———. *The Wheatstone English Concertina in Victorian England*. Oxford: Clarendon Press, 1996.
Barker, Andrew. *The Science of Harmonics in Classical Greece*. Cambridge: Cambridge University Press, 2007.

Barton, Ruth. "John Tyndall, Pantheist: A Rereading of the Belfast Address." *Osiris* 3 (1987): 111–34.

———. *The X Club: Power and Authority in Victorian Science.* Chicago: University of Chicago Press, 2018.

Bebbington, D. W. *Evangelicalism in Modern Britain: A History from the 1730s to the 1980s.* London: Unwin Hyman, 1989.

Beyer, Robert T. *Sounds of Our Times: Two Hundred Years of Acoustics,* New York: Springer-Verlag, 1999.

Biagioli, Mario, ed. *The Science Studies Reader.* New York: Routledge, 1999.

Bijsterveld, Karin. *Mechanical Sound: Technology, Culture, and Public Problems of Noise in the Twentieth Century.* Cambridge, MA: MIT Press, 2008.

Black, William B., and David Clarke. "Glasgow Time Signals." *Journal for the History of Astronomy* 47, no. 3 (2016): 256–93.

Blair, Kirstie. *Form and Faith in Victorian Poetry and Religion.* (Oxford: Oxford University Press, 2012.

Boase, G. C. "Collard, Frederick William (bap. 1772, d. 1860)." *Oxford Dictionary of National Biography.* Oxford: Oxford University Press, 2004; http://www.oxforddnb.com/view/10.1093/ref:odnb/9780198614128.001.0001/odnb-978019 8614128-e-5903.

Bowers, Brian. "Faraday, Wheatstone and Electrical Engineering." In David Gooding and Frank A. J. L. James, eds., *Faraday Rediscovered: Essays on the Life and Work of Michael Faraday, 1791–1867,* 163–73. Basingstoke: Macmillan, 1985.

———. *Sir Charles Wheatstone FRS, 1802–1875.* London: Institution of Electrical Engineers, 2001.

Bowers, Brian, and Lenore Symons. *Curiosity Perfectly Satisfyed: Faraday's Travels in Europe, 1813–1815.* London: Peter Peregrinus, 1991.

Brain, Robert Michael, and Kelly Joan Whitmer. "Dissecting Vision in Early Science and Medicine." *Perspectives in Biology and Medicine* 52, no. 3 (Summer 2009): 448–53.

Brent, Richard. "Copleston, Edward (1776–1849)." *Oxford Dictionary of National Biography.* Oxford: Oxford University Press, 2004; http://www.oxforddnb.com/view/10.1093/ref:odnb/9780198614128.001.0001/odnb-9780198614128-e-6267.

Brooke, John Hedley. *Science and Religion: Some Historical Perspectives.* Cambridge: Cambridge University Press, 1991.

Brück, Hermann A. "Smyth, Charles Piazzi (1819–1900)." *Oxford Dictionary of National Biography.* Oxford: Oxford University Press, 2004; http://www.oxforddnb.com/view/article/25948.

Buchanan, Alexandrina. *Robert Willis (1800–1875): And the Foundation of Architectural History.* Woodbridge: Boydell, 2013.

Buchanan, Alexandrina, James W. P. Campbell, Javier Giron, and Santiago Huerta, eds. *Robert Willis: Science, Technology and Architecture in the Nineteenth Century. Proceedings of the International Symposium Held in Gonville and Caius College, Cambridge UK, 16th–17th September 2016.* Madrid: Instituto Juan De Herrera, 2016.

Buttmann, Günther. *The Shadow of the Telescope: A Biography of John Herschel.* London: Lutterworth, 1974.

Caesar, Michael, and Franco D'Intino, eds. *Zibaldone: The Notebooks of Leopardi.* London: Penguin, 2013.

Cahan, David, ed. *Hermann Von Helmholtz and the Foundations of Nineteenth-Century Science*. Berkeley: University of California Press, 1993.

Cannon, Walter F. "John Herschel and the Idea of Science." *Journal of the History of Ideas* 22, no. 2 (April–June 1961): 215–39.

Cantor, Geoffrey. "John Tyndall's Religion: A Fragment." *Notes and Records of the Royal Society* 69 (2015): 419–36.

———. *Michael Faraday: Sandemanian and Scientist: A Study of Science and Religion in the Nineteenth Century*. Basingstoke: Macmillan, 1991.

———. *Optics after Newton: Theories of Light in Britain and Ireland, 1704–1840*. Manchester: Manchester University Press, 1984.

———. "Reading the Book of Nature: The Relation between Faraday's Religion and His Science." In David Gooding and Frank A. J. L. James, eds., *Faraday Rediscovered: Essays on the Life and Work of Michael Faraday, 1791–1867*, 69–81. Basingstoke: Macmillan, 1985.

———. *Religion and the Great Exhibition of 1851*. Oxford: Oxford University Press, 2011.

Cantor, Geoffrey, David Gooding, and Frank A. J. L. James. *Michael Faraday*. New York: Humanity Books, 1991.

Carlyle, E. I. "Bickersteth, Edward (1814–1892)." *Oxford Dictionary of National Biography*. Oxford: Oxford University Press, 2004; http://www.oxforddnb.com/view/10.1093/ref:odnb/9780198614128.001.0001/odnb-9780198614128-e-2346.

———. "Cowie, Benjamin Morgan (1816–1900)." *Oxford Dictionary of National Biography*. Oxford: Oxford University Press, 2004; http://www.oxforddnb.com/view/10.1093/ref:odnb/9780198614128.001.0001/odnb-9780198614128-e-6497.

Case, Stephen. "'Land-marks of the Universe': John Herschel against the Background of Positional Astronomy." *Annals of Science* 72, no. 4 (2015): 417–38.

———. *Making Stars Physical: The Astronomy of Sir John Herschel*. Pittsburgh: University of Pittsburgh Press, 2018.

Chadwick, Owen. *The Spirit of the Oxford Movement: Tractarian Essays*. Cambridge: Cambridge University Press, 1990.

Chandler, Michael. "Liddon, Henry Parry (1829–1890)." *Oxford Dictionary of National Biography*. Oxford: Oxford University Press, 2004; http://www.oxforddnb.com/view/10.1093/ref:odnb/9780198614128.001.0001/odnb-9780198614128-e-16644.

———. *The Life and Work of Henry Parry Liddon (1829–1890)*. Leominster: Gracewing, 2000.

Chapman, Allan. "The Astronomical Revolution." In John Fauvel, Raymond Flood, and Robin Wilson, eds., *Mobius and His Band: Mathematics and Astronomy in Nineteenth-Century Germany*, 35–77. Oxford: Oxford University Press, 1993.

———. "George Biddell Airy, F.R.S. (1801–1892): A Centenary Commemoration." *Notes and Records of the Royal Society of London* 46, no. 1 (January 1992): 103–10.

———. "An Occupation for an Independent Gentleman: Astronomy in the Life of John Herschel." *Vistas in Astronomy* 36 (1993): 71–116.

Charlton, David, ed. *E. T. A. Hoffmann's Musical Writings: Kreisleriana, the Poet and the Composer, Music Criticism*. Cambridge: Cambridge University Press, 1989.

Chua, Daniel K. L. "Vincenzo Galileli, Modernity and the Division of Nature." In Suzannah Clark and Alexander Rehding, eds., *Music Theory and Natural Order from the Renaissance to the Early Twentieth Century*, 17–29. Cambridge: Cambridge University Press, 2001.

Clark, Suzannah, and Alexander Rehding, eds. *Music Theory and Natural Order from the Renaissance to the Early Twentieth Century*. Cambridge: Cambridge University Press, 2001.

Clarke, Martin V., "'Meet and Right It Is to Sing': Nineteenth-Century Hymnals and the Reasons for Singing." In Martin V. Clarke, ed., *Music and Theology in Nineteenth-Century Britain*, 21–36. Farnham: Ashgate, 2012.

——, ed. *Music and Theology in Nineteenth-Century Britain*. Farnham: Ashgate, 2012.

Cobb, Aaron D. "Inductivism in Practice: Experiment in John Herschel's Philosophy of Science." *Journal of the International Society for the History of Philosophy of Science* 2, no. 1 (Spring 2012): 21–54.

Corbin, Alain. *Village Bells: Sound and Meaning in the Nineteenth-Century French Countryside*. Basingstoke: Papermac, 1999.

Corsi, Petro. *Science and Religion: Baden Powell and the Anglican Debate, 1800–1860*. Cambridge: Cambridge University Press, 1988.

Creese, David. *The Monochord in Ancient Greece Harmonic Science*. Cambridge: Cambridge University Press, 2010.

Crowe, Michael J. "Herschel, Sir John Frederick William, First Baronet (1792–1871)." *Oxford Dictionary of National Biography*. Oxford: Oxford University Press, 2004; http://www.oxforddnb.com/view/article/13101.

Curthoys, Judith. *The Cardinal's College: Christ Church, Chapter and Verse*. London: Profile Books, 2012.

Daunton, Martin. *Progress and Poverty: An Economic and Social History of Britain, 1700–1850*. Oxford: Oxford University Press, 1995.

——. *State and Market in Victorian Britain*. Woodbridge: Boydell, 2008.

Davies, James Q. "Instruments of Empire." In James Q. Davies and Ellen Lockhart, eds., *Sound Knowledge: Music and Science in London, 1789–1851*, 145–74. Chicago: University of Chicago Press, 2016.

Davies, James Q., and Ellen Lockhart. "Fantasies of Total Description." In James Q. Davies and Ellen Lockhart, eds., *Sound Knowledge: Music and Science in London, 1789–1851*, 1–26. Chicago: University of Chicago Press, 2016.

——, eds. *Sound Knowledge: Music and Science in London, 1789–1851* Chicago: University of Chicago Press, 2016.

Davis, John R. "Dilke, Sir (Charles) Wentworth." *Oxford Dictionary of National Biography*. Oxford: Oxford University Press, 2004; http://www.oxforddnb.com/view/10.1093/ref:odnb/9780198614128.001.0001/odnb-9780198614128-e-7646.

Desmond, Adrian. *The Politics of Evolution: Morphology, Medicine, and Reform in Radical London*. Chicago: University of Chicago Press, 1989.

DeYoung, Ursula. *A Vision of Modern Science: John Tyndall and the Role of the Scientist in Victorian Culture*. Basingstoke: Palgrave Macmillan, 2011.

Dickson, Melissa. "Charles Wheatstone's Enchanted Lyre and the Spectacle of Sound." In James Q. Davies and Ellen Lockhart, eds. *Sound Knowledge: Music and Science in London, 1789–1851*, 125–44. Chicago: University of Chicago Press, 2016.

Dolan, Emily I. "Music as an Object of Natural History." In James Q. Davies and Ellen Lockhart, eds. *Sound Knowledge: Music and Science in London, 1789–1851*, 27–46. Chicago: University of Chicago Press, 2016.

——. *The Orchestral Revolution: Haydn and the Technologies of Timbre*. Cambridge: Cambridge University Press, 2013.

Dupree, Mary Helen, and Sean B. Franzel, eds. *Performing Knowledge, 1750–1850*. Berlin: De Gruyter, 2015.
Eagles, Stuart. "Political Legacies." In Francis O'Gorman, ed., *The Cambridge Companion to John Ruskin*, 249–62. Cambridge: Cambridge University Press, 2015.
Erlman, Veit. *Reason and Resonance: A History of Modern Aurality*. New York: Zone Books, 2014.
Eve, A. S., and C. H. Creasey. *Life and Work of John Tyndall*. London: Macmillan, 1945.
Fauvel, John, Raymond Flood, and Robin Wilson, eds. *Mobius and His Band: Mathematics and Astronomy in Nineteenth-Century Germany*. Oxford: Oxford University Press, 1993.
Feiereisen, Florence, and Alexandra Merley Hill, eds. *Germany in the Loud Twentieth Century: An Introduction*. Oxford: Oxford University Press, 2012.
Fisch, Menachem. "A Philosopher's Coming of Age: A Study in Erotetic Intellectual History." In Menachem Fisch and Simon Schaffer, eds., *William Whewell: A Composite Portrait*, 31–66. Oxford: Oxford University Press, 1991.
Fisch, Menachem, and Simon Schaffer, eds. *William Whewell: A Composite Portrait*. Oxford: Oxford University Press, 1991.
Flood, Raymond, Adrian Rice, and Robin Wilson, eds. *Mathematics in Victorian Britain*. Oxford: Oxford University Press, 2011.
Forgan, Sophie. "Faraday—from Servant to Savant: The Institutional Context." In David Gooding and Frank A. J. L. James, eds., *Faraday Rediscovered: Essays on the Life and Work of Michael Faraday, 1791–1867*, 51–67. Basingstoke: Macmillan, 1985.
Forgan, Sophie, and Graeme Gooday. "'A Fungoid Assemblage of Buildings': Diversity and Adversity in the Development of College Architecture and Scientific Education in Nineteenth-Century South Kensington," *History of Universities* 13 (1994): 153–92.
Foucault, Michel. *Discipline and Punish: The Birth of the Prison*. Trans. Alan Sheridan Harmondsworth: Penguin, 1977.
Fox, Celina, ed. *London—World City, 1800–1840*. New Haven, CT: Yale University Press, 1992.
Fox, Leonard. "Introduction." In Leonard Fox, ed., *The Jew's Harp: A Comprehensive Anthology*, 15–39. Lewisburg, PA: Bucknell University Press, 1988.
———, ed. *The Jew's Harp: A Comprehensive Anthology*. Lewisburg, PA: Bucknell University Press, 1988.
Francis, Keith A. "Paley to Darwin: Natural Theology versus Science in Victorian Sermons." In Keith A. Francis and William Gibson, eds., *Oxford Handbook of the British Sermon, 1689–1901*, 444–62. Oxford: Oxford University Press, 2012.
Francis, Keith A., and William Gibson, eds. *Oxford Handbook of the British Sermon, 1689–1901*. Oxford: Oxford University Press, 2012.
Fussell, Paul. *The Great War and Modern Memory*. Oxford: Oxford University Press, 1977.
Galison, Peter, and Emily Thompson, eds. *The Architecture of Science*. Cambridge, MA: MIT Press, 1999.
Gant, Andrew. *O Sing unto the Lord: A History of English Church Music*. London: Profile Books, 2015.
Garber, Elizabeth. "Reading Mathematics, Constructing Physics: Fourier and His Readers, 1822–1850." In A. J. Kox and Daniel M. Siegel, eds., *No Truth Except in*

the Details: Essays in Honor of Martin J. Klein, 31–54. Dordrecht: Kluwer Academic Publishers, 1995.

Gatens, William J. *Victorian Music in Theory and Practice*. Cambridge: Cambridge University Press, 1986.

Gauld, Alan. "Gurney, Edmund (1847–1888)." *Oxford Dictionary of National Biography*. Oxford: Oxford University Press, 2004; http://www.oxforddnb.com/view/10.1093/ref:odnb/9780198614128.001.0001/odnb-9780198614128-e-11763.

Gavroglu, Kostas. "Strutt, John William, Third Baron Rayleigh (1842–1919)." *Oxford Dictionary of National Biography*. Oxford: Oxford University Press, 2004; http://www.oxforddnb.com/view/10.1093/ref:odnb/9780198614128.001.0001/odnb-9780198614128-e-36359.

Gieryn, Thomas F. "Boundary-work and the Demarcation of Science from Nonscience: Stains and Interests in Professional Ideologies of Scientists." *American Sociological Review* 48, no. 6 (December 1983): 781–95.

Gibson, William, "The British Sermon, 1689–1901: Quantities, Performance, and Culture." In Keith A. Francis and William Gibson (eds.), *Oxford Handbook of The British Sermon, 1689–1901*, (Oxford University Press: Oxford, 2012), pp. 3–30.

Gillin, Edward J. "Prophets of Progress: Authority in the Scientific Projections and Religious Realizations of the *Great Eastern* Steamship." *Technology and Culture* 56, no. 4 (October 2015): 928–56.

———. "Seismology's Acoustic Debt: Robert Mallet, Chladni's Figures, and the Victorian Science of Earthquakes." *Sound Studies* 6, no. 1 (2020): 65–82.

———. *The Victorian Palace of Science: Scientific Knowledge and the Building of the Houses of Parliament*. Cambridge: Cambridge University Press, 2017.

Gillin, Edward, and H. Horatio Joyce, eds. *Experiencing Architecture in the Nineteenth Century: Buildings and Society in the Modern Age*. London: Bloomsbury, 2019.

Gooday, Graeme. "Architectural Acoustics: Thomas Roger Smith and the Science of Hearing Buildings in Nineteenth-Century Britain." In Edward Gillin and H. Horatio Joyce, eds., *Experiencing Architecture in the Nineteenth Century: Buildings and Society in the Modern Age*, 101–14. London: Bloomsbury, 2019.

Gooday, Graeme J. N. *The Morals of Measurement: Accuracy, Irony, and Trust in Late Victorian Electrical Practice*. Cambridge: Cambridge University Press, 2004.

Gooding, David. "From Phenomenology to Field Theory: Faraday's Visual Reasoning." *Perspectives on Science* 14, no. 1 (2006): 40–65.

———. "'In Nature's Schools': Faraday as an Experimentalist." In David Gooding and Frank A. J. L. James, eds. *Faraday Rediscovered: Essays on the Life and Work of Michael Faraday, 1791–1867*, 105–35. Basingstoke: Macmillan, 1985.

Gooding, David, and Frank A. J. L. James, eds. *Faraday Rediscovered: Essays on the Life and Work of Michael Faraday, 1791–1867*. Basingstoke: Macmillan, 1985.

Gouk, Penelope. *Music, Science and Natural Magic in Seventeenth-Century England*. New Haven, CT: Yale University Press, 1999.

Grande, James. "On Tongues and Ears: Divine Voices in the Modern Metropolis." In Roger Parker and Susan Rutherford, eds. *London Voices, 1820–1840: Vocal Performers, Practices, Histories*, 137–58. Chicago: University of Chicago Press, 2019.

Grattan-Guinness, I. "Mathematics and Mathematical Physics from Cambridge, 1815–1840: A Survey of the Achievements and of the French Influences." In P. M. Harman, ed., *Wranglers and Physicists: Studies on Cambridge Physics in the Nineteenth Century*, 84–111. Manchester: Manchester University Press, 1985.

Gribenski, Fanny. "Negotiating the Pitch: For a Diplomatic History of A, at the Crossroads of Politics, Music, Science and Industry." In Frédéric Ramel and Cécile Prévost-Thomas, eds. *International Relations, Music and Diplomacy: Sounds and Voices on the International Stage*, 173–92. Cham: Palgrave Macmillan, 2018.

———. *Tuning the World: Acoustics, Aesthetics, Industry, and Global Politics*. (Chicago: University of Chicago Press, forthcoming).

Gribenski, Fanny, and Edward Gillin. "The Politics of Musical Standardization in Nineteenth-Century France and Britain." *Past and Present* no. 251 (May 2021): 153–87.

Hahn, Roger. *Pierre Simon Laplace, 1749–1827: A Determined Scientist*. Cambridge, MA: Harvard University Press, 2005.

Hamilton, James. *Faraday: The Life*. London: Harper Collins, 2002.

Hankins, Thomas L., and Robert J. Silverman. *Instruments and the Imagination*. Princeton, NJ: Princeton University Press, 1995.

Harman, P. M., ed. *The Scientific Letters and Papers of James Clerk Maxwell*. 3 vols. Cambridge: Cambridge University Press, 1990–2002.

———, ed. *Wranglers and Physicists: Studies on Cambridge Physics in the Nineteenth Century*. Manchester: Manchester University Press, 1985.

Haynes, Bruce. *History of Performing Pitch: The Story of "A."* Lanham, MD: Scarecrow, 2002.

Helmstadter, Richard J., and Bernard Lightman, eds. *Victorian Faith in Crisis: Essays on Continuity and Change in Nineteenth-Century Religious Belief*. Basingstoke: Macmillan:, 1990.

Herbert, Trevor, and Helen Barlow. *Music and the British Military in the Long Nineteenth Century*. Oxford: Oxford University Press, 2013.

Hilton, Boyd. "Moral Disciplines." In Peter Mandler, ed., *Liberty and Authority in Victorian Britain*, 224–46. Oxford: Oxford University Press, 2006.

Hirst, Paul. "Foucault and Architecture." *AA Files* 26 (Autumn 1993): 52–60.

Holmes, Caitlin. "The Astronomer Royal, the Hydrographer and the Time Ball: Collaborations in Time Signalling 1850–1910." *British Journal for the History of Science* 42, no. 3 (September 2009): 381–406.

Hoskin, Michael. *Discoverers of the Universe: William and Caroline Herschel*. Princeton, NJ: Princeton University Press, 2011.

Howse, Derek. *Greenwich Time and the Discovery of the Longitude*. Oxford: Oxford University Press, 1980.

———. *Greenwich Time and the Longitude*. London: Oxford University Press, 1997.

Hui, Alexandra. *The Psychophysical Ear: Musical Experiments, Experimental Sounds, 1840–1910*. Cambridge, MA: MIT Press, 2013.

Hui, Alexandra, Julia Kursell, and Myles W. Jackson. "Music, Sound, and the Laboratory from 1750 to 1980." *Osiris* 28, no. 1 (January 2013): 1–11.

Hunt, Bruce J. "The Ohm Is Where the Art Is: British Telegraph Engineers and the Development of Electrical Standards." *Osiris* 9, Instruments (1994): 48–63.

Hynes, Samuel. *A War Imagined: The First World War and English Culture*. London: Bodley Head, 1990.

Jackson, Myles W., "Charles Wheatstone: Musical Instrument Making, Natural Philosophy, and Acoustics in Early Nineteenth-Century London." In James Q. Davies and Ellen Lockhart, eds., *Sound Knowledge: Music and Science in London, 1789–1851*, 101–24. Chicago: University of Chicago Press, 2016.

———. "From Scientific Instruments to Musical Instruments: The Tuning Fork, the Metronome, and the Siren." In Trevor Pinch and Karin Bijsterveld, eds., *The Oxford Handbook of Sound Studies*, 201–23. Oxford: Oxford University Press, 2012.

———. *Harmonious Triads: Physicists, Musicians, and Instrument Makers in Nineteenth-Century Germany*. Cambridge, MA: MIT Press, 2006.

James, Frank A. J. L., ed. *The Correspondence of Michael Faraday. Vol. 1: 1811–December 1831. Letters 1–524*. London: Institution of Electrical Engineers, 1991.

———, ed. *The Correspondence of Michael Faraday. Vol. 2: 1832–December 1840. Letters 525–1333*. London: Institution of Electrical Engineers, 1993.

———, ed. *The Correspondence of Michael Faraday. Vol. 3: 1841–December 1848. Letters 1334–2145*. London: Institution of Electrical Engineers, 1996.

———, ed. *The Correspondence of Michael Faraday. Vol. 4: January 1849–October 1855. Letters 2146–3032*. London: Institution of Electrical Engineers, 1996.

———, ed. *The Correspondence of Michael Faraday. Vol. 5: November 1855–October 1860, Letters 3033–3873*. London: Institution of Engineering and Technology, 2008.

———, ed. *The Correspondence of Michael Faraday. Vol. 6: November 1860–August 1867, Undated Letters, Additional Letters for Volumes 1–5. Letters 3874–5053*. London: Institution of Engineering and Technology, 2012.

James, Jamie. *The Music of the Spheres: Music, Science, and the Natural Order of the Universe*. London: Little, Brown, 1993.

Jean, Yaron. "The Sonic Mindedness of the Great War." In Florence Feiereisen and Alexandra Merley Hill, eds., *Germany in the Loud Twentieth Century: An Introduction*, 51–62. Oxford: Oxford University Press, 2012.

Jenkins, Richard. "Disenchantment, Enchantment and Re-enchantment: Max Weber at the Millennium." *Max Weber Studies* (2000): 11–32.

Jenson, J. Vernon, "Return to the Wilberforce-Huxley Debate." *British Journal for the History of Science* 21 (1988): 161–79.

Kemp, Martin. "Wheatstone's Waves." *Nature* 394 (July 23, 1998): 327.

Kennedy, Michael. *The Autobiography of Charles Hallé: With Correspondence and Diaries*. London: Elek, 1972.

———. *The Hallé Tradition: A Century of Music*. Manchester: Manchester University Press, 1960.

Kent, Christopher. "Willis, Henry (1821–1901)." *Oxford Dictionary of National Biography*. Oxford: Oxford University Press, 2004; http://www.oxforddnb.com/view/10.1093/ref:odnb/9780198614128.001.0001/odnb-9780198614128-e-36935.

Kim, Stephen S. *John Tyndall's Transcendental Materialism and the Conflict between Religion and Science in Victorian England*. Lewiston, NY: Mellen University Press, 1996.

Kinns, Roger. "The Early History of the Edinburgh Time Ball and Time Gun." *The International Journal for the History of Engineering and Technology* 81, no. 2 (2011): 264–90.

Knight, Frances. "Parish Preaching in the Victorian Era: The Village Sermon." In Keith A. Francis and William Gibson, eds., *Oxford Handbook of the British Sermon, 1689–1901*, 63–78. Oxford: Oxford University Press, 2012.

Kox, A. J., and Daniel M. Siegel, eds. *No Truth Except in the Details: Essays in Honor of Martin J. Klein*. Dordrecht: Kluwer Academic Publishers, 1995.

Kromhout, Melle Jan. "'Antennas Have Long Since Invaded Our Brains': Listening to the 'Other Music' in Friedrich Kittler." In Sander van Maas, ed., *Thresholds of Listening: Sound, Technics, Space*, 89–104. New York: Fordham University Press, 2015.

Kursell, Julia. "'A Gray Box': The Phonograph in Laboratory Experiments and Fieldwork, 1900–1920." In Trevor Pinch and Karin Bijsterveld, eds., *The Oxford Handbook of Sound Studies*, 176–97. Oxford: Oxford University Press, 2012.

———, ed. *Sounds of Science—Schall im Labor (1800–1930)*. Berlin: Max Planck Institute for the History of Science, 2008.

Lajosi, Krisztina, and Andreas Stynen, eds. *Choral Societies and Nationalism in Europe*. Leiden: Brill, 2015.

Larrue, Jean-Marc, and Marie-Madeleine Mervant-Roux, eds. *Le son du theatre, XIXe-XXIe siècle: Histoire intermédiale d'un lieu d'écoute moderne*. Paris CNRS Éditions, 2016.

Latour, Bruno. "Visualization and Cognition: Thinking with Eyes and Hands." *Knowledge and Society: Studies in the Sociology of Culture Past and Present* 6 (1986): 1–40.

Latour, Bruno, and Steve Woolgar. *Laboratory Life: The Construction of Scientific Facts*. Princeton, NJ: Princeton University Press, 1986.

Lenoir, Timothy. "Helmholtz and the Materialities of Communication." *Osiris* 9, Instruments (1994): 184–207.

Lightman, Bernard. "Refashioning the Spaces of London Science: Elite Epistemes in the Nineteenth Century." In David N. Livingstone and Charles W. J. Withers, eds., *Geographies of Nineteenth-Century Science*, 25–50. Chicago: University of Chicago Press, 2011.

———. *Victorian Popularizers of Science: Designing Nature for New Audiences*. Chicago: University of Chicago Press, 2007.

———, ed. *Victorian Science in Context*. Chicago: University of Chicago Press, 1997.

Lightman, Bernard, and Michael S. Reidy. *The Age of Scientific Naturalism: Tyndall and His Contemporaries*. London: Pickering and Chatto, 2014.

Lightman, Bernard, and Bennett Zon, eds. *Evolution and Victorian Culture*. Cambridge: Cambridge University Press, 2014.

Livingstone, David N., and Charles W. J. Withers, eds. *Geographies of Nineteenth-Century Science*. Chicago: University of Chicago Press, 2011.

Loughridge, Deirdre. "Celestial Mechanisms: Adam Walker's Eidouranion, Celestina, and the Advancement of Knowledge." In James Q. Davies and Ellen Lockhart, eds., *Sound Knowledge: Music and Science in London, 1789–1851*, 47–76. Chicago: University of Chicago Press, 2016.

Lubbock, Constance A. *The Herschel Chronicle: The Life-Story of William Herschel and His Sister Caroline Herschel*. Cambridge: Cambridge University Press, 1933.

Maas, Sander van, ed. *Thresholds of Listening: Sound, Technics, Space*. New York: Fordham University Press, 2015.

Mackay, A. J. G. "Ramsay [formerly Burnett], Edward Bannerman (1793–1872)." *Oxford Dictionary of National Biography*. Oxford: Oxford University Press, 2004; http://www.oxforddnb.com/view/10.1093/ref:odnb/9780198614128.001.0001/odnb-9780198614128-e-23081.

Macleod, Roy. "The Support of Victorian Science: The Endowment of Research Movement in Great Britain, 1868–1900." *Minerva* 9, no. 2 (April 1971): 197–230.

MacMahon, M. K. C. "Ellis [formerly Sharpe], Alexander John (1814–1890)." *Oxford Dictionary of National Biography*. Oxford: Oxford University Press, 2004; http://

www.oxforddnb.com/view/10.1093/ref:odnb/9780198614128.001.0001/odnb
-9780198614128-e-8683.
Maconie, Robin. "Willis, Speech, Sound, and Music." In Alexandrina Buchanan, James W. P. Campbell, Javier Giron, and Santiago Huerta, eds., *Robert Willis: Science, Technology and Architecture in the Nineteenth Century. Proceedings of the International Symposium Held in Gonville and Caius College, Cambridge UK, 16th–17th September 2016*, 75–84. Madrid: Instituto Juan De Herrera, 2016.
Maitland, J. A. F. "Hullah, John Pyke (1812–1884)." *Oxford Dictionary of National Biography*. Oxford: Oxford University Press, 2004; http://www.oxforddnb.com/view/10.1093/ref:odnb/9780198614128.001.0001/odnb-9780198614128-e-14112.
Mandler, Peter. "Introduction: State and Society in Victorian Britain." In Peter Mandler, ed., *Liberty and Authority in Victorian Britain*, 1–21. Oxford: Oxford University Press, 2006.
———, ed. *Liberty and Authority in Victorian Britain*. Oxford: Oxford University Press, 2006.
Manstan, Roy R. *The Listeners: U-boat Hunters during the Great War*. Middletown, CT: Wesleyan University Press, 2018.
Marsden, Ben, and Crosbie Smith. *Engineering Empires: A Cultural History of Technology in Nineteenth-Century Britain*. Basingstoke: Palgrave Macmillan, 2005.
Martin, David. "Music and the Aesthetic in Worship and Collective Singing: England since 1840." *Culture and Society* 53 (2016): 647–55.
McKay, Charles. *Big Ben: The Great Clock and the Bells at the Palace of Westminster*. Oxford: Oxford University Press, 2010.
Metz, Don. "The Scientific Instruments of Charles Wheatstone and the Blending of Science, Art, and Culture." *Interchange* 46 (2015): 19–29.
Meulders, Michel. *Helmholtz: From Enlightenment to Neuroscience*. Cambridge, MA: MIT Press, 2010.
Mills, Mara. "Deafening: Noise and the Engineering of Communication in the Telephone System." *Grey Room* 43 (Spring 2011): 118–43.
Milner, Greg. *Perfecting Sound Forever: The Story of Recorded Music*. London: Granta, 2009.
Morgan, Benjamin. *The Outward Mind: Materialist Aesthetic in Victorian Science and Literature*. Chicago: University of Chicago Press, 2017.
Morrell, Jack, and Arnold Thackray. *Gentlemen of Science: Early Years of the British Association for the Advancement of Science*. Oxford: Oxford University Press, 1981.
Morris, Jeremey. "Preaching the Oxford Movement." In Keith A. Francis and William Gibson, eds. *Oxford Handbook of the British Sermon, 1689–1901*, 406–27. Oxford: Oxford University Press, 2012.
Morrison-Low, A. D. "Brewster, Sir David (1781–1868)." *Oxford Dictionary of National Biography*. Oxford: Oxford University Press, 2004; http://www.oxforddnb.com/view/article/3371.
Morus, Iwan Rhys. *Frankenstein's Children: Electricity, Exhibition, and Experiment in Early-Nineteenth-Century London*. Princeton, NJ: Princeton University Press, 1998.
———. "Illuminating Illusions, or, the Victorian Art of Seeing Things." *Early Popular Visual Culture* 10, no. 1 (February 2012): 37–50.
———. "'The Nervous System of Britain': Space, Time and the Electric Telegraph

in the Victorian Age." *British Journal for the History of Science* 33, no. 4, On Time: History, Science and Commemoration (December 2000): 455–75.

———. "Seeing and Believing Science." *Isis* 97, no. 1 (March 2006): 101–10.

———. *When Physics Became King*. Chicago: University of Chicago Press, 2005.

———. "Worlds of Wonder: Sensation and the Victorian Scientific Performance." *Isis* 101, no. 4 (December 2010): 806–16.

Morus, Iwan, Simon Schaffer, and Jim Secord, "Scientific London." In Celina Fox, ed., *London—World City, 1800–1840*, 129–42. New Haven, CT: Yale University Press, 1992.

Mowitt, John. "The Sound of Music in the Era of Its Electronic Reproducibility." In Jonathan Stern, ed., *The Sound Studies Reader*, 213–24. London: Routledge, 2012.

Muir, T. E. "Sacred Sound for a Holy Space: Dogma, Worship and Music at Solemn Mass during the Victorian Era, 1829–1903." In Martin V. Clarke, ed., *Music and Theology in Nineteenth-Century Britain*, 37–60. Farnham: Ashgate, 2012.

Murphy, G. Martin. "White, Joseph Blanco (1775–1841)." *Oxford Dictionary of National Biography*. Oxford: Oxford University Press, 2004; http://www.oxforddnb.com/view/article/29260.

Murphy, Martin. *Blanco White: Self-banished Spaniard*. New Haven, CT: Yale University Press, 1989.

Musselman, Elizabeth Green. *Nervous Conditions: Science and the Body Politic in Early Industrial Britain*. New York: State University of New York Press, 2006.

Neeley, Kathryn. *Mary Somerville: Science, Illumination, and the Female Mind*. Cambridge: Cambridge University Press, 2001.

Neswald, Elizabeth. "Saving the World in the Age of Entropy: John Tyndall and the Second Law of Thermodynamics." In Bernard Lightman and Michael S. Reidy, *The Age of Scientific Naturalism: Tyndall and His Contemporaries*, 15–31. London: Pickering and Chatto, 2014.

Neuffer. "To Eulenstein, Musician on the Jew's Harp." In Leonard Fox, ed., *The Jew's Harp: A Comprehensive Anthology*, 162–63. Lewisburg, PA: Bucknell University Press, 1988.

Newmarch, R. H. "Elvey, Sir George Job (1816–1893)." *Oxford Dictionary of National Biography*. Oxford: Oxford University Press, 2004; http://www.oxforddnb.com/view/10.1093/ref:odnb/9780198614128.001.0001/odnb-9780198614128-e-8772.

Nockles, Peter Benedict. *The Oxford Movement in Context: Anglican High Churchmanship, 1760–1857*. Cambridge: Cambridge University Press, 1994.

Nye, Mary Jo, ed. *The Cambridge History of Science. Vol. 5: The Modern Physical and Mathematical Sciences*. Cambridge: Cambridge University Press, 2003.

O'Gorman, Francis, ed. *The Cambridge Companion to John Ruskin*. Cambridge: Cambridge University Press, 2015.

Olleson, Philip. "Busby, Thomas (1754–1838)." *Oxford Dictionary of National Biography*. Oxford: Oxford University Press, 2004; http://www.oxforddnb.com/view/article/4158.

———. "Potter, (Philip) Cipriani Hambley (1792–1871)." *Oxford Dictionary of National Biography*. Oxford: Oxford University Press, 2004; http://www.oxforddnb.com/view/10.1093/ref:odnb/9780198614128.001.0001/odnb-9780198614128-e-22615.

Otis, Laura. *Networking: Communicating with Bodies and Machines in the Nineteenth Century*. Ann Arbor: University of Michigan Press, 2001.

Otter, Chris. *The Victorian Eye: A Political History of Light and Vision in Britain, 1800–1910*. Chicago: University of Chicago Press, 2008.
Pantalony, David. *Altered Sensations: Rudolph Koenig's Acoustical Workshop in Nineteenth-Century Paris*. New York: Springer, 2009.
Parker, Roger, and Susan Rutherford. "Introduction. London Voices, 1820–1840: A 'Luminous Guide.'" In Roger Parker and Susan Rutherford, eds., *London Voices, 1820–1840: Vocal Performers, Practices, Histories*, 1–14. Chicago: University of Chicago Press, 2019.
———, eds. *London Voices, 1820–1840: Vocal Performers, Practices, Histories*. Chicago: University of Chicago Press, 2019.
Parry, J. P. "Liberalism and Liberty." In Peter Mandler, ed., *Liberty and Authority in Victorian Britain*, 71–100. Oxford: Oxford University Press, 2006.
Patterson, Elizabeth Chambers. *Mary Somerville and the Cultivation of Science, 1815–1840*. Boston: Martinus Nijhoff, 1983.
Pawley, Margaret. "Wordsworth, Christopher (1807–1885)." *Oxford Dictionary of National Biography*. Oxford: Oxford University Press, 2004; http://www.oxforddnb.com/view/10.1093/ref:odnb/9780198614128.001.0001/odnb-9780198614128-e-29971.
Pesic, Peter. *Music and the Making of Modern Science*. Cambridge, MA: MIT Press, 2014.
Petersen, Sonja. "Craftsmen-Turned-Scientists? The Circulation of Explicit and Working Knowledge in Musical-Instrument Making, 1880–1960." *Osiris* 28, no. 1 (January 2013): 212–31.
Picker, John M. *Victorian Soundscapes*. Oxford: Oxford University Press, 2003.
Pinch, Trevor, and Karin Bijsterveld. "New Keys to the World of Sound." In Trevor Pinch and Karin Bijsterveld, eds., *The Oxford Handbook of Sound Studies*, 3–35. Oxford: Oxford University Press, 2012.
———. "'Should One Applaud?': Branches and Boundaries in the Reception of New Technology in Music." *Technology and Culture* 44, no. 3 (July 2003): 536–59.
———, eds. *The Oxford Handbook of Sound Studies*. Oxford: Oxford University Press, 2012.
Pink, Andrew. "Order and Uniformity, Decorum, and Taste: Sermons Preached at the Anniversary Meeting of the Three Choirs, 1720–1800." In Keith A. Francis and William Gibson, eds., *Oxford Handbook of the British Sermon, 1689–1901*, 215–26. Oxford: Oxford University Press, 2012.
Porter, Theodore M. *Trust in Numbers: The Pursuit of Objectivity in Science and Public Life*. Princeton, NJ: Princeton University Press, 1996.
Rainbow, Bernarr. *The Choral Revival in the Anglican Church, 1839–1872*. Woodbridge: Boydell, 2001.
Ramel, Frédéric, and Cécile Prévost-Thomas, eds. *International Relations, Music and Diplomacy: Sounds and Voices on the International Stage*. Cham: Palgrave Macmillan, 2018.
Reidy, Michael S. "Introduction: John Tyndall, Scientific Naturalism and Modes of Communication." In Bernard Lightman and Michael S. Reidy, *The Age of Scientific Naturalism: Tyndall and His Contemporaries*, 1–13. London: Pickering and Chatto, 2014.
Richards, Joan L. "*Mathematics in Victorian Britain* by Raymond Flood; Adrian Rice; Robin Wilson. Review." *Isis* 104, no. 4 (December 2013): 853–55.

Rigg, J. M. "Leopold, Prince, First Duke of Albany (1853–1884)." *Oxford Dictionary of National Biography*. Oxford: Oxford University Press, 2004; http://www.oxforddnb.com/view/10.1093/ref:odnb/9780198614128.001.0001/odnb-9780198614128-e-16475.

Roach, John. "Vaughan, Charles John (1816–1897)." *Oxford Dictionary of National Biography*. Oxford: Oxford University Press, 2004; http://www.oxforddnb.com/view/10.1093/ref:odnb/9780198614128.001.0001/odnb-9780198614128-e-28124.

Rubinstein, William D. "The Social Origins and Career Patterns of Oxford and Cambridge Matriculants, 1840–1900." *Historical Research* 82, no. 218 (November 2009): 715–30.

Schaffer, Simon. "Astronomers Mark Time: Discipline and the Personal Equation." *Science in Context* 2, no. 1 (March 1988): 115–45.

———. "The History and Geography of the Intellectual World: Whewell's Politics of Language." In Menachem Fisch and Simon Schaffer, eds., *William Whewell: A Composite Portrait*, 201–31. Oxford: Oxford University Press, 1991.

———. "Late Victorian Metrology and Its Instrumentation: A Manufactory of Ohms." In Mario Biagioli, ed., *The Science Studies Reader*, 457–78. New York: Routledge, 1999.

———. "Metrology, Metrication, and Victorian Values." In Bernard Lightman, ed., *Victorian Science in Context*, 438–74. Chicago: University of Chicago Press, 1997.

———. "Natural Philosophy and Public Spectacle in the Eighteenth Century." *History of Science* 21, no. 1 (1983): 1–43.

———. "Physics Laboratories and the Victorian Country House." In Crosbie Smith and Jon Agar, eds., *Making Space for Science: Territorial Themes in the Shaping of Knowledge*, 149–80. Basingstoke: Macmillan, 1998.

———. "Transport Phenomena: Space and Visibility in Victorian Physics." *Early Popular Visual Culture* 10, no. 1 (February 2012): 71–91.

Schickore, Jutta. *The Microscope and the Eye: A History of Reflections, 1740–1870*. Chicago: University of Chicago Press, 2007.

Schmidgen, Henning. "Silence in the Laboratory: The History of Soundproof Rooms." In Julia Kursell, ed., *Sounds of Science—Schall im Labor (1800–1930)*, 47–61. Berlin: Max Planck Institute for the History of Science, 2008.

Seccombe, Thomas. "Jebb, John (1805–1886)." *Oxford Dictionary of National Biography*. Oxford: Oxford University Press, 2004; http://www.oxforddnb.com/view/10.1093/ref:odnb/9780198614128.001.0001/odnb-9780198614128-e-14682.

Secord, James A. *Victorian Sensation: The Extraordinary Publication, Reception, and Secret Authorship of "Vestiges of the Natural History of Creation."* Chicago: University of Chicago Press, 2000.

———. *Visions of Science: Books and Readers at the Dawn of the Victorian Age*. Oxford: Oxford University Press, 2014.

Shapin, Steven. "The House of Experiment in Seventeenth-Century England." *Isis* 79, no. 3 (September 1988): 373–404.

———. *Never Pure: Historical Studies of Science As If It Was Produced by People with Bodies, Situated in Time, Space, Culture, and Society, and Struggling for Credibility and Authority*. Baltimore: Johns Hopkins University Press, 2010.

Shapin, Steven, and Simon Schaffer. *Leviathan and the Air-Pump: Hobbes, Boyle, and the Experimental Life*. Princeton, NJ: Princeton University Press, 1985.

Sharp, R. F. "Horsley, Charles Edward (1821–1876)." *Oxford Dictionary of National*

Biography. Oxford: Oxford University Press, 2004; http://www.oxforddnb.com/view/10.1093/ref:odnb/9780198614128.001.0001/odnb-9780198614128-e-13818.

Shaw, Watkins. "Ouseley, Sir Frederick Arthur Gore, Second Baronet (1825–1889)." *Oxford Dictionary of National Biography*. Oxford: Oxford University Press, 2004; http://www.oxforddnb.com/view/10.1093/ref:odnb/9780198614128.001.0001/odnb-9780198614128-e-20953.

Sibum, Heinz Otto. "Reworking the Mechanical Value of Heat: Instruments of Precision and Gestures of Accuracy in Early-Victorian England," *Studies in the History and Philosophy of Science* 26, no. 1 (1995): 73–106.

Sloterdijk, Peter. *Terror from the Air*. Los Angeles: Semiotext, 2009.

Smart, R. N. "Forbes, James David (1809–1868)." *Oxford Dictionary of National Biography*. Oxford: Oxford University Press, 2004; http://www.oxforddnb.com/view/article/9832.

Smith, Crosbie. *The Science of Energy: A Cultural History of Energy Physics in Victorian Britain*. London: Athlone, 1998.

Smith, Crosbie, and Jon Agar, eds. *Making Space for Science: Territorial Themes in the Shaping of Knowledge*. Basingstoke: Macmillan, 1998.

Smith, Mark M. *Sensing the Past: Seeing, Hearing, Smelling, Tasting, and Touching in History*. Berkeley: University of California Press, 2007.

Smith, Robert W. "Remaking Astronomy: Instruments and Practice in the Nineteenth and Twentieth Centuries." In Mary Jo Nye, ed., *The Cambridge History of Science*. Vol. 5: *The Modern Physical and Mathematical Sciences*, 154–73. Cambridge: Cambridge University Press, 2003.

Smith, Stanley. "Pole, William (1814–1900)." *Oxford Dictionary of National Biography*. Oxford: Oxford University Press, 2004; http://www.oxforddnb.com/view/article/22463.

Snyder, Laura J. *Reforming Philosophy: A Victorian Debate on Science and Society*. Chicago: University of Chicago Press, 2006.

Squire, W. B. "Donaldson, John (1789/90–1865)." *Oxford Dictionary of National Biography*. Oxford: Oxford University Press, 2004; http://www.oxforddnb.com/view/10.1093/ref:odnb/9780198614128.001.0001/odnb-9780198614128-e-7801.

Stanley, Matthew. *Huxley's Church and Maxwell's Demon: From Theistic Science to Naturalistic Science*. Chicago: University of Chicago Press, 2015.

Steege, Benjamin. *Helmholtz and the Modern Listener*. Cambridge: Cambridge University Press, 2012.

Stephen, Leslie. "Morgan, Augustus De (1806–1871)." Revised by I. Grattan-Guinness, *Oxford Dictionary of National Biography*. Oxford: Oxford University Press, 2004; http://www.oxforddnb.com/view/article/7470.

Sterne, Jonathan. *The Audible Past: Cultural Origins of Sound Reproduction*. Durham, NC: Duke University Press, 2003.

———, ed. *The Sound Studies Reader*. London: Routledge, 2012.

Strudwick, Vincent. *Christopher Wordsworth: Bishop of Lincoln, 1869–1885*. Lincoln: Honywood, 1987.

Sumner, W. L. *Father Henry Willis: Organ Builder*. London: Musical Opinion, 1955.

Swade, Doron. "Babbage, Charles (1791–1871)." *Oxford Dictionary of National Biography*. Oxford: Oxford University Press, 2004; http://www.oxforddnb.com/view/article/962.

Terraine, John. *White Heat: The New Warfare, 1914–18*. London: Sidgwick and Jackson, 1982.

Thistlewaite, Nicholas. *The Making of the Victorian Organ*. Cambridge: Cambridge University Press, 1990.

Thompson, Emily. "Listening to/for Modernity: Architectural Acoustics and the Development of Modern Spaces in America." In Peter Galison and Emily Thompson, eds., *The Architecture of Science*, 253–80. Cambridge, MA: MIT Press, 1999.

———. *The Soundscape of Modernity: Architectural Acoustics and the Culture of Listening in America, 1900–1933*. Cambridge, MA: MIT Press, 2002.

Thompson, E. P. "Time, Work-Discipline, and Industrial Capitalism." *Past and Present* 38 (December 1967): 56–97.

Thormählen, Wiebke. "From Dissent to Community: The Sacred Harmonic Society and Amateur Choral Singing in London." In Roger Parker and Susan Rutherford, eds., *London Voices, 1820–1840: Vocal Performers, Practices, Histories*, 159–78. Chicago: University of Chicago Press, 2019.

Tkaczyk, Viktoria. "Listening in Circles: Spoken Drama and the Architects of Sound, 1750–1830." *Annals of Science* 71, no. 3 (2014): 299–334.

———. "The Making of Acoustics around 1800, or How To Do Science with Words." In Mary Helen Dupree and Sean B. Franzel, eds., *Performing Knowledge, 1750–1850*, 27–55. Berlin: De Gruyter, 2015.

Tompson, Julia. "Salmon, George (1819–1904)." *Oxford Dictionary of National Biography*. Oxford: Oxford University Press, 2004; http://www.oxforddnb.com/view/10.1093/ref:odnb/9780198614128.001.0001/odnb-9780198614128-e-35914.

Tresch, John. *The Romantic Machine: Utopian Science and Technology after Napoleon*. Chicago: University of Chicago Press, 2012.

Tresch, John, and Emily I. Dolan. "Towards a New Organology: Instruments of Music and Science." *Osiris* 28, no. 1 (January 2013): 278–98.

Trippett, David. *Wagner's Melodies: Aesthetics and Materialism in German Musical Identity*. Cambridge: Cambridge University Press, 2013.

Turner, R. Steven. "The Ohm-Seebeck Dispute, Hermann Von Helmholtz, and the Origins of Physiological Acoustics." *British Journal for the History of Science* 10, no. 34 (1977): 1–24.

Tweney, Ryan D. "Faraday's Discovery of Induction: A Cognitive Approach." In David Gooding and Frank A. J. L. James, eds., *Faraday Rediscovered: Essays on the Life and Work of Michael Faraday, 1791–1867*, 189–209. Basingstoke: Macmillan, 1985.

———. "Inventing the Field: Michael Faraday and the Creative 'Engineering' of Electromagnetic Field Theory." In Robert J. Weber and David N. Perkins, eds., *Inventive Minds: Creativity in Technology*, 31–47. Oxford: Oxford University Press, 1992.

———. "Representing the Electromagnetic Field: How Maxwell's Mathematics Empowered Faraday's Field Theory." *Science and Education* 20 (2011): 687–700.

———. "Stopping Time: Faraday and the Scientific Creation of Perceptual Order." *Physics* 29 (1992): 149–64.

Van Drie, Melissa. "Hearing through the Théâtrophone: Sonically Constructed Spaces and Embodied Listening in Late Nineteenth-Century French Theatre." *Sound Effects—An Interdisciplinary Journal of Sound and Sound Experience* 5, no. 1 (2015): 74–90.

Warrack, Guy. *Royal College of Music: The First Eighty-five Years, 1883–1968 and Beyond*. London: Royal College of Music, 1977.
Warrack, John. "Smart, Sir George Thomas (1776–1867)." *Oxford Dictionary of National Biography*. Oxford: Oxford University Press, 2004; http://www.oxforddnb.com/view/10.1093/ref:odnb/9780198614128.001.0001/odnb-9780198614128-e-25740.
Warwick, Andrew. *Masters of Theory: Cambridge and the Rise of Mathematical Physics*. Chicago: University of Chicago Press, 2003.
Weber, Robert J., and David N. Perkins, eds. *Inventive Minds: Creativity in Technology*. Oxford: Oxford University Press, 1992.
Weinstein, Jerry L. "Musical Pitch and International Agreement." *American Journal of International Law* 46, no. 2 (April 1952): 341–43.
Werrett, Simon. "Disciplinary Culture: Artillery, Sound, and Science in Woolwich, 1800–1850." *19th Century Music* 39, no. 2 (2015): 87–98.
Whyte, William. *Unlocking the Church: The Lost Secrets of Victorian Sacred Space*. Oxford: Oxford University Press, 2017.
Williams, L. Pearce. *Michael Faraday: A Biography*. London: Chapman and Hall, 1965.
Wise, M. Norton. "Introduction." In M. Norton Wise, ed., *The Values of Precision*, 3–13. Princeton, NJ: Princeton University Press, 1995.
———, ed. *The Values of Precision*. Princeton, NJ: Princeton University Press, 1995.
Wise, M. Norton, and Crosbie Smith. "Measurement, Work and Industry in Lord Kelvin's Britain." *Historical Studies in the Physical and Biological Sciences* 17, no. 1 (1986): 147–73.
Wittje, Roland. *The Age of Electroacoustics: Transforming Science and Sound*. Cambridge, MA: MIT Press, 2016.
———. "The Electrical Imagination: Sound Analogies, Equivalent Circuits, and the Rise of Electro Acoustics, 1863–1939." *Osiris* 28, no. 1 (January 2013): 40–63.
Wright, Michael. *The Jews-Harp in Britain and Ireland*. Farnham: Ashgate, 2015.
Yates, Nigel. *The Oxford Movement and Anglican Ritualism*. London: Historical Association, 1983.
Yeo, Richard. *Defining Science: William Whewell, Natural Knowledge, and Public Debate in Early Victorian Britain*. Cambridge: Cambridge University Press, 1993.
———. "An Idol of the Market-place: Baconianism in Nineteenth-Century Britain." *History of Science* 23, no. 3 (September 1985): 251–98.
———. "Whewell, William (1794–1866)." *Oxford Dictionary of National Biography*. Oxford: Oxford University Press, 2004; http://www.oxforddnb.com/view/article/29200.
Zon, Bennett. "The 'Non-Darwinian' Revolution and the Great Chain of Musical Being." In Bernard Lightman and Bennett Zon, eds., *Evolution and Victorian Culture*, 196–226. Cambridge: Cambridge University Press, 2014.
———. "'Spiritual' Selection: Joseph Goddard and the Music Theology of Evolution." In Martin V. Clarke, ed., *Music and Theology in Nineteenth-Century Britain*, 215–35. Farnham: Ashgate, 2012.

Index

Page numbers in italics refer to figures and tables.

Académie des sciences, 62
acoustics, 2–4, 81, 88, 90, 92–93, 103, 107, 139, 197, 201–2, 213, 234, 239; acoustical experiments, and musical instruments, 61–62; acoustic experiments, 68; acoustic figures, 39–40, 53, 57, 59–63, 67, 69–70, 74, 76, 80, 101–2; acoustic phenomena, 44, 67; acoustic plate experiments, 27, 64–66, 89, 91; acoustic science, 138; acoustic vibrations, 65–67; acoustic time signals, as popular 196; architectural practices, 16; electroacoustics, 237; and hearing, 61; Houses of Parliament, 123; and measurement, 77–78; religious significance, 119–20; secondary mechanical sciences, 105–6
Adams, William Grylls, 80
Adelaide Gallery, 14
Africa, 47, 160
Airy, George Biddell, 20, 44, 86, 150, 152–54, 156, 168, 175–76, *184*, 202, 235; accuracy, calls for, 197; acoustic signaling, as unreliable, 193; acoustic time signaling, as popular, 196; audible signaling, 193–94; aural signaling, contempt toward, 197; Big Ben, 173–74, *176*, 180, 182; mathematical theory, 172–73, 183, 185–86; reliability of ear, distrust of, 165–66; semaphore, preference for, 189; sonorous problems, interest in, 167; sound waves, 170–71,

171; visual signaling, as superior, 193–94, 197; Westminster bells, 178–84
Airy, Wilfrid, 172–73, 186
Albert, Prince, 129–30
Ampére, André-Marie, 69
Analytical Society, 86–87
Anderson, Charles, 207
Anderson, Henry James, 85
Anglicanism, 203–4; Anglican evangelicalism, 116; fading philosophical authority, 227
Annual Festival of Church Choirs, 117
Aristotle, 106
Arnott, Neil, 133, 146
Arts and Crafts Movement, 238
Ashmolean Society, 58
Asia, 47
astronomy, 84, 86, 96, 105, 156, 169–70, 239; and music, 87–88; and sound, 85
Astronomy and General Physics (Whewell), 105
atheism, 3
audience, 18
Australasia, 160
Austria, 161
Aveling, Thomas, 7–8

Babbage, Charles, 86–89, 91–93
Bacon, Francis, 50
Bain, Alexander, 224
Barry, Charles, 123, 173
Barton, Ruth, 219

Bath Orchestra, 87
Battle of Waterloo, 21
Beethoven, 136, 155
Belgium, 159
Bell, Alexander Graham, 80, 233
Benedict, Julius, 131
Bennett, William Sterndale, 130, 133
Bentham, Jeremy, Panopticon design, 13
Berlin Observatory, 86
Berlioz, Hector, 128, 134–35, 146
Bernoulli, Daniel, 3, 57
Big Ben, 173, 180, 182–83, 194; accuracy, ensuring of, 174–75, 178; quarter chime bells, 175
Bijsterveld, Karin, 233–34
Binns, William, 219
Blagrove, Henry Gamble, 131
Blaikley, David James, 161
Blakesley, Joseph, 184–85
Bombay (India), 186
Bonaparte, Napoléon, 57–58, 123
Bowley, Robert Kanzow, 131
Boys, Charles Vernon, 231, 232, 233
Brewster, David, 27, 33, 101, 103, 232–33; "Invisible Girl," 30; kaleidoscope, invention of, 28, 30
Bristol (England), 34, 196
Britain, 3, 5, 6, 9, 12, 13, 18, 21, 26, 27, 63, 70–71, 73–74, 84–88, 92, 109, 116, 119, 123, 125, 131, 133, 135–37, 140, 154, 159, 170, 197, 201, 203, 207, 209, 229, 231–32, 237–38; church music revival, 132; *diapason normal*, 157, 160; Great Bell, 174; Greenwich time, 186; Houses of Parliament, 173; imperial system of weights, legalization of, 126, 141; laissez-faire tradition, 124, 128–29, 145, 163; mathematical standard pitch, rejection of, 162; mechanization in, 4; military, as employer of musicians, 143; musical pitch, 19–20, 235, 239; national standard time, 165; natural sciences in, 2; pitch reform, and global role, concerns over, 160–61; scientific culture in, 1–2; sonorous experiments, as part of scientific culture, 234; sound in, 81; time system,
visual character of, 166; urbanization in, 4. *See also* England; Scotland
British Army, 6
British Association for the Advancement of Science (BAAS), 9, 56–57, 61, 74, 80, 204, 214–15, 217–18, 232; Standards Committee, 126
British Empire, 186
British Isles, 2, 160
British Museum, 148–49
British science, sound in, 81
Broadwood, Walter Stewart, 132–33
Brown, Colin, 148
Brown, Robert William, 61, 146–47
Browning, Robert, 25
Brougham, Henry, 96
Brunel, Isambard Kingdom, 4, 129–30
Burdett-Coutts, Angela, 51
Burges, M., 136
Burney, Charles, 87
Busby, Thomas, 37

Calcutta (India), 196
Calicott, W. H. 148
Calvin, John, 6
Cambridge (England), 12, 86
Cambridge Camden Society, 112
Cambridge University, 110, 114–16; Cavendish Laboratory, 126
Canada, 160
Canterbury (England), 133
Cape of Good Hope, 186
Cape Town (South Africa), time-gun, 187
Carlisle (England), 133
Carlyle, Thomas, 216–17, 231
Carnot, Sadi, 73
Carpenter, William B., 218, 224
Cayley, Arthur, 173
Cazdet, W., 144–45
Challis, James, 111–12, 167
Chalmers, Thomas, 72, 218
Chambers, Robert, 9
Champneys, W. Weldon, 222–23
chanting, 28
Charles, M.: "Invisible Girl," 28–29, 29, 30

Charles X, 32
Chester, Harry, 129, 135, 140
Chladni, Ernst, 65, 68, 77, 90, 104, 107, 130–31; acoustic figures, 39, 53, 57, 60–63, 76, 101; acoustic investigations, 119; acoustic plates, 27, 29, 39, 58, 59, 62, 89, 211; experiments, 34, 94, 251n105; figure experiments, 16, 35, 40, 42, 58; sound waves, 4; vibrating plate experiments, 61–63, 102
Chorley, H. F., 144
Christian, Prince of Schleswig-Holstein, 226
Christianity, 213, 228
Christian Socialist movement, 132
Chua, Daniel, 11
Church of England, 28, 84, 106–7, 109, 116, 120, 132, 216; Catholic vision of, 115; High Church Anglicanism, 6, 110–11, 118; musical renaissance, 86; music and nature, 118
Church of Scotland, 28
Collard, Charles, 132
Collard, Frederick William, 132
Collard, Frederick William, Jr., 131
Collard & Collard, 132
collegiate music, 110, 117
Cooke, William Fothergill, 79
Copleston, Edward, 118–19
Cowie, Benjamin Morgan, 133
Cowper, William, 180–81
Crotch, William, 55
Curling, Edward, 227–28

Darlington (England), 196
Darwin, Charles, 9, 204–5, 217–18, 228, 233, 238–39
Davies, James, 16
Davison, Frederick, 133, 144
Davy, Humphry, 42–43
Davy, Jane, 43
de la Rive, Charles-Gaspard, 249n48
De Morgan, Augustus, 133, 141, 146, 162, 167–70, 175
Denison, Edmund Beckett, 173–74
Dent, Frederick, 176
Descent of Man, The (Darwin), 205, 233

DeYoung, Ursula, 216, 219
diapason normal, 127–28, 135, 140, 154, 157, 160–61, 183
Dickson, Melissa, 26
Dilke, Charles Wentworth, 129, 135, 144
Distin, Henry, 137, 137
Donaldson, John, 130–31
Donkin, William Fishburn, 139–40
Dover Castle, time-gun, 187
Driffield, G. T., 132, 140
Dublin (Ireland), 129
Duc d'Orléans, 32
Duke of Cambridge, 161
Duke of Edinburgh, 226
Dundee (Scotland), 196
Durham (England), 133
Dyson, J., 135

East India Trading Company, 51
Ecclesiological Movement, 111
Edinburgh (Scotland), 12, 95, 133; time-ball, *188*, 189; time cannon, 187; time-gun, 187, 189–90, *191*, *192*, 193–95
Edinburgh Observatory, 190
Edison, Thomas, 233
Elder Brethren of Trinity House, 207–8
electricity, 7, 14, 26–27, 42, 44–45, 62, 68–70, 73–75, 77–80, 88, 94–95, 98, 105, 119, 235, 252n134
Electric Telegraph Company, 187
electromagnetism, 27, 61, 63–64, 68
Eliot, George, 45, 83
Ellis, Alexander, 156–57, 159–61, 202
Elvey, George Job, 130
Enchanted Lyre, 35, 36, 37–40, 53, 74–75, 78–79, 81, 116
England, 37, 44, 92, 127, 159. See also Britain
Enlightenment, 15
Erard (manufacturer), 133
Erlman, Veit, 247n60
Eulenstein, Karl, 34; Jew's harp, 30–33, 47–49, 52, 81, 90; as musical celebrity, 32; scientific and nonscientific audiences, appeal to, 33
Euler, Leonhard, 3

Europe, 20, 47, 50, 62, 92, 127, 154, 156–57, 159–61, 201, 236
evolution, 20, 204–5, 217–20, 227, 239
"Explanations of the Musical Scale and Its Component Intervals" (Pole), 150

Faraday, Michael, 18–19, 25–27, 50, 55–56, 63–64, 71–73, 81, 84, 90, 103–4, 125, 206–8, 218–29, 233, 234, 249n48, 252n134; acoustic experiments, 68; acoustic figures, 67, 69, 70, 74; acoustic phenomena, as worthy of attention, 44; acoustic vibrations, experimenting on, 65–67; electromagnetism, 68–70, 74; experiments of, 45, 47–48; gender (Javanese musical instrument), 47; heat, investigation of, 70; hydrogen gas experiments, 44; kaleidophone, 49; music, importance of to, 42–43; psychology of, 69; sonorous phenomena, as crispations, 65–67, 69–70; sounds of nature, 44; theory of vibrations, 68, 70; Wheatstone, collaboration with, 42, 44–45, 47–49, 51–54, 59, 61–62, 119–20
Faraday, Sarah, 45
Farrar, F. W., 223
Fechner, Gustav, 5
Flanders (Belgium), 235–36
Fontenelle, Bernard Le Bovier de, 101
Forbes, James David, 72–73, 206, 207; second law, 207; third law, 207
Foster, Neve, 135–36
Foucault, Michel, 13
France, 3–4, 86, 125, 146, 159, 161, 163, 235–36; as center of mathematics, 87; *diapason normal* in, 127–28, 140, 154, 183; pitch, standardization of, 128; Republican Calendar, 8; tuning fork A435, 134–35
Frederick, William, 132
French Revolution, 3–4, 111
Fresnel, Augustin, 67
Froude, Richard, 111

geometry, 3, 107
George IV, 32

Germany, 16, 125, 161; German Romantic philosophy, 10
germ theory, 228
Gevaert, François-Auguste, 159
Gieryn, Thomas, 17
Gitelman, Lisa, 233
Gladstone, John Hall, 74
Gladstone, William, 114, 232
Glasgow (Scotland), 12, 133; time-ball, 194; time-gun, 195
Goddard, Joseph, 204
Goethe, Johann Wolfgang von, 10–11, 216
Goldschmidt, Otto Moritz David, 131
Gooday, Graeme, 16, 127
Goodeve, Thomas Minchin, 133, 146
Goodsir, Joseph Taylor, 215
Goss, John, 131
Gothic church architecture: revival of, 111–12
Gothic Revival, 117
Grajeda, Tony, 233
Gray & Davison, 133
Great Eastern (steamship), 4
Great Exhibition, 7, 129, 133–34, 181–82
Great First Cause, 97
Great Pyramid, 188–89
Great Western Railway (GWR), 79
Greenwich (England), 190, 194–95
Greenwich time, 186, 190, 193–94, 196; industrial connotations, 165; time-ball, 187
Gresley, William, 115–16
Gribenski, Fanny, 125
Griesbach, Henry, 131
Grisi, Giulia, 43
Gurney, Edmund, 202, 225–27, 234
Gye, Frederick, 43

Halévy, Fromental, 134–35, 146
Hallé, Charles, 131, 226
Hamilton, James, 68
Hamilton, Walter Kerr, 115
Handel, George Frideric, 128, 137–38; tuning fork A423, 132
Hanslick, Eduard, 11
harmony, 3, 10–11, 29–30, 146–50, 189, 201–2, 227; harmonic reform, hostility to, 153–54

Hartlepool (England), 196
Haweis, Hugh Reginald, 222
Hawes, Benjamin, 129–30
Hawes, William, 129–30
Haynes, Bruce, 125
heat, 61, 74, 126; caloric, 71; and electricity, 94; and sound, 70, 72–73, 97, 99; vibratory power, 72–73
Heat: A Mode of Motion (Tyndall), 207, 219
Hegel, Georg Wilhelm Friedrich, 10–11
Helmholtz, Hermann, 5, 15, 156–57, 201–3, 209, 213–14, 224–25, 234, 237
Herschel, John, 8, 18–20, 59–60, 63, 71, 75, 85–86, 92, 95, 98, 109–11, 116, 120, 125, 130, 133, 140, 151, 154, 168, 190, 201, 208, 215, 221, 229, 234; campaign for C_{512} pitch, 127, 142–45, 162, 235; cornfield analogy, 83, *84*, 89, 99, 103, 105, 107, 119; harmony, 93–94, 146–50, 153; mathematical theory, of sound's transmission, 89; mathematical works, broad readership of, 149; musical intervals, *148*, 149; musical standard, and nature, 141; natural laws, 93; nature, evoking of, 126; resonance, 90; Somerville, collaboration with, 101–2; sonorous phenomena, work on, 84; and sound, 104–5; sound waves, 83, 88–89; Wheatstone's acoustic contributions, ignoring of, 89–91
Herschel, Lady, 91, 95
Herschel, William, 86–88, 95–96
Hewat, John, 187–90, 193
Hewett, John William, 118
Hickson, William Edward, 55
Higgins, William Mullinger, 94–95
Hill, William, 133
Hipkins, Alfred, 157
Hirst, Thomas, 268n54
Hislop: time-gun map, *192*, 194
History of the Inductive Sciences (Whewell), 105
Hoffman, Ernst Theodor Amadeus, 10–11
Hogarth, George, 136
Holland, 161

Holland, W., 136
Holmes, Nathaniel, 194–95
Hook, Walter, 116–17
Hopkins, Edward John, 131
Hopkins, John Larkin, 178–79, *179*, 182–83
Horsley, Charles Edward, 130
Horsley, Mary Elizabeth, 130
Horsley, William, 130, 132
Hughes, Francis J., 149
Hui, Alexandra, 5, 16
Hullah, John Pyke, 55, 112, 116, 132–34, 136, 139–40, 143–45; ladder, 150–51, *151*
Hungary, 161
Huxley, Thomas, 9, 204–6, 216–19, 224, 228
Hymns Ancient and Modern, 111

imperialism, 1
India, 51, 160
industrialization, 3, 85, 235–36
Institution of Civil Engineers, 182
intelligent design, 119
Italy, 44, 161

Jackson, Myles, 16, 125, 251n105
James, Faber, 180–81
James, Jamie, 85
Java, 52
Jean, Yaron, 236
Jebb, John, 116–18, 131–32
Jenkin, Fleeming, 218
Jew's harp, 31–32, 47–49, 52, 81, 91; as children's toy, perception of, 30; magical influence of, 29; manipulation of nature, 71; as mysterious, 81; and resonance, 90; wonder of, 33–34
Joule, James, 126

kaleidophone, 49–50, 81
kaleidoscope, 28, 30
Kant, Immanuel, 219, 256n80
Keble, John, 111
Kempelen, Wolfgang von, 79–80
Kim, Stephen, 216
King's College London, 45, 75, 78, 80
Kingsland Congregational Church, 7

Kingsley, Charles, 132
Kittler, Friedrich, 245n3
knowledge production, 15–16, 107; within domestic spaces, 14; knowledge, visual representation, 13; and nature, 12; visual techniques in, 13–14
Koenig, Rudolph, 144
Köhler, Augustus Charles, 160
Kursell, Julia, 16

laboratory culture, 13
Lange, Friedrich, 11
Laplace, Pierre-Simon, 9, 78, 86, 96–97
Lardner, Dionysius, 93
Latour, Bruno, 13
law of gravity, 93
Leeds Parish Church, 116
Leopardi, Giacomo, 10–11
Leopold, Prince, 225–26
Leopold II, 157
Leslie, Henry David, 131, 144
Liddon, Henry, 115, 227
Lieber, Francis, 35
Lightman, Bernard, 14
Lincoln (England), 133
Lincoln Cathedral, 184
Lind, Jenny, 131
Lissajous, Jules, 138
Liverpool (England), 33, 133
Lloyd, Llewelyn S., 162
Lockhart, Ellen, 16
Lockyer, Norman, 232
London (England), 12, 14, 16, 19, 29–30, 33, 37–38, 40, 44, 48–51, 54, 60, 80, 82, 88, 92, 95–96, 187, 189, 194–95, 251n105; West End, 25–26, 34–35, 42, 53, 57, 81
London Philharmonic, 154
London Philharmonic Society, 124, 127
Lubbock, John, 103
Lunn, Henry Charles, 131, 146, 154–55, 223–25
Luther, Martin, 6

Macfarren, George Alexander, 130
Mach, Ernst, 5
Macnamara, Sophia, 129–30
Madras, 186
Mahillon, Charles, 159
Mahillon, Fernand, 159–61
Marchioness of Salisbury, 32
Marrian, J. P., 70
Martineau, James, 220–21
Mason, W., 135
materialism, 84; and Romanticism, 12
"Materialism and Its Opponents" (Tyndall), 221–22
mathematical theory, 89, 145, 165, 166, 168, 169, 172, 183, 186, 197
mathematics, 85–87, 107, 110, 114, 125, 127, 133–35, 146–47, 149, 165–67, 172–73, 184–85, 197, 214–15; musical tone, 4–5
Mather, James, 193
Mauritius, 186
May, Charles, 174
Maxwell, James Clerk, 9, 201, 218; "Maker," existence of, 217
Mears of Whitechapel, 174
measurement: and trust, 127
mechanical universe, 9
mechanism, 11–12, 29, 38, 95–96, 113, 128, 195, 216–17, 227, 238; industrial, 5; musical, 37; spiritual, 7
Mechanism of the Heavens, The (Somerville), 96–97, 104, 110
mechanization, 3–4
Mee, John H., 223
Mellon, Alfred, 131
Mendelssohn, Felix, 130, 136
Meyerbeer, Giacomo, 146
Middle Ages, 3
Middleborough (England), 196
Mill, John Stuart, 106
Miller, Edward, 113–14
Millington, John, 45
Mills, Marla, 233
Milner, Greg, 233
Mitscherlich, Eilhard, 69
modernity: vision, association with, 15
Monash, John, 236–37
Morris, William, 238
Morus, Iwan, 14–15
Mowitt, John, 233
Mozart, Wolfgang Amadeus, 10, 155
Municipal Corporations Act, 111

Murray, John, 204
music, 82, 203; and astronomy, 87–88; controversy over, 225; and dissonance, 5, 213; divine character, 7, 114, 205, 225–26; within ecclesiastical and theological reforming efforts, 111; emotional effects of, 209–10, 222, 225, 239; and faith, 118; as God-given gift, 7; as God's creation, 228; language, preceding of, 205; magical quality of, 154; manufacture of, 1; materiality of, 2, 224–25, 235, 238; mathematical laws, 94; mechanical principles, reduction to, 11; and mechanics, 11; melody and harmony, 37; metaphysical character of, 60, 120, 222, 225; moral value of, 111, 114, 119; musical entertainments, as philosophical demonstrations, 51; musical instruments, as scientific exhibitions, 49; "Music of the Spheres," 84–85; mystery in, 113; and nature, 6, 118, 223, 239; orchestral chorus, 227; Romantic understandings of, 10; as sacred, 203; and science, 5, 20–21, 113, 118–19, 162–63, 224, 234–35, 239; scientific explanation, transcendence of, 7, 227; and sound, 1–2, 10, 12–13, 20–21, 85, 109, 120, 213, 227, 235, 238; and speech, 205; spiritual quality, 202, 223, 225, 229; study of nature, 3; as universal language, 154; in worship, 6
Music and Morals (Haweis), 222
musicology, 238
Musselman, Elizabeth, 13–14

Napoléon, 4, 9, 128
Napoléon III, 125, 163
Nasmyth, Alexander, 95
Nathan, Isaac, 85
National Training School for Music, 225–26
natural philosophy, 10, 12, 18–21, 27, 33, 50, 83–86, 114, 204, 233, 235, 239; intelligent design, 119; thunderstorms, as acoustic spectacles, 44
natural sciences, 2, 12, 19; and sound, 3
natural selection: and nature, 217

nature, 9, 81, 93–94, 100–101, 126, 204–5, 214, 219, 224, 229; God, invoking of, 8, 20; knowledge production, 12; materialist interpretation of, 206, 235; and music, 6, 118, 223, 239; as mystery, 203; natural selection, 217; and science, 61–62, 216; song church worship, relationship between, 117; and sound, 19, 95, 109, 119–20, 232, 234
Neale, John, 111–12
Neeley, Kathryn, 100
Neptune, 86
Newcastle (England), 193–95; time-gun, 187
Newman, Francis William, 60
Newman, John Henry, 58, 60, 111
Newton, Isaac, 3, 78, 86–87, 107, 171–72; Newtonian calculus, 4
New York, 129
Nightingale, Florence, 51
North Shields (England), 187
Norwich Cathedral, 117

On the Connexion of the Physical Sciences (Somerville), 97–98, 100–104
"On Musical Scales" (J. Herschel), 147–49
On the Origin of Species (Darwin), 9, 204, 239
On Sound and Atmospheric Vibrations (Airy), 150, 152–53, 170, 172, 183, 185, 202
Orsted, Christian, 35, 42, 62, 68–69
Otter, Chris, 13
Ouseley, Frederick Gore, 126, 131–34, 141, 150–51, 153
Ouseley, Harriot, 141
Oxford (England), 12, 58, 86, 232
Oxford Movement, 111, 115, 118; medievalism, cultivation of, 117
Oxford University, 110–11, 114–16
Oxford University Museum, 204

Paley, William, 227
Pantalony, David, 202
Paris (France), 32, 57, 129; time-gun, 189
Parker, John, 143
Parr, Alfred, 235

Patti, Adelina, 155
Peacock, George, 86–87, 167, 170
Peel, Robert, 105, 114
Percy, John, 70
Pesic, Peter, 16
Philharmonic Society, 137–39
Phillips, Thomas, 134, 142, 144
philosophical knowledge, 13
Philosophical Transactions of the Royal Society (journal), 18, 26, 42, 62–63, 103
Philosophy of the Inductive Sciences (Whewell), 107
physics, 12
"Physics and Metaphysics" (Tyndall), 213
pitch, 17, 19, 123, 151, 153, 163, 197, 235, 239; British Army pitch ("Kneller Hall Pitch"), support for, 160; campaign for reform, 154–55; controversy over, 20, 201; frequency, question over, 138; national standard, queries over, 135, 138, 160; pitch committee (1859–60), 126–27, 130–37, 137, 139, 139, 140–41, 145, 154–55; problem of, as instrumental, 156; regulation of, 124–25, 128–29; standardizing, attempts to, 125–28, 137, 140, 155–57, 161–62, 196; and trust, 127, 162; tuning forks, 136–38, 139–40, 145–46; unifying attempts, 126–27, 136–37, 157, 161
Pole, William, 133–34, 141, 150, *150*, 151–52, *152*, 153–55, 162, 181–83
Poor Law Amendment Act, 111
popular medievalism, 111
popular science, 19
Portsmouth (England), 186
Potter, Cipriani Hambley, 131, 134, 144–45
Pouillet, Claude, 103
Preliminary Discourse on the Study of Natural Philosophy, A (Herschel), 63, 98, 105
Presbyterianism, 218
Principia (Newton), 97, 171–72
Protestantism, 6
Prussia, 161

Ptolemy, 3, 108
Public Worship Regulation Act, 115
Pugin, Augustus, 111
Pusey, Edward, 111, 227
Pythagoras, 3, 211, 213; heavenly bodies and musical harmony, linking of, 84–85

Quarry, John, 219–20

Raffles, Lady Sophia, 51
Raffles, Sir Stamford, 51
Rankine, Macquorn, 218
Rayleigh, 3rd Baron, 201–3, 234, 237
Rea, William, 158
Reeves, Sims, 154–55
Reform Act, 111
Reid, David Boswell, 71–72, 123
Renaissance, 3
Ringer, Sydney, 134
Ritchie, Jamie, 195
Ritchie, William, 54–55
Ritter, Johann Wilhelm, 62
River Tyne, 190, 193
River Tyne Commission, 193–194
Robison, John, 71
Roman Catholicism, 6, 8–9, 28, 111; song worship, 117
Romanticism, 10–11, 246n37; and materialism, 12; and mechanism, 12
Rose, Rendell, 137
Rossini, Gioachino, 128, 134–35, 146
Royal Academy of Sciences, 89, 223
Royal Albert Hall, 155
Royal Commission on Scientific Instruction and the Advancement of Science, 232
Royal Institution, 14, 18, 25–26, 33, 42, 44, 47, 48, 50–53, 67, 76, 80–81, 202, 209, 222; Friday Evening Discourses, 45, 54–55, 71, 73, 206–7; lectures of, and female attendance, 55–56; lecture seasons (1829 and 1831), 55, 56
Royal Military School of Music, 160
Royal Navy, 6, 186
Royal Observatory, 20, 86, 156, 165, 186–87, 235

Royal Scottish Society of Arts, 190
Royal Society, 18, 26, 63, 65, 67, 69, 87, 91–92, 103, 195, 206–7, 209, 232
Royal Society of Arts, 129, 161–62. *See also* Society of Arts
Royal Society of Edinburgh, 71–72
Ruskin, John, 111, 165; popularity, loss of, 237–38
Russell, John Scott, 4–5, 129
Russia, 159, 161

Sacred Harmonic Society, 6, 154
sacred music, 117–18
Saint-Saëns, Camille, 235
Saldman, Charles, 135
Salisbury (England), 133
Salmon, George, 227
Sandemanianism, 63
Sauveur, Joseph, 124
Savart, Felix, 62, 65, 68, 89, 101, 103, 130–31
Saxony, 161
Schaffer, Simon, 14
science, 107, 120, 218; art, as mutually beneficial, 223; boundary work, 17; and British state, 232; church, undermining of, 84; conservation of energy, 220; as form of entertainment, 26, 39; and harmony, 118; and music, 5, 20–21, 113, 118–19, 162–63, 224, 234–35, 239; and musical instruments, 49; and nature, 61–62, 216; as politically dangerous, 84; and religion, 239; social authority of, 12–13; social distinctions, 17; and sound, 16–17, 74, 83–84, 166, 226–27; visual cultures of, 13–15
scientific revolution, 13
Scotland, 33. *See also* Britain
Scottish Telegraph Construction and Maintenance Company, 194–95
Secord, James, 84
7th Duke of Devonshire, 232
Seven Years' War, 87
Shepherd, E., 135
Shore, John, 132
Siemens, Carl Wilhelm, 80

slavery, 106
Sloterdijk, Peter, 236
Smart, Benjamin Humphrey, 54
Smart, George Thomas, 130, 145
Smith, Crosbie, 218
Smith, Richard, 167
Smith, Robert, 87, 168
Smith, Sidney, 32
Smyth, Charles Piazzi, 187, 190, 194–96; "Pyramidology" of, 188–89
social elitism, 106
Society of Arts, 125, 126, 129, 131, 134–35, 140, 145, 154, 157, 158, 160, 162, 183, 196, 204. *See also* Royal Society of Arts
Society of British Musicians, 136
Society for the Diffusion of Useful Knowledge, 96
Society for the Improvement of Church Music, 118
Society of Telegraphic Engineers, 80
Somerville, Mary, 8, 19, 84–86, 95, 97, 98, 105, 109, 111, 115–16, 119–20, 125, 143, 221, 223, 229, 234; acoustics, 103; criticism of, 101; and heat, 99; Herschel, collaboration with, 101–2; light, study of, 99; nature, 100–101; scientific sublime, 100; and sound, 104; sound and heat, relationship between, 100; sunlight, power of, 96
Somerville, William, 95
sonorous, 12; divine design of, 113; metaphysical character of, 27–28; sonorous experiments, 19, 58, 62, 102, 232–34; sonorous investigations, 201; sonorous knowledge, 222, 238–39; sonorous measurements, 88; sonorous time signaling, accuracy of, 196
sonorous phenomena, 21, 83–84, 202, 203, 206, 207, 223, 229, 233, 238; mathematics, applying to, 165, 166; social control, as means of, 27
Sopwith, Thomas, 129
sound, 26, 28–30, 34–35, 45, 80, 101, 103, 105, 166, 209–10, 214, 228–29; accuracy of, 165; and astronomy, 85; divine creation, 119; divine nature of, 222; ear, as instrument of war, 236;

sound (*continued*)
  ear, mathematical precision of, 15; ear, as wondrous mechanism, 227; earthly and astronomical phenomena, intellectual link between, 119; experiential quality of, 27; harmonious universe, 85; harmony, 93–94; and hearing, 233; and heat, 70–73, 97, 99–100, 221; keeping time, 165; and light, 3, 93–94, 98–100, 216, 221; measuring of, as subjective, 186–87, 189–90, 193–94; militaristic applications of, 237; moral implications of, 95; moral teachings, 86; as movement of waves, 40, 42, 215–16; and music, 1–2, 10, 12–13, 20–21, 85, 109, 120, 213, 227, 235, 238; natural phenomena, linking of, 6, 84; in natural sciences, 25; and nature, 19, 95, 119–20, 232, 234; as object of entertainment, 82; philosophical inquiries into, 201–4; philosophical knowledge of, 120; popular and elite science, transcending boundaries of, 27; production of knowledge, 108; religious controversy over, 213; as sacred, 203; and science, 16–17, 74, 83–84, 166, 226–27; scientific authority, 3; scientific discourse, 82; as scientific object, 61, 70; sensory perceptions of, 5; sound waves, 2, 4, 83, 170–71, 209, 222–23; as spiritual, 229; theory of, 20; variations in, and temperature changes, 161; velocity of, 172; and vibrations, 89; weapon, mobilized as, 237
"Sound" (J. Herschel), 88–92, 98, 103, 146, 201; footnote, as damning, 92–93
Sound: A Course of Eight Lectures Delivered at the Royal Institution of Great Britain (Tyndall), 209
South Eastern Railway Company, 186
South Kensington Physical Laboratory: "Albertopolis," scientific and cultural center of, 232
Spain, 159
Spencer, Herbert, 9, 204–6, 216–17, 224, 228, 234
Stainer, John, 133, 160–61

steam engines, 73
Steilbelt, Daniel, 38
Stephenson, Robert, 129
Sterne, Jonathan, 15–16, 166
St. Helena, 186
Stone, William Henry, 156
Stones of Venice (Ruskin), 238
St. Paul's Cathedral, 75–76, 115; choir, 129
Strutt, John William, 201
Stumpf, Carl, 5
Sunderland (England), 194–95
Surman, J., 136
Sweden, 161
Switzerland, 161
Symington, Robert, 194–95
System of Logic, A (Mill), 106

Tait, Peter Guthrie, 218
Taylor, Edward, 56
Taylor, William, 181
Taylor, William Cooke, 101
Tempest, The (Shakespeare), 1
Test Act, 111
theories of evolution, 9
thermodynamics, 9, 73; divine architect, 218; first law of, 219; second law of, 219
Thompson, Thomas, 143
Thomson, William, 218
time: accuracy of, 196–97; acoustic time signaling, as popular, 196; disseminating of, 196–97; dropping of balls, 186; seeing and hearing, dichotomy between, 196; time-balls, 186, *188*, 189–90, 194, 197; time cannon, 187; time-gun maps, *192*, 194; time-guns, 187, 189–90, *191*, *192*, 193, 195–97; time-guns, southern skepticism toward, 195
Time Machine, The (Wells), 201
Tkaczyk, Viktoria, 16, 57
Tractarianism, 111
Trasch, John, 12
Treatise on Harmony (Ouseley), 150–51, 153
Treaty of Versailles, 161
Trevelyan, Arthur, 70–73, 206–7

Trinity College, 112, 132
Truro (England), 133
Turle, James, 181
Turner, Steven, 203
Tweney, Ryan, 68
Tyndall, John, 5, 9, 14, 20–21, 44, 73, 202–3, 205, 207, 226–27, 229, 234, 238; acorn comparison, 220–22, 233; acoustic knowledge, 215; Catholicism, dislike of, 216; Christianity, criticism of, 213; Church of England, rejection of, 216; cyclical account of nature, preference for, 219; demonstrations, of, 209–11; infinite universe, 219; inorganic and organic, uniting of, 221; invisible acoustic clouds, 209; hostility toward, 221–22; light and sound, 216; and nature, 217; nature, materialist interpretation of, 206, 213, 217–20, 223, 235; nature, and natural selection, 217; scientific materialism, 268n54; solitaire, 209, *210*; sonorous trials of, 208–9; sound, movement of, 215–16; sound, repopularizing of, 209–10; sound waves, 209–10; soul, overlooking of, 222; transcendental materialism, 268n54; vibrations, *211*, 211; vibrations, and consciousness, effect on, 214; Wheatstone's Enchanted Lyre, evoking of, 220–21; Wheatstone's kaleidophone, 211, *212*

Unitarianism, 95, 203
University College London, 45
Uranus, 86–87
urbanization, 4
United States, 127, 161

Van Drie, Melissa, 233
Verdi, Giuseppe, 157
vibratory theory, 70
Victoria, Queen, 226
Vitruvius, 106
Volta, Alessandro, 34

Waddell, James, 131
Wagner, Richard, 11, 155
Walker, Adam, 87–88

Walker, Charles, 186
Walmisley, Thomas, 130
Warner and Sons, 174
Washington Conference, 165
Webb, Benjamin, 111–12
Webb, Jane, 51
Weber, Max, 11, 202
Wells (England), 133
Wells, H. G., 201
Werner, Robert, 176, 177, 178, 180
Wesley, Charles, 88, 116
Wesley, Samuel, 116, 133–34
Wesley, Samuel Sebastian, 116
Wheatstone, Charles, 17–19, 25–26, 34, 50, 55–58, 65, 67–68, 70, 81, 84, 88, 90, 101, 103–4, 125, 130–31, 133, 141, 146, 162, 206, 233–35, 251n105; acoustic figures, 59, 61, 64, 80, 101, 102; acoustic knowledge, 27, 80; acoustics, and measurement, 77–78; architectural acoustics, 75; electricity, 78–80; Enchanted Lyre, 35, 36, 37–40, 53, 74–75, 78–79, 210–11, 220–21; Faraday, collaboration with, 42, 44–45, 47–49, 51–54, 59, 61–62, 119–20; "Invisible Girl," 35; Jew's harp, 91; kaleidophone, 39–40, *40*, *41*, 60, 66, 78, 80, 91, 211; kaleidoscope of, 30; music and natural philosophy, combining of, 39; scientific knowledge, and communication, 53–54; sonorous communication, 79; telegraphic work, 78–80, 118; vibrations, 37, 75–78, 80, 102
Whewell, William, 8, 19, 72, 85–86, 92–93, 104, 123, 130, 132, 134, 141, 145–46, 162, 167, 180, 215, 221, 223, 229, 234–35, 256n80; acoustics, 105–7, 110–13, 116, 119–20; acoustic vibrations, 108–9; a priori philosophy, champion of, 106; evidence of design, 105; hearing, exactness of, 108; as idealist, 106; intuitionist philosophy, as leader of, 106; and nature, 109; physicist and scientist, coining of terms, 104–5; pitch, 123; secondary mechanical sciences, 105, 107–8; senses and ideas, epistemological balance between,

Whewell, William (*continued*)
    107; and sound, 108; sound and music, 109
White, Joseph Blanco, 58–61
Wilberforce, Samuel, 204
Williams, Charles James Blasius, 74–75
Williamson, Alexander, 218
Willis, Henry, 133, 137, 141, 146, 155
Willis, Robert, 55, 89–90, 162; speech-producing machine of, 79–80
Willis and Hill, 133
Wilson, Francis, 88
Wittje, Roland, 237

Wordsworth, Christopher, 112–13
World War I, 2, 10, 21, 235, 237; Allies victory, hearing at center of, 236; ear, as instrument of war, 236; idealism, demise of, 238
World War II, 161
Wundt, Wilhelm, 5
Württemberg, 161
Wylde, Henry, 134

X Club, 205–6, 217, 219, 268n54

Young, Thomas, 39, 88, 92